AF598121

METHODS IN MOLECULAR BIOLOGY™

For further volumes:
http://www.springer.com/series/7651

DNA Electrophoresis

Methods and Protocols

Edited by

Svetlana Makovets

Institute of Cell Biology, University of Edinburgh, Edinburgh, UK

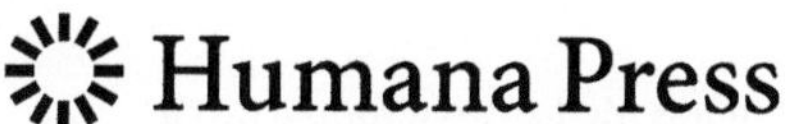

Editor
Svetlana Makovets
Institute of Cell Biology
University of Edinburgh
Edinburgh, UK

ISSN 1064-3745 ISSN 1940-6029 (electronic)
ISBN 978-1-62703-564-4 ISBN 978-1-62703-565-1 (eBook)
DOI 10.1007/978-1-62703-565-1
Springer New York Heidelberg Dordrecht London

Library of Congress Control Number: 2013945081

Printed on acid-free paper

Humana Press is a brand of Springer
Springer is part of Springer Science+Business Media (www.springer.com)

Preface

Since the discovery of structure of DNA by Watson and Crick, electrophoresis has been established as the most commonly used approach to analyze size, shape, and structure of DNA molecules. DNA molecules vary greatly in size, from a few nucleotides one can easily synthesize in vitro to long chains of millions of base pairs such as eukaryotic chromosomes. DNA comes in a variety of forms: it can be single stranded or double stranded, linear or circular, or supercoiled or relaxed; it can be bound, bent, or modified by proteins; it forms sophisticated branched structures during replication and recombination. DNA electrophoresis is a powerful tool that allows separating DNA molecules according to their size and shape.

Beginning with a historic overview on DNA electrophoresis and its principles as an approach, each of the follow-up chapters is devoted to a single protocol and consists of a brief introduction, a list of materials and reagents, a step-by-step protocol, and a list of notes presenting the author's unique know-hows to help readers successfully navigate through the protocol. Chapter 2 is aimed at early career researchers, to introduce them to the basics of the most commonly used DNA electrophoresis in agarose gels to analyze samples of linear DNA such as plasmid digests and PCR products. Chapters 3–7 are devoted to 2-dimensional gel electrophoresis which allows resolving more complex DNA molecules: branched replication and recombination intermediates and supercoiled plasmids. Chapters 8–11 describe DNA electrophoresis under conditions in which DNA molecules are completely or partially denatured during the runs, allowing either analysis of ssDNA based on its mobility or comparison of sequences of dsDNA molecules derived from their mobility change due to partial in-gel melting. Chapters 12 and 13 are dedicated to pulse field gel electrophoresis employed to analyze very large DNA molecules such as full-length eukaryotic chromosomes or bacterial chromosomes cut into a relatively small number of fragments. Single-cell DNA electrophoresis (comet assay) used to analyze DNA damage and repair is described in Chapter 15. Chapter 16 adds another level of resolution to this approach by coupling the electrophoresis to fluorescence in situ hybridization which allows following the dynamics of damage repair at a specific locus in a genome. Finally, methods for studying protein–DNA interactions using electrophoreses are presented in Chapters 16–19.

DNA Electrophoresis: Methods and Protocols is written by expert scientists with hands-on experience. I would like to thank all of them for sharing their invaluable experience with the scientific community by contributing to this book.

Edinburgh, UK *Svetlana Makovets*

Contents

Contributors

DIANA ANDERSON • *School of Life Sciences, University of Bradford, Bradford, UK*
DAVID T.F. DRYDEN • *School of Chemistry, University of Edinburgh, Edinburgh, UK*
AZIZ EL HAGE • *Wellcome Trust Centre for Cell Biology, University of Edinburgh, Edinburgh, UK*
FIONA FLETT • *Institute of Cell Biology, University of Edinburgh, Edinburgh, UK*
ANDREW FREE • *Institute of Cell Biology, University of Edinburgh, Edinburgh, UK*
LAURA GARDANO • *UFR SMBH/University Paris 13, Paris, France*
PABLO HERNÁNDEZ • *Department of Cell and Molecular Biology, Centro de Investigaciones Biológicas (CSIC), Madrid, Spain*
JONATHAN HOUSELEY • *The Babraham Institute, Cambridge, UK*
NEIL HUNTER • *Howard Hughes Medical Institute and the Departments of Microbiology & Molecular Genetics, Molecular & Cellular Biology and Cell Biology & Human Anatomy, University of California Davis, Davis, CA, USA*
HEIDRUN INTERTHAL • *Institute of Cell Biology, University of Edinburgh, Edinburgh, UK*
ANDREAS S. IVESSA • *Department of Cell Biology and Molecular Medicine, University of Medicine and Dentistry of New Jersey, Newark, NJ, USA*
DORA B. KRIMER • *Department of Cell and Molecular Biology, Centro de Investigaciones Biológicas (CSIC), Madrid, Spain*
JESSICA P. LAO • *Howard Hughes Medical Institute and the Departments of Microbiology & Molecular Genetics, Molecular & Cellular Biology and Cell Biology & Human Anatomy, University of California Davis, Davis, CA, USA*
JULIAN LAUBENTHAL • *School of Life Sciences, University of Bradford, Bradford, UK*
DAVID R.F. LEACH • *Institute of Cell Biology, University of Edinburgh, Edinburgh, UK*
FENFEI LENG • *Department of Chemistry and Biochemistry, Florida International University, Miami, FL, USA*
DANIEL L. LEVY • *Department of Molecular Biology, University of Wyoming, Laramie, WY, USA*
SVETLANA MAKOVETS • *Institute of Cell Biology, University of Edinburgh, Edinburgh, UK*
MARÍA-LUISA MARTÍNEZ-ROBLES • *Department of Cell and Molecular Biology, Centro de Investigaciones Biológicas (CSIC), Madrid, Spain*
JULIA S.P. MAWER • *Institute of Cell Biology, University of Edinburgh, Edinburgh, UK*
JUN NAKAMURA • *Department of Environmental Sciences and Engineering, University of North Carolina, Chapel Hill, NC, USA*
BRIAN PACHKOWSKI • *Department of Environmental Sciences and Engineering, University of North Carolina, Chapel Hill, NC, USA*
LYNN POWELL • *Centre for Integrative Physiology, School of Biomedical Sciences, University of Edinburgh, Edinburgh, UK*
GARETH A. ROBERTS • *School of Chemistry, University of Edinburgh, Edinburgh, UK*

NICHOLAS P. ROBINSON • *Department of Biochemistry, University of Cambridge, Cambridge, UK*
JORGE B. SCHVARTZMAN • *Department of Cell and Molecular Biology, Centro de Investigaciones Biológicas (CSIC), Madrid, Spain*
FIONA STRATHDEE • *Institute of Cell Biology, University of Edinburgh, Edinburgh, UK*
SHANGMING TANG • *Howard Hughes Medical Institute and the Departments of Microbiology & Molecular Genetics, Molecular & Cellular Biology and Cell Biology & Human Anatomy, University of California Davis, Davis, CA, USA*
VIKTOR VIGLASKY • *Faculty of Sciences, Department of Biochemistry, Institute of Chemistry, Kosice, Slovakia*
TANYA L. WILLIAMS • *Department of Biochemistry and Biophysics, University of California, San Francisco, CA, USA*

Chapter 1

DNA Electrophoresis: Historical and Theoretical Perspectives

Gareth A. Roberts and David T.F. Dryden

Abstract

The technique of gel electrophoresis is now firmly established as a routine laboratory method for analyzing DNA. Here, we describe the development of the methodology as well as a brief explanation of how the technique works. There is a short introduction to pulsed-field agarose gel electrophoresis, which represents a critical advancement in the method that facilitates the analysis of very large fragments of DNA. Finally, theoretical considerations are included.

Key words Electrophoresis, DNA separation, Agarose, Polyacrylamide, Pulsed-field, Reptation

1 Historical Perspective

The idea of separating charged particles by applying an electric field is a very old concept. For example, Tiselius in 1937 [1] used zone-boundary electrophoresis to resolve proteins. There were, however, a number of technical problems that needed to be overcome before the technique could be regarded as a routine procedure. For example, the generation of convection currents leads to mass transport processes that can be greater in magnitude than the electrophoresis unless drastic measures are taken to ameliorate this problem. An elegant solution to the difficulties arising from convection currents is to perform the electrophoresis in a gel matrix [2]. The analysis of DNA by electrophoresis through a supported matrix was originally developed by Vin Thorne in the mid 1960s while working at the Institute of Virology in Glasgow, Scotland [3]. The study involved the separation of different DNA species extracted from polyoma virus particles. Thorne was able to separate the superhelical, nicked, and linear forms of DNA by electrophoresis through an agar gel. However, the DNA first had to be radiolabeled in order to facilitate detection of the resulting bands. Although the original investigation did not invoke a huge response

Svetlana Makovets (ed.), *DNA Electrophoresis: Methods and Protocols*, Methods in Molecular Biology, vol. 1054, DOI 10.1007/978-1-62703-565-1_1, © Springer Science+Business Media New York 2013

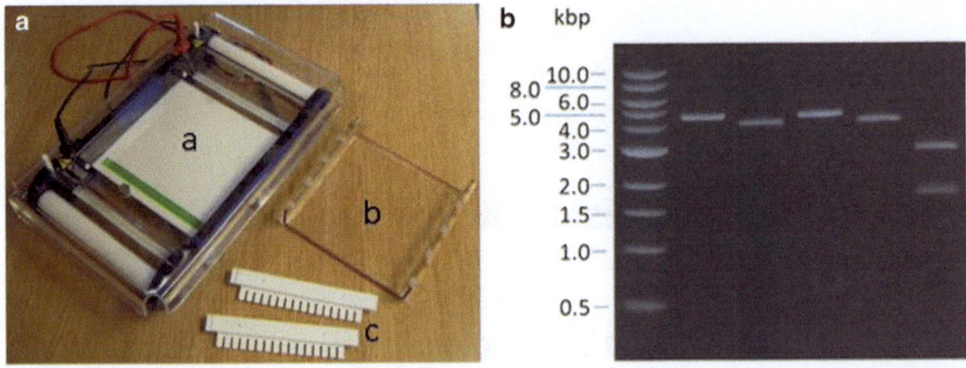

Fig. 1 Agarose gel electrophoresis is a standard laboratory procedure for the analysis of DNA. *Panel A*: Standard laboratory apparatus for performing agarose gel electrophoresis. (*a*) Horizontal gel electrophoresis tank. (*b*) UV transparent gel acrylic casting tray. (*c*) Set of combs. *Panel B*: Analysis of DNA fragments resolved on a 0.9 % agarose gel. The gel was stained with ethidium bromide and viewed on a UV transilluminator. The size of the DNA fragments in the marker lane is indicated

from the scientific community, the advent of recombinant DNA technology in the early 1970s brought about a sudden need to readily analyze DNA fragments. This led to the rapid development of DNA electrophoresis as an established technique to map the structure of DNA. The visualization of the DNA bands with ethidium bromide, a fluorescent intercalating dye, was originally reported by Aarj and Borst [4]. The subsequent development of horizontal gel electrophoresis in the form of a submerged ("submarine") gel constituted a further critical advance in the technique over the tube-gel method in which the gel was formed in a thin glass tube which could only take a single sample. Another important refinement took place in 1984 with the development of pulsed-field gel electrophoresis by Schwartz and Cantor [5] as a revolutionary method for the electrophoretic separation of extremely large DNA fragments (up to 10 Mbp).

The electrophoresis of DNA is now an established procedure in molecular biology for routinely analyzing, separating, and purifying DNA samples. A typical modern-day laboratory agarose gel electrophoresis kit is shown in Fig. 1 alongside a gel stained with ethidium bromide and viewed on a transilluminator. Specific examples in the field of genetics and medicine where this technique is used include the sequencing of genomes, DNA fingerprinting as well as the molecular identification of pathogens and genetic disorders.

2 General Background

DNA electrophoresis is an analytical technique that facilitates the separation of DNA molecules on the basis of their size or conformation [6]. The general principle behind this technique is that the sample to be analyzed is subjected to an electric field that induces the analyte to migrate towards the anode due to the negative charge on the sugar-phosphate backbone of the DNA molecule. Separation is achieved because migration of the DNA molecules occurs in a viscous sieving medium in the form of a gel. The rate of migration varies according to the length of the linear DNA molecules within the size range of the technique (typically <50 kbp). Longer DNA fragments migrate more slowly through the gel than shorter DNA fragments because they encounter greater resistance to their movement.

There are, however, a number of factors other than the length of the DNA molecule that determine their rate of migration. For example, the overall conformation of the DNA plays a role in their separation. In particular, the amount of supercoiling or degree of single-stranded segments also affects the migration of the molecule through the gel. DNA electrophoresis can be used to separate circular DNA with different supercoiling topologies. Plasmid DNA is easily resolved into covalent closed circular, linear, and nicked forms by this technique. The amount of DNA damage resulting in cross-linked species may also be conveniently monitored by DNA electrophoresis.

In order to maximize the resolution of the technique, various parameters can be changed. These parameters include the sieving medium, voltage applied, and the length of the running time. Typically, DNA electrophoresis is carried out in a gel, usually either agarose or polyacrylamide. The composition and strength of the gel is carefully chosen to enhance the resolution of the technique.

Polyacrylamide gels, which have a pore size of 5–100 nm, are used to separate small DNA fragments (<500 bp). The polyacrylamide itself is a flexible neutral polymer that can be cross-linked by incorporation of a bifunctional crosslinker (bisacrylamide). Polyacrylamide gels are normally cast vertically between plates or in tubes, thereby allowing the exclusion of oxygen that would otherwise prevent polymerization. These gels are particularly useful for resolving single-stranded DNA fragments. As such, this technique is well suited to the sequencing of DNA in which DNA molecules differing in length by just a single base need to be separated.

By contrast, agarose gels with pore sizes in the range of 200–500 nm have a lower resolution but can be used to resolve much larger DNA fragments (100–50,000 bp). Agarose is a rigid polysaccharide derived from seaweed made up of alternating residues of D- and L-galactose joined via α-(1-3) and β-(1-4) glycosidic bonds. The agarose is soluble in water at high temperature but forms a gel as the solution cools by lateral association of the chains (each made up of ~800 galactose residues) into helical bundles.

Highly purified preparations of agarose are commercially available that display the following properties: (1) very low background fluorescence, (2) essentially DNase-free, and (3) low inhibition of enzymes used in molecular biology (e.g., restriction endonucleases, DNA ligases, DNA kinases). The greater the voltage the faster the DNA molecules migrate. At low voltages, the rate of migration of linear DNA fragments is directly proportional to the applied voltage. However, this relationship breaks down at higher voltages, which can also induce unwanted heating in the gel matrix. Therefore, in practice, DNA of greater than about 2 kbp should be separated on an agarose gel at 5–8 V/cm.

3 Pulsed-Field Gel Electrophoresis

In continuous field electrophoresis, DNA molecules larger than 50 kbp migrate at the same rate making it impossible to achieve resolution. However, pulsed-field gel electrophoresis (PFGE) facilitates the separation of very large pieces of DNA, effectively increasing the upper limit for this technique from 50 to 10,000 kbp. Increased resolution is achieved by alternating the direction of the electric field relative to the gel during electrophoresis so that the DNA molecules are forced to change direction [7]. With each reorientation of the electric field, the smaller DNA molecules are able to respond more quickly than the larger DNA molecules. Specifically, the smaller DNA molecules unravel and "snake" through the gel more rapidly than their longer counterparts. Thus, there is a time-associated size-dependent reorientation of DNA migration. In this fashion, the DNA fragments separate and resolve themselves into discrete bands. Reorientation of the electric field is achieved using a number of instrumentation approaches, including orthogonal field agarose gel electrophoresis (OFAGE) [8], transverse alternating field electrophoresis (TAFE) [9], field inversion gel electrophoresis (FIGE) [10], and contour-clamped homogeneous electric field electrophoresis (CHEF) [11] (Fig. 2). While all of these methods are variations on the pulsed-field theme, the overwhelming majority of laboratories use CHEF, which has now become essentially synonymous with the term PFGE. CHEF comprises a hexagonal array of 24 electrodes that generates a highly uniform electrophoresis gradient causing the DNA molecules to reorient, most commonly over an angle of 120°. This configuration of electrodes produces equidistant DNA migration to the left and right of the agarose gel center resulting in straight vertical lanes of separation. As with all electrophoretic methods, CHEF is influenced by factors such as agarose gel concentration and thickness, buffer composition, temperature, and strength of the electric field (V/cm).

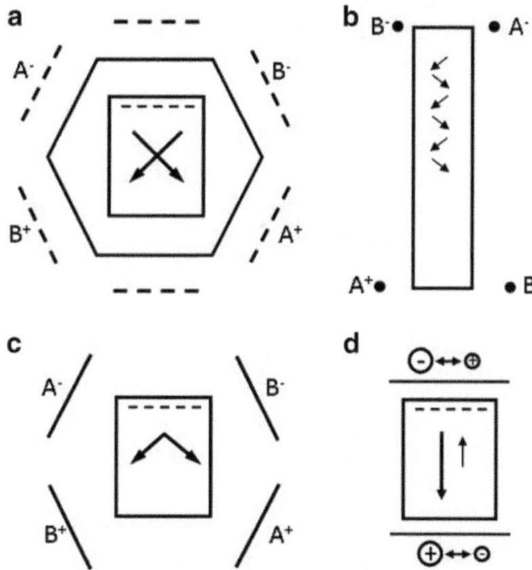

Fig. 2 Examples of pulsed-field gel electrophoresis (PFGE) arrangements that are in common use. *Panel A*: Contour-clamped homogeneous electric field electrophoresis (CHEF), *Panel B*: transverse alternating field electrophoresis (TAFE), and *Panel C*: orthogonal field agarose gel electrophoresis (OFAGE), *Panel D* field inversion gel electrophoresis (FIGE)

Traditional methods of sample preparation are not applicable to PFGE because large pieces of DNA are highly susceptible to shearing forces, leading to strand breakage. However, the DNA samples can be prepared by encapsulation in ultrapure low-melting-temperature agarose plugs. The precise method of sample preparation varies according to the source of DNA. Perhaps the simplest method is the preparation of DNA from the mammalian cells. Initially, the cells are washed several times in buffer and resuspended at 2×10^7 cells/mL and mixed with an equal volume of cool low-melting-temperature agarose. The mixture is then immediately dispensed into appropriate molds. The resulting agarose plugs are carefully removed from the molds and incubated for 2 days at 50 °C in a solution containing 450 mM ethylenediaminetetraacetic acid (EDTA) pH 9.5, 1 % sodium lauroyl sarcosine, and 1 mg/mL proteinase K. The detergent and proteinase K remove all cellular components to liberate the DNA, and the EDTA maintains a high-chelating environment to minimize nuclease activity. Moreover, the plugs may be stored in this EDTA solution for extended periods of time at 4 °C. The preparation procedure is slightly more involved when preparing DNA from yeast because the cell wall first needs to be removed. For example, *Saccharomyces cerevisiae* is pretreated with zymolyase or lyticase, whereas fission yeast *Schizosaccharomyces pombe* requires treatment with novozym. Similarly, bacteria, such as *Escherichia coli*, require pretreatment with lysozyme for cell wall digestion.

4 Theoretical Considerations

The so-called biased reptation model (BRM) provides a good theoretical framework for explaining how DNA migrates through a gel under the influence of an electric field. The precursor of this model, known as the reptation mechanism, was originally devised in 1971 by Pierre DeGennes in order to clarify the dynamics of polymer melts of high molecular weight [12]. According to the model proposed by DeGennes, the dynamics of such systems is influenced by entanglement effects between the long polymer chains. Specifically, each polymer is constrained to move within a topological tube caused by the presence of surrounding polymers that restrict its movement. Within this tube the polymer undergoes a snake-like motion, hence the name reptation. In such a fashion, the polymer advances through the melt by diffusion of stored length along its own contour. Analyzing the motion of the chains in such a system is extremely complex. However, the model focuses on the much simpler dynamics of uniformly charged polymers migrating through a network of fixed obstacles. It must be assumed that long chains do not affect the physical properties of the system as a whole.

The application of an external electric field has a big influence on the reptation picture of how an extended molecule like DNA migrates through a gel. A thorough theoretical investigation of the behavior of DNA in a gel during electrophoresis was independently carried out by two groups in 1982 [13, 14]. This led to the development of BRM, which is currently the most popular mathematical model for explaining the electrophoresis of DNA [15]. In BRM, the DNA is confined within a theoretical tube of length L_{tube} similar to the model proposed by DeGennes for a polymer melt (Fig. 3). Specifically, the gel matrix can be considered to constitute a frozen network of obstacles in which the polymer migrates through the diffusion of stored length. The reptating motion of DNA through a gel has also been invoked experimentally to explain the electrophoretic mobility of linear DNA molecules through a gel where the anticipated radius of gyration exceeds the size of the gel pores [16].

The DNA molecule is a semiflexible polymer that adopts a negative charge in aqueous solution due to dissociation of H^+ from the phosphate backbone. To a first approximation, the total charge on a fragment of double-stranded DNA could be thought to equal $2eN_{\text{base}}$, where e is the charge of an electron and N_{base} is the number of base pairs. However, Manning [17] pointed out that this approximation neglects the effect of counterions on the charge. Counterions, such as Na^+, condense the DNA so that the effective charge spacing (l) equals the Bjerrum length:

$$l = \frac{e^2}{4\pi\varepsilon_0\varepsilon_b k_B T}$$

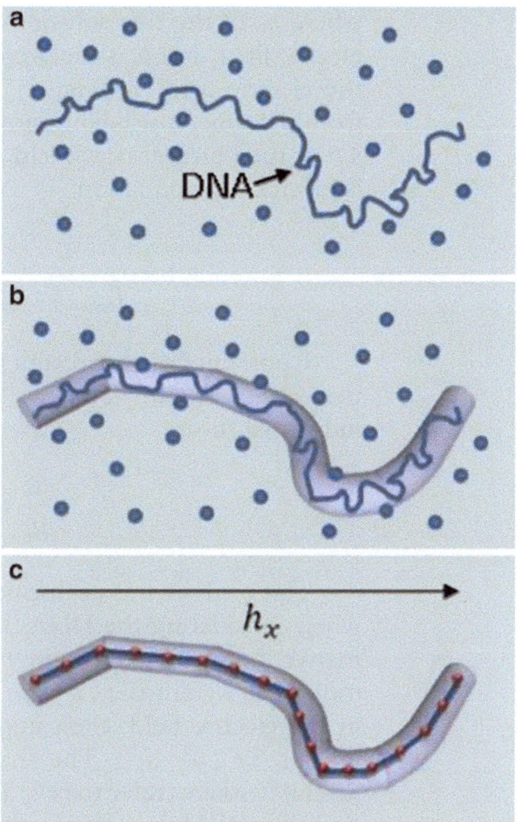

Fig. 3 Schematic representation of the reptation model for the migration of DNA through a gel matrix during agarose gel electrophoresis. *Panel A*: DNA molecule is confined by gel fibers (denoted by *solid circles*) with the matrix, *Panel B*: the DNA can be considered to migrate through the matrix by being confined within a theoretical tube, and *Panel C*: the DNA can be simplified in the form of a set of rigid links moving along the tube axis. The length h_x represents the projection of the tube in the direction of the electric field (indicated by the *arrow*)

where ε_0 is the permittivity of free space, ε_b is the bulk permittivity, and $k_B T$ is the Boltzmann factor. The net effect of this phenomenon is to considerably reduce the net charge on the DNA from the anticipated value if one assumes all the phosphate groups dissociate. Moreover, the DNA polymer can be considered to comprise a series of N_k Kuhn segments of Kuhn length l_k. Thus, the contour length (L) of the polymer is given by $L = N_k l_k$. By taking into account the electric force acting on the DNA and the frictional force opposing its movement through the gel, the resulting curvilinear velocity of the DNA in the tube (υ_{tube}) is given by the equation:

$$\upsilon_{tube} = \frac{\mu_0 E h_x}{L}$$

where μ_0 of the free solution mobility, E is the magnitude of the electric field, and h_x is the projection of the tube in the direction of the electric field (*see* Fig. 3). Once the DNA molecule migrates from the tube, it will have moved by a distance of h_x and generated a new tube further downfield. Hence, the net velocity of the DNA is given by the equation:

$$\upsilon = \frac{h_x}{L_{tube} / \upsilon_{tube}} = \mu_0 E\left(\frac{h_x^2}{LL_{tube}}\right)$$

Upon migrating through many tubes, the effective mobility of the chain is the average over the mobility through each individual tube;

$$\frac{\mu}{\mu_0} = \left(\frac{h_x^2}{LL_{tube}}\right)$$

This analysis was performed by Zimm [18] and is appropriate when considering the DNA that is loosely confined within the gel matrix. In this situation, the mobility of the DNA is a function of molecular weight (i.e., $\mu \sim N_k^{-1}$). However, if the DNA is oriented in the electric field, then migration is independent of molecular weight (i.e., $\mu \sim N_k^0$). The analysis has been further developed by several workers (refer to refs. 19, 20 for review articles). In conclusion the BRM has been extremely successful in describing the behavior of DNA during electrophoresis through a gel matrix. Indeed, the concept of reptation is now firmly embedded within a theoretical analysis of DNA electrophoresis.

References

1. Tiselius A (1937) Electrophoresis of serum globulin. Biochem J 31:313–317
2. Grabar P, Williams CA (1953) Method permitting the combined study of the electrophoretic and the immunochemical properties of protein mixtures; application to blood serum. Biochim Biophys Acta 10:193–194
3. Thorne HV (1966) Electrophoretic separation of polyoma virus DNA from host cell DNA. Virology 29:234–239
4. Aaij C, Borst P (1972) The gel electrophoresis of DNA. Biochim Biophys Acta 269:192–200
5. Schwartz DC, Cantor CR (1984) Separation of yeast chromosome-sized DNAs by pulsed field gradient gel electrophoresis. Cell 37:67–75
6. Brody JR, Kern SE (2004) History and principles of conductive media for standard DNA electrophoresis. Anal Biochem 333:1–13
7. Goering RV (2010) Pulsed field gel electrophoresis: a review of application and interpretation in the molecular epidemiology of infectious disease. Infect Genet Evol 10:866–875
8. Carle GF, Olson MV (1984) Separation of chromosomal DNA molecules from yeast by orthogonal-field-alternation gel electrophoresis. Nucleic Acids Res 12:5647–5664
9. Gardiner K, Laas W, Patterson D (1986) Fractionation of large mammalian DNA restriction fragments using vertical pulsed-field gradient gel electrophoresis. Somat Cell Mol Genet 12:185–195
10. Carle GF, Frank M, Olson MV (1986) Electrophoretic separations of large DNA molecules by periodic inversion of the electric field. Science 232:65–68

11. Chu G, Vollrath D, Davis RW (1986) Separation of large DNA molecules by contour-clamped homogeneous electric fields. Science 234:1582–1585
12. DeGennes PG (1971) Reptation of a polymer chain in the presence of fixed obstacles. J Chem Phys 55:572–579
13. Lumpkin OJ, Zimm BH (1982) Mobility of DNA in electrophoresis. Biopolymers 21: 2315–2316
14. Lerman LS, Frisch HL (1982) Why does the electrophoretic mobility of DNA in gels vary with the length of the molecule? Biopolymers 21:995–997
15. Slater GW, Noolandi J (1985) New biased-reptation model for charged polymers. Phys Rev Lett 55:1579–1582
16. Edmondson SP, Gray DM (1984) Analysis of the electrophoretic properties of double-stranded DNA and RNA in agarose gels at a finite voltage gradient. Biopolymers 23:2725–2742
17. Manning GS (1969) Limiting laws and counterion condensation in polyelectrolyte solutions I. Colligative properties. J Chem Phys 51:924–933
18. Zimm BH (1991) "Lakes-straits" model of field-inversion gel electrophoresis of DNA. J Chem Phys 94:2187–2206
19. Viovy J-L (2000) Electrophoresis of DNA and other polyelectrolytes: physical mechanisms. Rev Mod Phys 72:813–872
20. Slater GW (2009) DNA gel electrophoresis: the reptation model(s). Electrophoresis 30: S181–S187

Chapter 2

Basic DNA Electrophoresis in Molecular Cloning: A Comprehensive Guide for Beginners

Svetlana Makovets

Abstract

Presented here is a complete molecular cloning protocol consisting of a number of separate but interconnected methods such as preparation of *E. coli* competent cells; in vitro DNA digestion and ligation; PCR; DNA agarose gel electrophoresis and gel extraction; and screening transformants by colony PCR, analytical restriction digests, and sequencing. The method is described in a lot of details so that it can be easily followed by those with very little relevant knowledge and skills. It also contains many tips that even experienced researchers may find useful.

Key words Cloning, Agarose gel electrophoresis, DNA gel extraction, Competent cells, Clone screening, Restriction enzymes, Ligation

1 Introduction

A protocol for molecular cloning involves several basic molecular biology techniques such as in vitro enzymatic reactions, DNA agarose gel electrophoresis, bacterial transformations, and PCR which form a core set of skills for a qualified researcher. Gaining experience in any or all of these techniques is very common among early career scientists. The introduction below is a short summary of knowledge essential for using the method. Additional reading on the subject can be found in ref. 1.

1.1 Cloning: Vectors, Inserts, and Enzymes

Cloning is the construction of a DNA molecule of a novel nucleotide sequence generated by bringing together two or more existing DNA fragments. Like many construction projects, it involves cutting things at designed sites and gluing them together to assemble novel structures. Restriction endonucleases, the enzymes that cut DNA at specific nucleotide sequences, play the role of molecular scissors in cloning. Another enzyme, the T4 DNA ligase (originally isolated from the bacterophage T4) is used to "glue" DNA fragments together.

Svetlana Makovets (ed.), *DNA Electrophoresis: Methods and Protocols*, Methods in Molecular Biology, vol. 1054,
DOI 10.1007/978-1-62703-565-1_2, © Springer Science+Business Media New York 2013

In any cloning experiment there are two major DNA components—a vector and an insert—that are to be joined into a single DNA molecule (Fig. 1). The vector plays the role of a transport vehicle that is used to move its passengers—the cloned genes—between cells and cell-free environments in which they are manipulated by researchers. In most cases, the vector is a plasmid that can be selected for and maintained in the bacterium *Escherichia coli*. Vectors always contain a selective marker, a gene for resistance to an antibiotic such as ampicillin, kanamycin, or chloramphenicol, and a bacterial origin of replication required for plasmid propagation. The frequency of replication initiation varies in different vectors depending on the origin sequence (some fire more often than others) and results in a different plasmid copy number per bacterial cell, from 1 to ~500. High-copy-number vectors are good for plasmid DNA amplification but can be toxic to bacteria if a gene cloned in the vector is highly over-expressed as a result of plasmid amplification. To overcome this problem, genes are either cloned into low-medium-copy-number vectors or placed under the control of inducible promoters so as to keep bacteria viable until the gene expression is required.

If the expression of a cloned gene is designed to occur in an organism other than *E. coli*, so-called shuttle vectors are used. These can be maintained in *E. coli* during cloning or plasmid amplification but have additional features allowing them to be introduced into a different type of cell once the plasmid is constructed and isolated from *E. coli*. For example, budding yeast shuttle vectors contain yeast genetic markers, such as *URA3*, *LEU2*, *TRP1*, *HIS3*, and *KAN*, used to select for the plasmids upon transformation into yeast. In addition, the vectors need to be replicated and segregated inside a yeast cell. This can be achieved in several ways. Integration vectors are designed to insert the plasmid DNA, via homologous recombination, into a locus of the yeast genome that shares homology with the plasmid. The plasmid DNA is then replicated and segregated as a part of the chromosome in which it has been integrated. In contrast, CEN/ARS vectors contain a yeast centromere DNA sequence (CEN) and an origin of replication (ARS for autonomous replication sequence), which allow them to be maintained in vivo as a circular molecule of a single copy per cell. Another series of yeast vectors is based on the yeast endogenous 2 μ plasmid that replicates autonomously and has variable copy number (10–40 per cell) as it segregates through diffusion between daughter cells. It is mainly used for gene over-expression in yeast.

The source of a gene or DNA fragment you want to clone is, normally, either genomic DNA or cDNA. The DNA sequence of interest represents just a tiny fraction of the genomic DNA and therefore would be hard to clone using genomic DNA directly. PCR is used to enrich for the desired sequence to simplify cloning

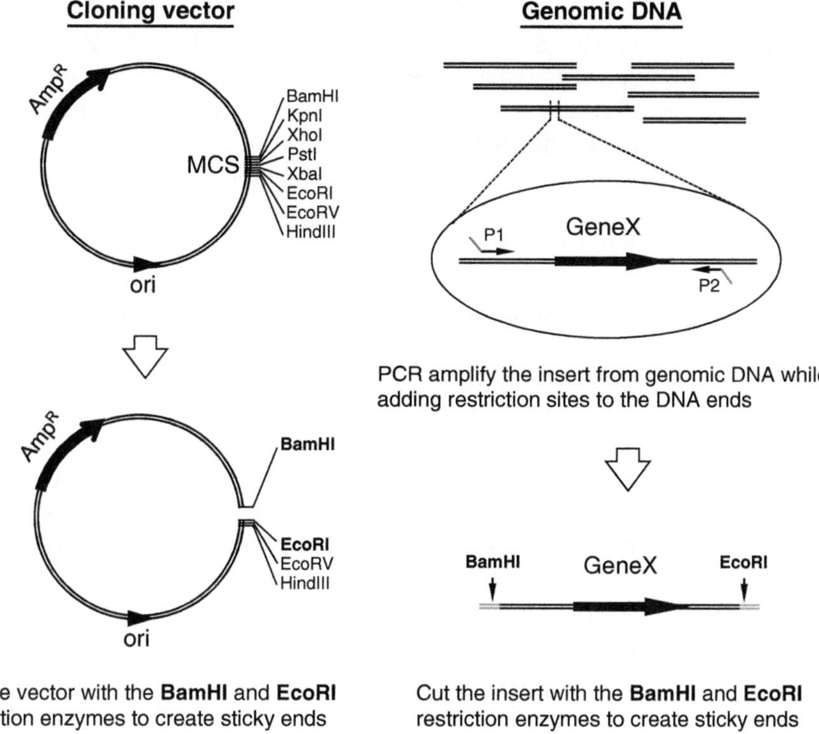

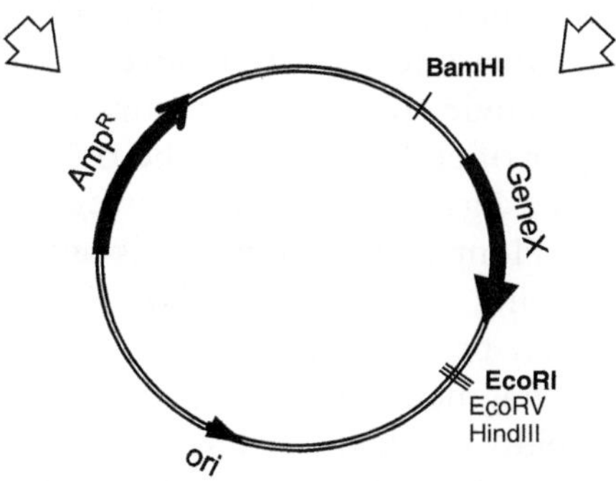

Ligate the linearized vector and insert together. GeneX has been cloned

Fig. 1 A schematic diagram of a cloning experiment in which the insert (GeneX) is PCR-amplified using genomic DNA. The PCR primers P1 and P2 are designed so that their 5′ ends contain sequences which, when double-stranded, are recognized by BamHI and EcoRI, respectively. When both vector and insert are cleaved using these two restriction enzymes, their ends can be joined together by DNA ligase, to produce a recombinant plasmid with GeneX cloned in to the vector. MSC, multiple cloning sequence (sites for eight different restriction enzymes are shown); *AmpR* β-lactamase gene coding for resistance to ampicillin, *ori* origin of replication

(Fig. 1). PCR products have either blunt ends or overhangs of a single adenosine at the 3′ end. Though such ends can be ligated, the ligation efficiency is higher if longer overhangs are present on both ends or, at least, on one end of a PCR fragment. This is easily achieved by placing recognition sequences for restriction enzymes at the 5′ ends of the primers used to PCR amplify the DNA sequence of interest. When the PCR product is generated, it can be digested with the appropriate restriction enzymes to generate the overhangs for the ligation step.

Once you have decided which gene you are cloning and into which vector, you can begin designing your cloning experiment. Each vector contains a multiple cloning sequence (MCS) sometimes called polylinker. This is a short ~100 bp region containing a cluster of unique restriction sites, the position of which is indicated on a plasmid restriction map. One or more often two of these sites are used to "open up" the vector, i.e., to convert the circular molecule into a linear DNA fragment, so that another DNA fragment, the one you want to clone, can be inserted into the created gap. The two fragments are "glued" together by ligase that essentially reverses the cleavage reactions performed by restriction enzymes. For joining of the end of the vector and the fragment to be cloned to happen, the DNA ends have to be compatible, i.e., the ssDNA overhangs have to be 100 % complementary to each other. The easiest way to achieve this is to use the same pair of restriction enzymes to cut both vector and insert DNA. When vector and insert cannot be cut with the same enzyme, the incompatible sticky ends generated by different enzymes can be converted into compatible, though harder to ligate, blunt ends. This is achieved by using either nucleases to remove overhangs or polymerases to convert 5′ single-stranded overhangs into blunt-end dsDNA. Having one pair of ends sticky and the other pair blunt works reasonably well in a ligation reaction. However, relying on ligation of completely blunt-ended fragments is not advisable.

During ligation, any DNA fragments with compatible DNA ends can be randomly joined by ligase, whether those are two vector fragments, two inserts, or one of each. To maximize the probability of the desired ligation, i.e., a vector fragment joined to an insert, the ends of a linearized vector are dephosphorylated by a phosphatase so that vector ends cannot be ligated back to each other or to the ends of another vector molecule. Ligation of multiple inserts to each other without a vector molecule involved is less of a problem as they do not result in transformants due to the lack of the vector-encoded drug marker in the recombinant molecule and the inability to be maintained in bacteria.

Ideally, only linearized, dephosphorylated vector molecules and insert fragments digested on both ends should be present in the ligation reaction. Using agarose gel electrophoresis to separate linearized vector from residual uncut supercoiled molecules is particularly important as even very small amounts of the latter would result in an extremely high level of unwanted transformants with the "empty" vector rather than a recombinant plasmid.

Having the right DNA fragments with intact compatible ends going into ligation is the most critical step in any cloning experiment.

Transformants containing putative recombinant plasmids are then screened for having the desired plasmid. The first screen employs colony PCR as it allows screening a larger number of colonies reasonably fast. Then, PCR-positive candidates are further screened by restriction digest of the plasmid DNA they contain. Finally, the clones that are positive after the first two screens are sequenced to confirm the sequence of the cloned fragment and to make sure no mutations were introduced during the cloning procedure.

1.2 DNA Agarose Gel Electrophoresis

DNA electrophoresis in agarose gel is one of the essential molecular biology techniques routinely used in many biological, medical, forensic, and other research laboratories dealing with the analysis of DNA samples from a variety of sources: PCR products, genomic DNA or DNA from mitochondria or chloroplasts, plasmid DNA, to name the most common ones.

DNA molecules are negatively charged due to phosphoric acid in the DNA sugar-phosphate backbone. Because of this charge, DNA moves in an electric field as an anion, from a cathode to an anode (marked black and red, respectively, on most gel electrophoresis tanks). The DNA electrophoresis technique separates DNA molecules of different sizes and shapes in agarose gels that are subjected to an electric field. This is possible due to the molecules' mobility in a gel being dependent on their size and shape. The agarose gel matrix works like a net, trapping DNA molecules as they move through the gel. Larger and/or extended molecules are affected by the trapping more than smaller and/or compact ones. The higher the agarose concentration is, the smaller the holes in the net are, and the slower the DNA molecules move.

Agarose gels normally run at a constant voltage that is stated in volts per centimeter (V/cm) of distance between the electrodes in a gel tank, to unify the running conditions for gel tanks of different designs. However, one has to remember that factors other than voltage and agarose gel concentration have an effect on gel-run progress. The temperature of the gel running buffer is one such factor. The buffer temperature may increase as the gel runs, particularly if a small tank is used and/or the voltage is high; this leads to faster DNA migration. Often the buffer warming up is uneven through the gel tank leading to gel "frowning" if the temperature is higher on the sides of the gel than in the middle or to gel "smiling" if the temperature is higher in the middle than on the sides of the gel. Using a pump to recirculate/mix the buffer or running your gel at a lower voltage can help if even mobility of all the samples is critically important for an experiment. Otherwise, loading DNA size marker into both side wells and the one in the middle helps estimate the size of the analyzed DNA fragments more accurately, even if the migration is somewhat uneven. Another factor that influences gel runs is the amount of buffer in the gel tank. One can have either just enough buffer to cover the gel on the top or quite a bit more. In the first case the gel would run considerably faster as additional buffer increases resistance in a system with set voltage and, as a result, the current and the rate of DNA migration decrease.

The DNA itself is not visible with the naked eye but when bound by ethidium bromide, it can be easily detected by visualizing the ethidium in the UV spectrum. Adding ethidium to agarose when casting a gel is sufficient for visualizing DNA; having ethidium in the running buffer is normally not required. However, ethidium will diffuse out of the gel left in running buffer for a few hours or longer. DNA also diffuses in agarose, both during gel runs and afterwards. Therefore, the best gel images are taken right after the run. For the same reason, DNA bands for gel extraction should be excised as soon as the run is complete. The gel slices can then be stored at −20 °C.

In the cloning protocol described in this chapter, agarose gel electrophoresis is used at several steps, each time for a different purpose: to purify an insert PCR product, to separate linearized vector molecules from the uncut circular ones, to assay PCR products when screening transformants by colony PCR, and to analyze the composition of recombinant plasmids which have been digested with restriction enzymes. All these are simple agarose gel electrophoresis experiments ideal for training early career molecular biologists.

2 Materials

All reagents should be made using deionized water (dH_2O). Using sterile double-deionized water (ddH_2O) is recommended for setting up PCR, restriction digests, and ligation reactions.

2.1 Cloning Design

1. New England Biolabs catalogue.
2. Software for DNA sequence analysis and manipulation, for example, Serial Cloner.

2.2 Preparing *E. coli* Competent Cells

1. L-broth: Dissolve 10 g Bacto Tryptone, 5 g Bacto yeast extract, and 10 g NaCl in 900 mL dH_2O, adjust pH to 7.0 using 1 M NaOH. Bring the volume to 1 L, aliquot in screw cap glass bottles, autoclave at 121 °C for 15 min, and store at room temperature, away from light.
2. LB-agar plates with antibiotics. Dissolve 10 g Bacto Tryptone, 5 g Bacto yeast extract, and 10 g NaCl in 900 mL dH_2O,

adjust pH to 7.0 using 1 M NaOH. Pour the broth into a 1 L graduated glass bottle with a stir bar inside. Add 20 g Bacto agar powder and bring the volume to 1 L. Autoclave at 121 °C for 20 min. Take the bottle out of the autoclave and place it on a stirring plate. Stir until the agar media cools down for you to hold the bottle in your bare hand (*see* **Note 1**). Add desired antibiotic, stir briefly, and pour agar into Petri dishes, ~25 mL per plate. Once the agar has set, turn the plates upside down, stack them in plastic bags (to prevent drying), and store in this position at 4 °C away from light for up to several months. On the day of the experiment, dry the number of plates to be used (*see* **Note 2**).

3. Antibiotic for selecting the plasmid vector used in the cloning experiment. Most commonly used antibiotics come in a powder form, which is then used to prepare stock solutions.
   - Ampicillin: prepare stock at 100 mg/mL in dH_2O, make 1 mL aliquots in microcentrifuge tubes, and store them at −20 °C. Use in media at a final concentration of 100 μg/mL. When transforming low-copy-number plasmids, decrease the concentration to 20–50 μg/mL for the initial plasmid selection.
   - Chloramphenicol: stock solution 10 mg/mL in 50 % ethanol, store at 4 °C; final concentration 20 μg/mL.
   - Kanamycin or streptomycin: stock solution 10 mg/mL in dH_2O, store at 4 °C; final concentration 20–50 μg/mL.
   - Rifampicin: stock solution 10 mg/mL in methanol, store at 4 °C; final concentration 50 μg/mL.
   - Tetracycline: stock solution 10 mg/mL in 50 % ethanol, wrap in foil, and store at 4 °C in the dark; final concentration 10 μg/mL. The solution should be light yellow. When the color turns dark yellow, a new stock solution should be made.
4. 1 M $CaCl_2$: Dissolve $CaCl_2$ in dH_2O, adjust the volume, filter sterilize or autoclave, and keep at room temperature.
5. 50 % (w/w) glycerol stock solution: To prepare a ~200 mL solution, pour 100 mL dH_2O into a 250 mL glass beaker. Place the beaker on a scale and press the TARE button. Slowly pour 100 g of glycerol while the beaker is on the scale. Put a stir bar into the beaker and stir on a stirring plate until the solution looks well mixed (no glycerol swirls in water should be visible). Filter-sterilize and keep in the dark at room temperature.
6. Sterile dH_2O.
7. 0.1 M $CaCl_2$, 20 % glycerol: In a 50 mL sterile conical tube, mix 5 mL $CaCl_2$, 20 mL of 50 % glycerol, and 25 mL sterile dH_2O.

8. Sterile flasks of various volumes: 50 or 100 mL and 2 L.
9. Sterile 50 mL conical tubes.
10. Shaking platform with flask holders, set at 37 °C.
11. Roller drum for microcentrifuge tubes or rocker set at 37 °C (*see* **Note 3**).
12. Spectrophotometer (visible light spectra) and cuvettes.
13. Refrigerated tabletop centrifuge.
14. Water bath at 42 °C (*see* **Note 4**).
15. Ice bucket large enough to accommodate a 2 L flask.
16. Sterile 1.5–1.7 mL microcentrifuge tubes (*see* **Note 5**).
17. Saran Wrap or any other cling film.
18. Dry ice.
19. *E. coli* DH5α or *E. coli* XL1-Blue (*see* **Note 6**).
20. Plasmid DNA sample of known DNA concentration.

2.3 Generation of Insert DNA Fragment by PCR Amplification

1. DNA template for PCR amplification (genomic DNA or cDNA).
2. Primers designed in Subheading 3.1.
3. High-fidelity DNA polymerase with a reaction buffer, such as Pfu DNA polymerase (Promega, M7741) or Herculase II Fusion Enzyme (Stratagene, 600677).
4. Stock of dNTPs, 10 mM each.
5. QIAquick Gel Extraction kit (Qiagen) or equivalent.
6. ddH_2O.

2.4 Agarose Gel Electrophoresis

1. Gel tank with a casting tray and a set of combs with different well sizes. For simple applications such as analyzing diagnostic PCRs, diagnostic restriction digests, and cloning experiments, I would recommend using a mini-gel electrophoresis apparatus which is basically a 10 cm × 10 cm gel box with two electrodes on opposite sides (Fig. 2). There are two gaskets that when inserted into the box generate a 10 cm × 8 cm compartment in the middle, used to cast an agarose gel. Therefore, there is no casting tray as such; the gel box is used both as a casting tray and a gel tank. Either one or two combs can be used to cast a single gel and up to 40 samples can be run at a time. Once the gel is polymerized, the gaskets are removed and the box is filled with 1× TBE and is ready to be used. Because the box is made of UV-transparent material, the gel can be photographed while in the box filled with buffer. This setting is very economical with respect to the amount of running buffer (50 mL) and gel volume (40–50 mL) required as well as very convenient for gel handling.

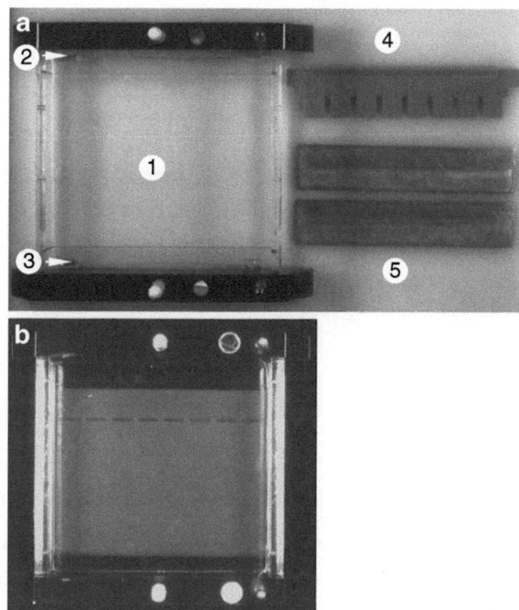

Fig. 2 Agarose gel electrophoresis equipment for a 10 × 8 cm mini-gel. (**a**) On the *left* is a 10 cm × 10 cm gel tank (1) with the electrodes, a cathode (2), and an anode (3), indicated by *white arrows*. On the *right* is a set of matching accessories for gel casting: an 8-well comb (4) and a pair of gaskets (5). (**b**) The same gel tank as in **a** but with an agarose gel inside

2. 10× TBE: 890 mM Tris base, 890 mM boric acid, and 20 mM EDTA. For 1 L, dissolve 108 g Tris and 55 g boric acid in 900 mL water, add 40 mL of 0.5 M EDTA, pH 8.0, and adjust the final volume to 1 L. Do not pH (*see* **Note 7**).
3. Agarose.
4. Ethidium bromide 10 mg/mL (*see* **Note 8**).
5. 6× sample buffer (gel-loading dye): 15 % Ficoll 400, 0.25 % bromophenol blue. Dissolve 1.5 g of Ficoll 400 in a total volume of 10 mL of dH_2O. Pinch in some bromophenol blue (*see* **Note 9**). Filter sterilize, aliquot into microcentrifuge tubes, and store at 4 °C (*see* **Note 10**).
6. DNA size markers: 1 kb and 100 bp DNA ladders (New England Biolabs).
7. Ready-to-load DNA size markers: mix 40 μL a DNA size marker, 40 μL 10× NEB 3 buffer (NEB), 320 μL dH_2O, and 80 μL 6× sample buffer (*see* **Note 11**). Mix well and store at −20 °C. When needed, defrost, mix, and load 3–5 μL per well.

2.5 Digestion and Ligation of DNA Fragments

1. Vector plasmid DNA of high purity.
2. Insert DNA generated by PCR and gel-purified (*see* Subheading 3.3).

3. Restriction enzymes with 10× reaction buffers and 100× BSA, provided by the enzyme suppliers (New England Biolabs, Roche, Promega, Fermatas, etc.).
4. Calf intestinal phosphatase (New England Biolabs) (*see* **Note 12**).
5. T4 DNA ligase with 10× reaction buffer (New England Biolabs) (*see* **Note 13**).
6. Agarose gel electrophoresis equipment and reagents (*see* Subheading 2.3).
7. Scalpels (preferred) or razor blades.
8. Long-wave (366 nm) UV box.
9. UV-protecting goggles or face mask.
10. Qiagen Gel Extraction kit or equivalent.
11. Qiagen MinElute PCR purification kit.
12. Water baths at 37 °C (possibly other temperature, depending on the choice of restriction enzymes) and 55 °C.

2.6 Recovery of Recombinant Plasmids in Bacteria: E. coli Transformation

1. *E. coli* competent cells (from Subheading 3.2).
2. Water bath at 42 °C (*see* **Note 4**).
3. L-broth and LB agar plates with an antibiotic to select for transformants (*see* Subheading 2.2 for media recipes).
4. Roller drum for microcentrifuge tubes or rocker set at 37 °C (*see* **Note 3**).

2.7 Colony Screening by PCR

1. 10 mM NaOH.
2. 0.2 mL thin wall PCR tubes, eight tube strips, and eight cap strips.
3. Thermocycler (PCR machine).
4. PCR reagents: Taq polymerase with 10× reaction buffer and 10 mM dNTPs.
5. Primers designed so that the PCR runs across one of the ligation sites (*see* Subheading 3.7 for more explanations).
6. L-broth with plasmid-selecting antibiotic (*see* Subheading 2.2 for recipes).
7. Agarose gel electrophoresis equipment and reagents (*see* Subheading 2.3).

2.8 Plasmid Diagnostics by Restriction Digests

1. Tabletop centrifuge.
2. Plasmid DNA mini-prep kit (Qiagen), Wizard Plasmid DNA Purification kit (Promega), or similar (*see* **Note 14**).
3. Restriction enzymes with 10× reaction buffers and 100× BSA, provided by the enzyme suppliers (New England Biolabs, Roche, Promega, Fermatas, etc.).
4. Agarose gel electrophoresis equipment and reagents (*see* Subheading 2.2).

2.9 Plasmid Verification by Sequencing

1. Primers for sequencing (*see* Subheading 3.9 for design).
2. Sequencing analysis software, for example FinchTV.

3 Methods

Your local Institution health and safety rules and regulations should be observed, particularly when working with potent carcinogens such as ethidium bromide and UV light. Wear gloves, lab coat, and UV protective goggles or a face mask.

3.1 Cloning Design

1. Download the complete sequence of your plasmid vector, and use it to create a file in one of the programs to manipulate and analyze DNA sequences, for example, Serial Cloner.
2. Using the software, generate and print out a restriction map of your vector in which only unique restriction sites are shown. Identify the multiple cloning site (MCS).
3. Download the sequence of the DNA fragment you want to clone, for example, the gene of interest, with its promoter and transcriptional terminator if desired. Create another DNA sequence file using the software for sequence analysis as above.
4. Scan the fragment sequence against the enzymes in the MCS and identify restriction sites that are present in MCS but absent in the fragment to be cloned.
5. Use a New England Biolabs catalogue to find out (a) which of the four reaction buffers are optimal for each of the enzymes selected in **step 4** and (b) how many base-pairs overhang they produce upon cutting. Identify a pair of enzymes that are active 75–100 % in the same buffer, preferably at the same temperature and generate 4 bp overhangs (*see* **Note 15**).
6. Decide if you want to clone the fragment in one orientation rather than the other. This will determine which restriction site will be placed upstream of the coding sequence and which one downstream.
7. Design a pair of primers to PCR amplify the fragment to be cloned, with the addition of four cytosines (*see* **Note 16**) and the corresponding restriction site at the 5′ end. For example, if in **step 5** you chose EagI and SalI to use for cloning your DNA fragment, then your oligos will look as follows:

   5′-ccccCGGCCGnnnnnnnnnnnnnnnnnn-3′,

   5′-ccccGTCGACnnnnnnnnnnnnnnnnnn-3′,

   where poly-n is the sequence specific for your fragment. Since the ten nucleotides at the 5′ ends are not homologous to the PCR template, exclude them when calculating annealing temperature for the PCR.

8. Order the oligos and the two restriction enzymes if you do not have them in your lab collection.
9. Using the software for sequence manipulation, create a file with a reconstituted sequence corresponding to the plasmid you are constructing. This will be useful in Subheadings 3.7–3.9 when performing clone screening.

3.2 *E. coli* Competent Cells

In most single-gene cloning experiments, competent cells with the efficiency of ~10^7 transformants per 1 μg of DNA allow to recover sufficient number of clones for further screening. The described protocol for preparing chemically competent *E .coli* cells is simple and very reproducible. In just a day, a batch of up to 100 aliquots can be easily generated and stored frozen for at least 1 year without loss of transformation efficiency.

The rules of sterility should be observed throughout the procedure to avoid culture contamination.

3.2.1 Preparing Competent Cells

1. Pick a single colony of a *recA E. coli* strain (*see* **Note 6**) and start a 10 mL L-broth culture in a 50 or 100 mL conical flask. Grow cells overnight on a shaker at 37 °C.
2. The next morning, dilute the overnight culture 1:100 in 200 mL of fresh L-broth in a 2 L flask. Grow the culture at 37 °C with vigorous aeration (*see* **Note 17**).
3. After 1 h of culture growth, start taking culture OD_{600} every 30 min. When the OD_{600} approaches 0.3–0.4, do the measurements more often.
4. While the culture is growing, precool a tabletop centrifuge to 4 °C. Also, prechill on ice four 50 mL conical tubes, 150 mL of sterile 0.1 M $CaCl_2$, and 15 mL of sterile 20 % glycerol in 0.1 M $CaCl_2$.
5. Place open microcentrifuge tubes (*see* **Note 5**) in a tube rack with 12 places × 8 rows using every other row. Normally, the tube tops from the rows filled later overlap with the tube bases from the previous row. When four rows are filled with 48 tubes, wrap the rack with the tubes with Saran Wrap and place it into a −80 °C freezer. Repeat this procedure with another rack, bringing the total number of tubes to 96.
6. When the bacterial culture OD_{600} reaches 0.4–0.5, chill the culture rapidly by placing the flask in an ice-water bath for 5 min. Swirl the flask every 10–20 s while it is in the bath to facilitate cooling.
7. Transfer the culture to the four prechilled 50 mL conical tubes and harvest cells in the prechilled tabletop centrifuge (from **step 4**) at 2,500 × *g* for 10 min.

8. Carefully pour off the supernatants and place the tubes on ice. Use a 1 mL pipetman to remove as much of the remaining media as possible (*see* **Note 18**).
9. Add 25 mL of cold 0.1 M $CaCl_2$ to each tube and resuspend the cells by vortexing. No cell clumps should be seen.
10. Incubate cells on ice for 1 h.
11. Combine the four 25 mL aliquots in two tubes and harvest the cells in the prechilled tabletop centrifuge at 2,500 × *g* for 10 min.
12. Carefully pour off the supernatant and place the tubes on ice. Use a 1 mL pipetman to remove as much of the remaining $CaCl_2$ as possible.
13. Add 5 mL of ice-cold 20 % glycerol in 0.1 M $CaCl_2$ to each tube and resuspend the cells by vortexing. No cell clumps should be visible.
14. Combine the two aliquots in one tube and place the cells on ice.
15. Pour some dry ice into an ice tray. Take one of the two racks with prechilled microcentrifuge tubes out of the –80 °C freezer, remove the Saran Wrap, and place the rack on dry ice.
16. Quickly pipette 100 μL of cells into each tube and close the tubes (*see* **Note 19**). Transfer the tubes to a –80 °C freezer.
17. Repeat the aliquoting with the rest of the tubes.
18. Store the competent cell aliquots at –80 °C until needed.

3.2.2 Testing Competent Cells

It is important to test each newly made batch of competent cells before using them in any sort of experiments. One should test the cells for competency, i.e., score the efficiency of transformation, as well as test for the presence of any contamination in the culture.

1. Take two aliquots out of the –80 °C freezer and place them on ice for ~5 min.
2. Gently flick the tubes to make sure that the aliquots are defrosted.
3. Label one of the aliquots with the name of the plasmid to be transformed and use the other as a negative, "no DNA" control.
4. Add 10 ng of plasmid DNA to the first aliquot. Flick the tube to mix. Incubate on ice for 30 min.
5. Heat-shock both aliquots of cells by placing the tubes into a 42 °C water bath.
6. Transfer the tubes back on ice for 1 min.
7. Add 0.9 mL L-broth to each tube.

8. Incubate the tubes at 37 °C with aeration (on a roller drum or a rocker) for 1 h or without aeration for 1.5 h.
9. Make tenfold culture dilutions in LB-broth, and plate 0.1 mL from each dilution onto plates with LB-agar with an antibiotic to select for the transformants. Incubate the plates at 37 °C overnight.
10. First, check the plates with the dilutions for the "no DNA" control aliquot. There should be no colonies as no plasmid was added to the cells. If there are any colonies, then either the culture was contaminated or your antibiotic is not working for some reason.
11. If the control plates from the previous step have no colonies but the ones for the plasmid transformation do, then your colonies are the expected transformants. Count colonies on a plate that has between 50 and 500 colonies and calculate the number of transformants per 1 μg of DNA (efficiency of transformation, *E*) using the following formula:

$$E = N \times A \times 1{,}000,$$

where *N* is the number of colonies on a plate and *A* is the dilution factor for that plate (*see* **Note 20**).

3.3 Generation of Insert DNA Fragment by PCR Amplification

Cloning a PCR amplified fragment has the advantage of having plenty of DNA to work with, as a result of PCR amplification. However, DNA synthesis during PCR is error-prone, and one has to address this issue by using high-fidelity polymerases and running as few amplification cycles as possible to generate sufficient DNA. For obvious reasons, amplifying longer fragments, mutation-free, presents a more difficult challenge than working with shorter ones.

1. Make threefold serial dilutions of your DNA template (normally genomic DNA or cDNA) and use them for three or four PCR reactions with various amounts of the template.
2. Set up PCR reactions, each in a total volume of 50 μL. Use a high-fidelity DNA polymerase and add the other components into the reactions according to the manufacturer's recommendations (*see* **Note 21**).
3. While the PCR is running, prepare a small agarose gel as described in Subheading 3.4. Choose a comb to generate wells holding ~60 μL (*see* **Note 22**).
4. When the PCR is finished, add 10 μL of sample buffer to each reaction and mix well (*see* **Note 23**). Load each reaction into a separate well. Load a DNA size marker on one or both sides of the gel, next to the PCR samples.

5. Run the gel at ~10 V/cm until the bromophenol blue migrates 6–8 cm from the wells (*see* **Note 24**).
6. Visualize the gel using a gel documentation system (*see* **Note 25**).
7. Identify the required PCR product and compare its amount in the samples with the different amount of the template used. Identify the sample with the highest amount of the template which generated a highly visible amount of product.
8. Place the gel in the casting tray on top of a long-wave (normally 366 nm) UV box. Use a clean scalpel or a razor blade to cut out a gel slice with the PCR product you chose in the previous step. Work fast to minimize the DNA exposure to UV.
9. Extract the DNA from the gel slice using QIAquick Gel Extraction kit or equivalent according to manufacturer's recommendations (*see* **Note 26**).
10. Keep the purified DNA fragment in the freezer until needed.

3.4 Agarose Gel DNA Electrophoresis: Casting, Loading, and Running a Gel

When casting a gel, one has to think of what is to be achieved by electrophoresing a given set of DNA samples through agarose. For example, if the purpose is to test samples from a PCR run for the presence/absence of a DNA product of the expected size, then the DNA migration distance does not have to be long, 2–3 cm would be sufficient in most cases. In contrast, when DNA fragments close in size are analyzed, long, up to 20 cm, migration may be required. The migration distance is one of the factors that should be taken into consideration when choosing the gel apparatus for your experiment. The other factor is the number of samples to be run on a gel. The gel comb should contain a sufficient number of teeth to provide wells for all the samples as well as a DNA size marker, preferably loaded on both sides of the sample set.

The agarose concentration in a gel depends on the size of DNA fragments electrophoresed. 0.4–0.5 % gels are used to separate longer, 4–12 kb fragments as well as for separating supercoiled and linear DNA of the same or very similar size in the range of 4–12 kb (*see* more on this in Subheading 3.5). For resolving short, 100–500 bp DNA fragments, 2 % agarose gels work very well. Even shorter dsDNA molecules (40–100 bp) can be resolved when run in a 3–4 % gel made with low melting point agarose.

When working in the presence of ultraviolet light, protect your skin from exposure by wearing gloves, long-sleeve clothing, and a UV protective face mask or goggles. When handling bottles with hot agarose, use heat protective gloves.

1. Assemble a gel-casting tray on a bench. Use a levelling table if casting a larger gel (15 cm × 20 cm or larger).
2. Choose an agarose concentration appropriate for the expected size of the DNA fragments to be analyzed. To prepare 100 mL of agarose gel, weigh the appropriate amount of agarose (e.g., 1 g

for 1 % gel) and place it into a 200–250 mL heat-resistant screw-top glass bottle (or a flask with a sponge inserted at the top) with 100 mL of 1× TBE and a small stirring bar.

3. Stir on a stirring plate on a lower setting for a few seconds to disperse the agarose.
4. Using a marker pen, mark the top of the liquid level on glass, screw the bottle top slightly loose, and place the bottle into a microwave. Heat on "high" with occasional swirling until the solution starts boiling. Then, lower the power and simmer until the agarose is completely dissolved (*see* **Note 27**).
5. Carefully take the bottle out of the microwave and place it inside a container with cold water. Place the container on a stirring plate and stir until the bottle is cool enough that you can hold it in your hand (*see* **Note 28**).
6. Remove the bottle from the container and place it directly on a stirring plate. Open the bottle and add ethidium bromide to a final concentration of 5 μg/mL.
7. Stop the stirrer and check if the volume of the TBE has dropped below the mark on the bottle and, if needed, add some water to bring it back to the original volume and stir briefly.
8. Pour the gel solution into the casting tray and use a plastic tip to move any bubbles to the sides of the casting tray. Let the gel solidify (*see* **Note 29**).
9. When the gel is set, place it into the gel tank and pour just enough 1× TBE running buffer to cover the gel with a 2–3 mm layer of liquid (*see* **Note 30**).
10. Carefully remove the comb. Place a strip of dark paper or plastic on the bench under the gel tank; the wells are much easier to see against a dark background.
11. For sample loading, choose a pipette with the lower volume range (e.g., to load 20 μL sample use P2–20 rather than P20–200) as its gentler spring allows more controlled, gentle loading. Pipette a sample into the pipette tip and while holding the pipette in one hand, rest the index finger of the other hand on the middle part of the pipette to keep it steady. Lower the tip into a well, about halfway through and slowly load the sample. If the well is going to be fully loaded, slowly pull the pipette out of the well while loading so that the tip is always above the loaded sample.
12. Repeat the loading with the rest of the samples and size marker(s).
13. Attach the leads to the gel tank and the power supply, set the voltage, and start the run.
14. Once the run has begun, look at the electrodes in the gel tank and make sure more bubbles comes from the electrode closest

to the wells (cathode). If this is not the case, stop the run and check if the leads are connected to the gel tank and the power supply in the right order.

3.5 Digestion and Ligation of DNA Fragments

Set up restriction digests of the vector plasmid and the PCR product to be cloned into the vector.

1. For the vector digest, mix 0.5–1 μg of DNA, 3 μL of 10× restriction buffer, and 5–20 U of each restriction enzyme (*see* **Note 31**) and 5 U of calf intestinal phosphatase (*see* **Note 32**) in a total volume of 30 μL.
2. In parallel, set up three control reactions in 10 μL each (*see* **Note 33**). The control reactions should be in the same reaction buffer and have the same concentration of vector DNA and the same concentration of each restriction enzyme as in the 30 μL digest. The first two should each have one of the two restriction enzymes used in the 30 μL digest (two single-enzyme digests), and the third one should have no restriction enzymes at all (undigested DNA control).
3. For the DNA fragment digest, mix 0.2–0.5 μg of DNA (*see* **Note 34**), 3 μL of 10× restriction buffer, and 5–20 U of each restriction enzyme in a total volume of 30 μL (*see* **Note 31**).
4. Incubate all the digests at 37 °C (*see* **Note 35**).
5. During the digestion, prepare a small or medium (8 cm or slightly longer) 0.5 % agarose gel in 1× TBE with at least six 10–12 mm wells (*see* **Notes 36** and **37**) as described in Subheading 3.4.
6. After the DNA has been digested for 1 h, remove all of the four plasmid reactions (three controls plus the 30 μL double digest) and leave the DNA fragment digest for further incubation while you perform **steps 7–12**.
7. Add 2 μL of the 6× sample buffer to each of the control reactions and 6 μL of the 6× sample buffer to the double digest, mix well, and load on the 0.5 % gel in the following order:
   (a) 1 kb DNA ladder size marker.
   (b) Single-enzyme digest 1.
   (c) Single-enzyme digest 2.
   (d) Undigested DNA.
   (e) Double-enzyme digest.
   (f) 1 kb DNA ladder size marker.
8. Run the gel at 5–6 V/cm until the bromophenol blue migrates at least 6–7 cm away from the wells.
9. Take a picture of the gel using a gel documentation system.

10. From the gel picture assess the following:
    (a) If each of the enzymes cut the vector close to completion. This can be found out from comparing the lanes with single digests against the lane with uncut DNA. Circular DNA is predominantly in a supercoiled form that runs faster than linear DNA molecules of the same size, but a minor fraction of nicked circular molecules is often present and runs slower than the linear molecules (Fig. 3). If each of the single digests ran as a single band corresponding to the linear form, then the digests are close to completion.
    (b) If the double digest and supercoiled vector are well separated during the run. Estimate the run distance between the supercoiled DNA in lane (d) and the linear DNA in lane (e). If the difference is less than 3 mm, continue running the gel until the two forms of DNA are separated clearly (*see* **Note 38**).
11. Place the gel in the casting tray on a long-wave UV box and use a clean scalpel or a razor blade to cut out a gel slice with the linear double-digested vector DNA from lane (e). Cut right around the band and stay away from lane (d) as well as from the area in lane (e) in which uncut vector molecules are expected to run.
12. Purify the DNA using Qiagen Gel Extraction kit. At the elution step, elute the DNA with 30 μL of pre-warmed (55–60 °C) deionized water and leave at room temperature for at least 15 min before spinning down the column in the centrifuge.
13. Measure the volume of the eluate and split the sample into two equal aliquots, normally 14–15 μL each.
14. Recover the DNA fragment digest from 37 °C and use Qiagen minElute PCR purification kit to purify the DNA from the restriction enzymes and buffer components. Follow the manufacturer's protocol until the elution step. At this point, instead of using water for eluting the DNA fragment, use one of the linearized vector aliquots from the previous step. Pipette 14–15 μL of the vector DNA solution right in the center of the purple column and let it stand for about 15 min until recovering the DNA by centrifugation. The tube contains 14–15 μL mixture of the vector and insert, while the other tube from **step 13** contains vector DNA only, at the same concentration and in the same volume as the tube with the vector/insert mix.
15. Set up ligation reactions. Add 2 μL of 10× T4 DNA ligase buffer to each of the two tubes from the previous step. Flick the tubes to mix. Add 200–400 U of T4 DNA ligase to each tube. Mix well by flicking the tubes. Spin briefly in the microcentrifuge to collect the liquid at the bottom of the tube. Incubate on the bench for 1–16 h (*see* **Note 39**).

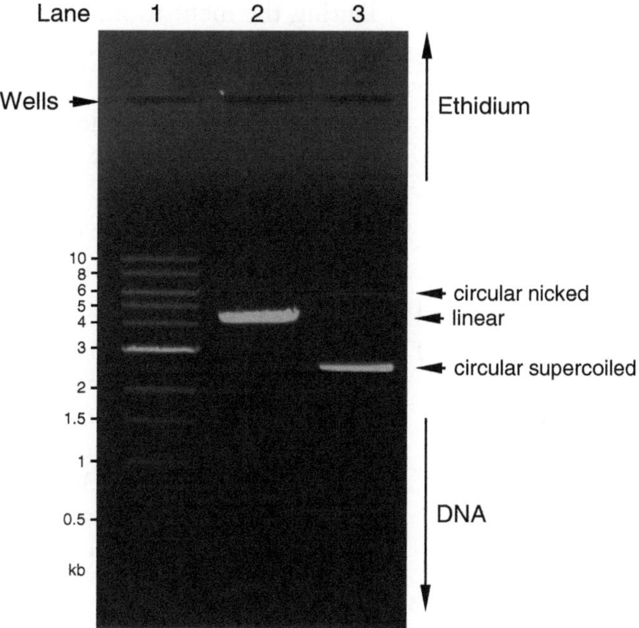

Fig. 3 Different isoforms of the plasmid vector pRS404 (4,274 bp) electrophoresed in 0.5 % agarose gel at 5 V/cm. *Lane 1* 1 kb DNA ladder (NEB). Sizes of individual bands are shown on the *left*. *Lane 2* pRS404 linearized with the EagI restriction enzyme. *Lane 3* undigested pRS404 contains two isoforms. The majority of plasmids in the sample are supercoiled circles but some nicked circles are present, as indicated by *horizontal arrows* on the *right*. Compare the *lanes 2* and *3* supercoiled circles move through the gel faster than linear DNA molecules, which in turn migrate faster than *nicked circles* of the same size. Notice that the band of linear DNA in *lane 2* looks brighter than the sum of bands in *lane 3* (the same amount of DNA was loaded in the *two lanes*) because ethidium binds relaxed DNA better than supercoiled DNA. On the *right*, *vertical arrows* indicate the directionalities of gel migration for DNA and ethidium. During electrophoresis, ethidium still remains in the *upper part* of the gel which looks brighter on the image

3.6 Recovery of Recombinant Plasmids in Bacteria: E. coli Transformation

1. Defrost on ice two 100 μL aliquots of *E. coli* competent cells.
2. Add the same volume of the "vector control" and "vector plus insert" ligation reactions, up to 10 μL, to the separate aliquots of competent cells.
3. Incubate on ice for 30 min or longer.
4. Place the tubes in a 42 °C water bath for exactly 2 min to heat-shock the bacteria and return the samples on ice for 1 min.
5. Add 900 μL of L-broth to each sample.
6. Place the tubes on a rotating weal at 37 °C for 1 h.

7. During the incubation, prepare (dry, pre-warm, or both) two plates with LB-agar and the antibiotic for plasmid selection (most commonly ampicillin at 100 μg/mL).
8. Spin the cells for 1–2 min at 2,000 × *g* in a microcentrifuge and remove ~900 μL of the supernatant by pouring it out of the open tubes.
9. Vortex to resuspend the cells in the remaining liquid and plate each sample onto a Petri dish prepared in **step** 7.
10. Incubate plates at 37 °C overnight.
11. Compare the number of colonies on the "vector control" plate and the "vector plus insert" plate. If the number of colonies on the two plates looks more or less similar or the "vector plus insert" plate has fewer colonies than the control plate, use colony PCR to screen 15–30 clones, as described in the next section. If the "vector plus insert" plate has at least three times more colonies than the control plate, then inoculate 6–10 random colonies from the "vector plus insertion" plate, each into 5 mL LB with the plasmid-selecting drug in a 15 mL conical tube and grow at 37 °C overnight with aeration. Then proceed to Subheading 3.8.

3.7 Colony Screening by PCR

It is reasonable to screen 15–30 colonies by colony PCR to identify candidate clones to be further analyzed by plasmid restriction digests as described in Subheading 3.8. In the colony PCR screen, one primer should hybridize to the vector sequence and the other one to the insert sequence so that a PCR product of 0.2–1 kb is generated when the expected plasmid is constructed. Use the plasmid sequence file you generated in Subheading 3.1 to design the PCR primers.

1. Use a Sharpie pen to number each colony to screen on the back of the plate.
2. Aliquot 10 mM NaOH to 0.2 mL PCR tubes in 8 tube strips, 3 μL per tube. The number of tubes should be equal to the number of colonies you are screening plus a tube for a negative control. Number the tubes.
3. Use a tip for 200 μL pipettes to pick a tiny bit of cells from each colony on the plate (*see* **Note 40**) and resuspend the bacteria in the NaOH in the PCR tube with the corresponding number.
4. Repeat the previous step for all the colonies to be screened. As a negative control, use a colony from the "vector only" plate or your bacterial strain without any plasmid transformed.
5. Close the PCR tubes with strips of caps and lyse the cells by heating the samples in the PCR machine for 10 min at 99 °C.

6. While lysing the cells, prepare a PCR master mix in a microcentrifuge tube. For *N* reactions (*see* **Note 41**), mix:

   $(N+1) \times 2.5$ μL of 10× Taq polymerase reaction buffer.

   $(N+1) \times 0.5$ μL of 10 mM dNTPs.

   $(N+1) \times 0.5$ μL of primer 1.

   $(N+1) \times 0.5$ μL of primer 2.

   $(N+1) \times 18$ μL of ddH_2O.

   Mix all the components well, using a 1 mL pipette, then add $(N+1) \times 0.25$ μL of Taq polymerase and mix again by pipetting (*see* **Note 42**).
7. Aliquot the master mix into fresh PCR tube caps, 22 μL per cap.
8. When the cell lysis (**step 4**) is finished, take the tubes out of the PCR machine, remove the caps, and replace them with the caps filled with the PCR master mix (**step 7**).
9. Spin the strips of tubes in a mini-centrifuge. Flick the samples to mix the lysed cells with the PCR mix and spin again.
10. Place the tubes in the PCR machine and run a colony PCR program:

1 cycle of	94 °C for 2 min
30–40 cycles of	94 °C for 30 s
	45–60 °C for 30 s (*see* **Note 43**)
	72 °C for 15–60 s (*see* **Note 44**)
1 cycle of	72 °C for 2 min

    Keep samples cold afterwards.
11. While the PCR is running, pour a 1.5–2 % agarose gel (as described in Subheading 3.4) to accommodate all the PCR samples along with a DNA size marker. We normally use a small (7 cm × 10 cm) gel with two rows of wells, 20 wells per row.
12. When the PCR run is finished, add 5 μL of 6× sample buffer to each sample (*see* **Note 23**).
13. Set your pipette at 15 μL, pipet up and down each sample and load 15 μL on a gel. When loading samples leave the first and the last lane in each row empty.
14. Load 5 μL of 100 bp ladder mix on either side of the set of lanes with the PCR samples.
15. Run the gel at 10 V/cm for 10–20 min, or until the bromophenol blue has migrated 2–3 cm into the gel.
16. Take an image of the gel and identify the lanes with the PCR product of the expected size.

17. Mark the corresponding *E. coli* colonies as PCR-positive on the plate from the cloning experiment.
18. Inoculate 4–8 positive colonies into L-broth with a drug selecting for the plasmid, each colony into 5 mL of media in a 15 mL conical tube.
19. Grow the cultures overnight at 37 °C with aeration.

3.8 Plasmid Diagnostics by Restriction Digests

Plasmid diagnostic by restriction digest allows testing if the transformants contain the desired plasmid by purifying and cleaving the analyzed plasmids with restriction enzymes and estimating the sizes of the generated fragments using agarose gel electrophoresis. The most straightforward analytical digest employs the same restriction enzymes that were used in the cloning. These kinds of digests should generate the linear fragments that were taken into the ligation reaction, i.e., one fragment corresponding to the linearized vector and the other one to the insert. As a control, cut vector plasmid DNA should be included in the analysis as its presence in the gel alongside the analyzed plasmids helps verifying the sizes of the DNA fragments.

When a DNA fragment is inserted into a vector using a single restriction enzyme, additional digests are required to find out the orientation of the insert in the vector. For this purpose, use the plasmid sequence you built earlier (Subheading 3.1, **step 9**) to choose a pair of unique site restriction enzymes so that one of them cuts within the insert and the other one in the plasmid backbone, both should cut away from the middle of the corresponding DNA regions, i.e., considerably closer to one side versus the other. Calculate the expected fragment sizes for each orientation and make sure that the two cases will be easily distinguishable in gel electrophoresis. For example, 5 + 1 kb is different enough from 4.5 + 1.5 kb, whereas 3.4 + 2.6 kb is not easy to distinguish from 3.8 + 2.2 kb, particularly if only one kind of clones is present on a gel.

1. Spin down the 5 mL bacterial cultures in a tabletop centrifuge and purify plasmid DNA from *E. coli* cells using a commercial kit, for example, Plasmid DNA mini-prep kit (Qiagen), Wizard Plasmid DNA Purification kit (Promega), or similar. At the elution step, use deionized water (*see* **Note 45**) pre-warmed to 55–60 °C and allow the elution step to proceed for 15 min or longer (*see* **Note 46**) before collecting the plasmid DNA solution by centrifugation. When not used, plasmid DNA samples should be kept at −20 °C.
2. Digest the analyzed plasmids as well as the vector plasmid used for the cloning using the restriction enzymes utilized in the cloning. For each digest, use 0.5–1 μL of the DNA mini-prep (*see* **Note 47**) in a total volume of 15 μL per digest (*see* **Note 48**). Incubate at the temperature optimal for the restriction enzymes for 1 h.

3. During the plasmid digestion, prepare a small (10 cm × 8 cm) 0.5–1 % agarose gel as described in Subheading 3.4. Use a comb with small-medium size wells (3–5 mm wide), enough wells for all the digests plus, at least, two wells for the size markers. The exact agarose concentration depends on the size of the smallest fragment expected for the positive colonies.
4. Once the 1 h digest incubation is finished, add 3 μL of 6× sample buffer to each reaction, mix, and load samples on the gel, starting from the second well. Load 5 μL of 1 kb DNA ladder into the first lane; into the last lane—either the same marker or 5 μL of 100 bp ladder mix if fragments smaller than 1 kb are expected.
5. Run the gel at 7–10 V/cm until the bromophenol blue is 1 cm from the bottom of the gel. You can monitor the gel running using a UV box.
6. Take an image of the gel using a gel documentation system.
7. Analyze each lane for the presence of the DNA fragments of the expected size and mark your positives. The corresponding plasmid samples can now be verified by sequencing.

3.9 Plasmid Verification by Sequencing

When cloning involves PCR, the cloned DNA should be thoroughly analyzed by sequencing as a single base pair mutation can result in a change of gene expression and/or protein structure and function. When designing primers for sequencing, make sure that you will be able to read the sequence from one restriction site that you used in your cloning all the way to the other one. Depending on the sequencing service available, 500–800 bases of DNA sequence can be read in one reaction. Design your first primer ~100 bp upstream of the cloned fragment with the 3′ end towards it. Then design more primers on the same strand, every 500 bp starting from the first one and until it is less than 500 bp to the end of the end of the cloned fragment.

In present days, sequencing is outsourced to private companies or university run services. The sequencing reactions should be set up/run according to the requirements of the sequencing service provider available to you.

When the sequencing data are received, the files usually can be opened by FinchTV, a freely available program. The sequencing data are presented as a linear set of colored peaks, each peak labelled with A, G, C, or T. Each of the four bases is color coded, and the base present at a given position in the DNA sequence results in a peak of density of the corresponding color.

1. For each sequencing reaction, check the density tracks to identify the region in which the sequencing went well and the data can be trusted. This region will have a clear single-color

peak per position. Normally, the readable sequence starts from ~40th base and, depending on the quality of sequencing, may extend all the way to 800–850 bases.

2. Select the DNA sequence from the "trusted" region, copy and paste it in a new window in Serial Cloner or another program for sequence analysis.
3. Use the Alignment tool to align the sequence against the expected plasmid sequence you generated in the Subheading 3.1.
4. Carefully scroll through the alignment and see if there are any mismatches. If there is a mismatch, go back to the sequencing file and find the color peak corresponding to the position. If this is a clear one-color peak, then there is a mismatch in the cloned sequence. If there are two overlapping peaks, then it could be a sequencing artifact and the sequencing information can be clarified by re-sequencing.
5. Mark the region of the plasmid that is covered by the sequencing reaction and has no mismatches.
6. Repeat **steps 1–5** for each sequencing reaction.
7. Check if all the sequence from one restriction site to the other is covered by the sequencing and if there are any mismatches in the sequence. Once you have identified a clone satisfying both requirements, the analysis of sequencing reactions you may have for other clones in the experiment is no longer required.

4 Notes

1. Most antibiotics are heat sensitive and therefore should be added when a medium has been cooled down after autoclaving. Measuring the temperature directly without breaking the sterility is somewhat problematic; instead, using your hand touch as a sensor is convenient and works well enough, assuming reasonable precautions are taken to avoid skin burns.
2. Drying plates can be done at any temperature in the range from room temperature (on the bench) to 50 °C (in special ovens or incubators), but the time required is longer at lower temperatures. It should be done in clean environment to preserve the media sterility. While drying, the plates should stay open or half-open to allow moisture evaporation. Turn all the plates upside down and slide the agar-containing bases halfway over the lids and rest them on the plate lids. Leave plates in this position until you can see a marble-like pattern on the agar surface.
3. These pieces of equipment are not essential. They provide aeration for bacteria to speed up their recovery after transformation as well as the expression of drug resistance genes from

the plasmid transformed. The latter is required prior to the cells being plated on drug-containing agar plates. The same can be achieved by incubating cultures in 37 °C water bath for additional 15–30 min.

4. For the 42 °C heat-shock, a water bath, not a heat block, must be used as the heat transfer properties of water and the metal in the block are different. The time of the heat-shock in this protocol is optimized for a water bath.
5. Autoclaving of microcentrifuge tubes is not necessary. For making competent cells, I just open a brand new box of RNase/DNase-free tubes but handle them with special care by avoiding touching the edges of the tube as well as the inner side of the lids, i.e., the parts that are likely to be in contact with bacterial cultures.
6. *E.coli recA*-mutant strains should be used for cloning or plasmid amplification experiments to avoid plasmid dimers, trimers, tetramers, etc., produced via homologous recombination in a RecA-dependent manner.
7. 10× TBE tends to precipitate, particularly during cold nights in labs with no overnight heating. Once precipitated, it is practically impossible to get back into solution, even at a lower concentration. 5× TBE is less prone to precipitation. Filtering the stock solution after it has been prepared lowers the chances of precipitation as it removes tiny particles of undissolved chemicals that seed precipitation.
8. Ethidium bromide, as any other DNA-binding agent, is highly toxic and carcinogenic. Be particularly careful and use gloves when handling it. I prefer using it over the less toxic new generation type of DNA-binding agents (SYBR Green and similar) as ethidium can be present in the gel during the electrophoresis and therefore allows monitoring of the DNA migration during the gel run. Please notice that ethidium is a cation, and during electrophoresis it migrates in the direction opposite to that for DNA and therefore the bottom part of the gel may be depleted of ethidium (*see* Fig. 3). If proper staining of this part of the gel is essential, the gel can be stained in 1× TBE with ethidium bromide after electrophoresis.
9. Bromophenol blue and often xylene cyanol are added to the sample buffer to monitor the progress of electrophoresis. The mobility of the dyes in agarose gels depends on agarose concentration, but bromophenol blue always runs faster than xylene cyanol. Having too much of a dye in a sample has a negative effect on the visibility of DNA fragments that have the same mobility as the dye. To avoid this problem, I pinch just enough bromophenol blue to make the 6× sample buffer greenish and do not use xylene cyanol at all.

10. In cloning experiments, it is very important to preserve ligatable DNA ends after they have been generated by restriction enzymes. Even a trace of nuclease or phosphatase activities will make the ends unligatable. Therefore, the 6× sample buffer should be made and handled as a sterile solution. For the same reason, I recommend using Ficoll as a density-increasing agent rather than the commonly used glycerol and sucrose, which are good carbon sources for microbes, should any contamination of the buffer occur. Bromophenol blue is also a pH indicator and turning the color of your buffer from greenish to bluish-purplish is an indication of contamination. Avoid using commercially made sample buffers that come with some kits as they are not sterile, normally kept at room temperature (along with the kits) for extended amount of time, and in my experience have led to DNA degradation.
11. DNA size markers come as DNA solutions in water, whereas most DNA samples we load in a gel are in some kind of reaction buffer, such as a restriction digest and PCR. Having some salt in the reaction buffers helps loading samples into agarose gel wells; otherwise the samples may float out of the wells despite the presence of the sample buffer. To prevent this problem, I add 10× NEB3 to the size marker samples, to a final concentration of 1×. In general, if your samples float out of the wells, it means that the density of your sample is lower than that of the buffer in the well. The possible reasons are (a) the presence of residual ethanol in your sample (normally after DNA purification), (b) not enough salt in your sample, (c) not enough sample buffer or the concentration of Ficoll in your sample buffer is lower than it should be, or (d) the concentration of salt in the wells is too high. This often happens when a gel tank with a gel in it is left open for hours and water from the running buffer surface evaporates leading to an increase in the concentration of TBE on the surface of the gel and the wells. Mixing the buffer and washing the wells easily solve this problem.
12. Other phosphatases, such as Antarctic phosphatase or shrimp alkaline phosphatase, can be used instead. These alternative phosphatases can be inactivated by heat but since my protocol does not require enzyme inactivation, calf intestinal phosphatase has no disadvantage in this case.
13. The 10× ligase buffer contains ATP, which is required for the ligation reaction to occur. ATP is sensitive to thaw-freeze cycle and therefore the original aliquot of 10× buffer should be defrosted on ice, pipetted into 0.2 mL PCR tubes as 10 μL aliquots, and stored at −20 °C. When needed, an aliquot should be defrosted on ice, the amount of buffer needed taken out of the tube, and the rest should be disposed of.

14. There is a great variety of kits on the market for purifying plasmid DNA from *E. coli*. While most of them rely on the same alkaline lysis protocol, the purity of the DNA after using kits from different suppliers can vary dramatically. DNA purity is particularly important for the sequencing step. Therefore, before you commit to a given supplier, test if plasmids purified using their kit can be sequenced.

15. It is very important to pick the "right" pair of restriction enzymes for cloning. Carefully study the profiles for each potential enzyme using the New England Biolabs (NEB) catalogue and take into consideration the following aspects.

    Firstly, ligation and recutting, or what fold over-digestion, leaves 90–95 % of ends ligatable and re-cleavable. The higher the allowed over-digestion is, the lower the presence of contaminating nucleases that affect the intactness of the cleaved ends.

    Secondly, if the enzyme activity is affected by the Dam or Dcm methylation. Dam and Dcm are two *E. coli* resident methylases, and if a recognition sequence for a restriction enzyme contains a methylated base(s), some of the enzymes will be inhibited but others will not be. Only DNA purified from *E. coli* cells, such as plasmids, are methylated while in vitro synthesized PCR products are not. Use the NEB catalogue to find out if the enzymes you are considering are sensitive to methylation and, if so, check the target sequence in the plasmid to see if there are sequences that would be methylated.

    Thirdly, if there is a common reaction buffer for the pair of enzymes you plan to use. Sometimes, when such buffer is not available from the list of four NEB buffers, it is worth checking if such buffer can be found among the Roche five buffers. For example, NEB recommends using a sequential digest for BamHI and HindIII while both enzymes can be used in the Buffer B from Roche. Enzymes from different manufacturers can be used in the same reaction but keep in mind that the enzyme unit definitions can be different for different companies. Also, an enzyme from one manufacturer can be used in a buffer from another one if the activity of this particular enzyme (not of its Isoschizomers!) is known.

    Fourthly, give the priority to the enzymes that generate longer, normally four base, overhangs as this will increase the efficiency of ligation. Be creative with your cloning design by taking advantage of some restriction enzymes generating compatible ends. Use Cross Index of Recognition Sequences in the NEB catalogue to find out which enzymes produce compatible ends. For example, the enzymes SpeI, AvrII, NheI, and XbaI each recognize a different 6 bp sequence (ACTAGT, CCTAGG,

GCTAGC, and TCTAGA, respectively) but all generate the same four base 5′ overhang 3′-CTAG-5′. Therefore, if you use any of the four enzymes to cut the vector you can use any of the four to cut you insert.

And finally, check if the enzymes cut at the same temperature. Most restriction enzymes are active at 37 °C, but some nucleases from thermophilic bacteria require higher temperatures (50–65 °C), while an enzyme such as SmaI is quickly inactivated at temperatures above 25 °C. If you need to digest DNA with two enzymes that cut at different temperatures, first add the one that requires higher temperature and proceed with the reaction. Then add the second enzyme and incubate the digest at the lower temperature.

Here are some pairs of enzymes that worked well in my hands in cloning experiments: EagI and SalI in NEB3, PstI and SalI in NEB3, PstI and BglII in NEB3, Acc65I and SalI in NEB3, BamHI and HindIII in Roche Buffer B, and XhoI and HindIII in NEB2.

16. The cccc is added at the 5′ end to facilitate the cleavage by restriction enzymes as the enzyme activity often goes down if the recognition sequence is too close to the DNA end. Having strong C–G base pairing is also advantageous over A–T as it is less prone to DNA end breathing (strand unpairing) and therefore keeps the DNA end double-stranded. While the same can be achieved by having four Gs in the primers, it is not recommended as G-quartet DNA can be formed by primer molecules with stretches of guanines.

17. The rate of aeration/rate of cell growth is very important for cell competency and therefore the volume of the culture should not exceed 1/10 of the volume of the flask.

18. The cells become competent due to their starvation in calcium and therefore it is important to remove as much of the remaining medium as possible.

19. Aliquoting the cells fast is much easier when done by two people. One person starts pipetting cells into tubes in the order reverse to filling up the rack, i.e., the last row of tubes is used first. The other person closes the tubes with cells right away.

20. Count the tube with 1 mL bacterial culture as undiluted (dilution factor of 1), the first tenfold dilution should have the dilution factor of 10, the second tenfold dilution the factor of 100, and so on. The 1,000 in the equation comes from 100× and 10× adjustments that should be taken into account. Because only 100 μL out of 1 mL (i.e., 1/10) is plated, multiplying by 10 corrects for the plating volume. Also, 10 ng of DNA is used per transformation whereas the Efficiency E is

calculated per 1 μg and therefore the difference of 100-fold requires additional multiplication by 100.

21. If running more than two reactions, use tubes linked into strips as those are more convenient to handle. It is recommended to make a master mix as it simplifies pipetting of small volumes and ensures the uniformity of the PCR reactions within the experiment. When making a master mix, always make a bit more than needed. For four reactions, make a master mix sufficient for 4.5 reactions. When mixing the PCR components, add everything except the enzyme and mix well. Then add the enzyme and mix by pipetting up and down with a pipetman set at least half the volume of the mix. Never vortex enzyme-containing solutions as vortexing causes proteins to denature.
22. To find out how much you can load into a well before casting the actual gel, measure the base of a tooth in a comb in mm and multiply the comb thickness by the tooth width. Multiply the resulting number by 5 as a well in a 6 mm thick agarose gel is about 5 mm deep. The resultant number is the well volume. For example, a 4 mm tooth of a 1 mm thick comb will generate a well of 20 mm loading capacity ($4 \times 1 \times 5$). If you want to load more you either need to use a different comb or pour a thicker gel.
23. I keep an 8 strip of PCR tubes filled with 6× sample buffer and use a multichannel pipette to add sample buffer to any samples in PCR tubes, such as PCR reactions and restriction digests.
24. The purpose of this gel run is (a) to evaluate the results of the PCR runs, i.e., to see if a PCR product of the expected size is present, and (b) to separate the PCR product from any nonspecific products and PCR components (primers, dNTPs, buffer, DNA polymerase). Therefore, if a nonspecific PCR product of a size close to the desired product is present, make sure you resolve them well from each other on the gel.
25. DNA exposure to UV causes DNA damage that, in turn, increases the chance of mutations in cloned DNA. Minimize the gel exposure to UV while taking a gel image. It is advantageous to have long-wave UV bulbs in your gel documentation system.
26. Often there is an optional wash step in the DNA purification protocols from a gel to get rid of residual agarose. Include this step as residual agarose may inhibit restriction enzymes later in the experiment. At the very last step, when the DNA is eluted from the DNA binding matrix, use water rather than elution buffer. Pre-warm water to 55–60 °C and let the elution proceed for 15 min, particularly if the DNA fragment is longer than 1 kb. This will increase the DNA recovery.

27. Hold the bottle by the cap and look at it against light while gently swirling the agarose. There should be no particles visible when agarose is melted completely.
28. A handful of ice can be added to the water in the container to speed up agarose cooling.
29. Casting the gel in a cold room decreases the time needed for the agarose to solidify.
30. The amount of buffer in the gel tank affects the speed of DNA migration. At a constant voltage, increasing the volume of the buffer will increase resistance in the system and lower the current, resulting in slower migration of DNA samples.
31. Many restriction enzymes show a loose-site recognition specificity (also called star activity) when glycerol concentration in the reaction is higher than 5 %. Because enzyme stocks normally come in 50 % glycerol, one has to be particularly careful when calculating the total volume of enzyme stocks that can be added to a reaction. Not more than 1/10 of the total volume of a reaction should come from enzyme stocks, e.g., if a restriction digest is set up in 30 μL, the combined volume of restriction enzymes and phosphatases added should not exceed 3 μL.

    When adding two (or more) restriction enzymes to a reaction, their relative amounts should be balanced. Each enzyme stock comes at a certain concentration in U/μL, written on a stock tube. For most restriction enzymes from NEB, 1 U is defined as the amount of enzyme required to digest 1 μg of phage λ DNA in 1 h, and therefore the number of target sites in λ for each enzyme (can be found in the NEB catalogue) should be taken into account when calculating the relative amount of enzymes for your digest. The formula reflecting this balance for a pair of enzymes A and B as follows:

    $$C_A \times N(\lambda)_A \times V_A / N(S)_A = C_B \times N(\lambda)_B \times V_B / N(S)_B,$$

    where C is the enzyme stock concentration; $N(\lambda)$ and $N(S)$ are the numbers of restriction sites for a given enzyme in phage λ and your sample DNA, respectively; and V is a volume of the enzyme to be added to digest the sample DNA. To calculate the unknown V_A and V_B, a second equation is required:

    $$V_A + V_B = V_T,$$

    where V_T is the total volume of the two enzyme stocks to be added to the reaction.

    Let's consider an example in which a plasmid with single sites for BamHI and EagI is to be digested and dephosphorylated in a reaction volume of 30 μL. Because of the "not more than 1/10" rule, and to be on a safe side, 2.5 μL will be allowed for all the enzymes. 0.5 μL of Phosphatase will be added, leaving 2 μL for the two restriction enzymes. BamHI has five sites in λ and comes at 20 U/μL, while EagI has two sites in λ and comes at 10 U/μL. Therefore, our first equation will look as:

$$20 \times 5 \times V_{\mathrm{BamHI}}/1 = 10 \times 2 \times V_{\mathrm{EagI}}/1, \text{ or } 5 \times V_{\mathrm{BamHI}} = V_{\mathrm{EagI}}.$$

Because the total volume of the two restriction enzymes in the reaction is 2 μL,

$$V_{\mathrm{BamHI}} + V_{\mathrm{EagI}} = 2.$$

Solving the system of two simple equations, $V_{\mathrm{BamHI}} = 0.33$ μL and $V_{\mathrm{EagI}} = 1.67$ μL.

32. If your digest requires NEB1 buffer you have to use Antarctic phosphatase as CIP is not active in this buffer.
33. When both restriction sites are close to each other, the products of a single and a double digest are indistinguishable from each other on a gel. To assay the activity of each enzyme two single digests are required. The uncut control is useful to monitor the mobility of undigested DNA during electrophoresis.
34. It is very important not to overload the digest with DNA as under-digestion will result in most of the DNA fragments being cut on one end or the other while only the fragments cleaved on both sides will contribute to the productive ligation at the next stage. In contrast, single-end digested fragments will have an inhibitory effect on the productive ligation by competing for the vector molecules. When deciding how much DNA to take into a digest, remember that 1 μg of 100 bp DNA fragment has ten times as many DNA ends (and therefore restriction sites) as 1 μg of 1 kb DNA fragment. Therefore, using DNA amounts in moles rather than DNA mass will produce more accurate calculations.
35. Special care should be taken to prevent water evaporation from the reaction to tube walls or lids as this will increase the amount of salt and glycerol in the reaction and may result in loss of enzyme activity or encouragement of star activity. Reactions set up in small volumes are particularly sensitive to the problem. Use water bath with a lid or an incubator so that tube lid are heated too and do not stimulate condensation. I avoid sample evaporation by setting up digests in a thermocycler, with the lid heated to at least 70 °C.
36. The agarose concentration of 0.5 % is suitable for separating supercoiled and linear molecules of 4 kb and larger. If your vector is smaller, increase the agarose concentration to 0.6–0.7 %.
37. Loading DNA into larger wells helps keeping bands thinner and sharper, which results in better separation of species of similar mobility in a gel.

38. The gel can be run further even if the bromophenol blue migrates out of the gel completely but the run progress should be monitored more carefully not to lose the DNA due to overrunning of the gel.
39. Sticky ends are ligated rather fast and the ligation reactions transformed into bacteria after 1 h normally produce enough colonies to screen. However, when a blunt end is involved, ligating for longer time periods might be needed for the ligation to occur. A sensible approach is to transform half of the reaction after 1–2 h ligation and leave the rest to proceed overnight. The next morning, if the number of transformants is not satisfactory the rest of the ligation reactions can be transformed.
40. It is important not to use too many cells in the colony PCR reaction as some cell lysate components, such as cell wall and lipids, inhibit amplification. Touch a side of a colony with a pipette tip to pick just enough cells to see with the naked eye.
41. When making a master mix, always make some extra to ensure sufficient amount of the mix is made for all the samples.
42. Always add enzymes last to master mixes to ensure they get diluted into the appropriate buffer conditions to prevent enzyme denaturation or precipitation.
43. The exact temperature depends on the primer melting temperature. Calculate the melting temperature of your primers in the PCR buffer and use a temperature 1–2 °C lower than the lower of the two melting temperatures of the primers.
44. The amplification time is dependent on the length of the expected PCR product. Allow 1 min per 1 kb of DNA to be amplified by the Taq polymerase.
45. Using water for elution allows using the plasmid samples for sequencing. If TE is used, EDTA will inhibit the sequencing reactions.
46. Pre-warming water and increasing elution time result in higher plasmid yields as they both promote DNA release from the column matrix. The larger the plasmids are, the harder it is to elute them.
47. If the insert is very small (shorter than 1 kb) more DNA may need to be digested to visualize the short DNA fragment. Also, because ethidium migrates in the direction opposite to DNA, the bottom of the gels in which smaller fragments run become depleted of the stain by the end of the run. Stain the gel in 1× TBE with 5 μg/mL ethidium for 30 min before visualizing the gel.
48. Making a master mix and setting up the digests in eight PCR tube strips help pipetting multiple digests faster and more accurately.

Acknowledgements

I would like to thank Julia S.P. Mawer for the critical reading of the manuscript and helpful suggestions.

Reference

1. Green MR, Sambrook J (2012) Molecular cloning: a laboratory manual. CSHL Press, New York

Chapter 3

Analysis of Branched DNA Replication and Recombination Intermediates from Prokaryotic Cells by Two-Dimensional (2D) Native–Native Agarose Gel Electrophoresis

Nicholas P. Robinson

Abstract

Branched DNA molecules are generated by the essential processes of replication and recombination. Owing to their distinctive extended shapes, these intermediates migrate differently from linear double-stranded DNA under certain electrophoretic conditions. However, these branched species exist in the cell at much low abundance than the bulk linear DNA. Consequently, branched molecules cannot be visualized by conventional electrophoresis and ethidium bromide staining. Two-dimensional native–native agarose electrophoresis has therefore been developed as a method to facilitate the separation and visualization of branched replication and recombination intermediates. A wide variety of studies have employed this technique to examine branched molecules in eukaryotic, archaeal, and bacterial cells, providing valuable insights into how DNA is duplicated and repaired in all three domains of life.

Key words 2D electrophoresis, Replication, Recombination, Branched DNA intermediates, Archaea

1 Introduction

Native–native (or neutral–neutral) two-dimensional DNA agarose gel electrophoresis is a versatile technique that enables the visualization of DNA intermediates that are generated during DNA replication and recombination. In 1982, Byers and Bell first described a method whereby branched DNA molecules, such as recombination intermediates, could be separated from linear double-stranded DNA [1]. This partitioning required the DNA substrates to be run sequentially through two gels of differing agarose concentration (the two dimensions). Early adaptations to this technique by Brewer and Fangman have led to robust protocols for the visualization of both branched replication and recombination intermediates [2]. This classic protocol has undergone various modifications and refinements, and the technique has since been utilized to study genomic DNA isolated from a variety of different organisms.

Svetlana Makovets (ed.), *DNA Electrophoresis: Methods and Protocols*, Methods in Molecular Biology, vol. 1054,
DOI 10.1007/978-1-62703-565-1_3, © Springer Science+Business Media New York 2013

The method described in this chapter has been optimized to visualize replication intermediates from the hyperthermophilic archaeon *Sulfolobus solfataricus* [3, 4], but, with minor modifications, the technique has also been used to separate branched intermediates from other archaeal species. Similar protocols have been used to visualize branched intermediates in the closely related *Sulfolobus acidocaldarius* and also other archaea such as *Aeropyrum pernix* [5]. The method is similar to the procedure first described by the Forterre laboratory, during their characterization of the *Pyrococcus furiosus* origin of replication [6], which was the first archaeal initiation site to be verified experimentally. The protocol outlined below has also been adapted to study replication intermediates from *Escherichia coli* [7] and, therefore, with minor modifications, will be suitable for separating branched DNA molecules from a range of prokaryotic organisms.

The technique begins with the restriction digestion of genomic DNA, to produce suitably sized fragments. These are then run slowly at low voltage, on a low-percentage agarose gel, to partition the DNA molecules purely on the basis of size, following standard gel electrophoresis principles. The separation of the branched intermediates from the bulk linear DNA does not occur until the second dimension. Once the first dimension has been run, the entire sample lane is excised from the first dimension, rotated through 90°, and placed at the top of a new high-percentage second dimension agarose gel, which is poured around the first dimension gel slice (Fig. 1). Ethidium bromide is incorporated at high concentration in the second dimension gel, and during electrophoresis this intercalating agent locks the DNA molecules into rigid conformations. As a result, when the second dimension gel is run at high voltage, the DNA migrates not only on the basis of size but also separates according to the shape of the molecule. Under these conditions, the branched intermediates are retarded in the gel, becoming distributed predominantly on the basis of shape, and are partitioned from the more streamlined linear DNA. However, the branched replication and recombination intermediates will only represent a small proportion of the total genomic DNA, and these molecules are therefore generally not abundant enough to visualize with UV fluorescence, following ethidium bromide staining. It is therefore necessary to transfer the DNA to a nylon membrane by Southern blotting and hybridize with a specific radiolabelled or fluorescent probe to detect the minor subpopulation of branched intermediates, particular to the region of the genome under analysis. The protocol described in this chapter details the entire 2D process, from the isolation and preparation of the genomic DNA to the two-dimensional electrophoresis conditions, and finally describes the Southern transfer, probe hybridization, and visualization of the intermediates.

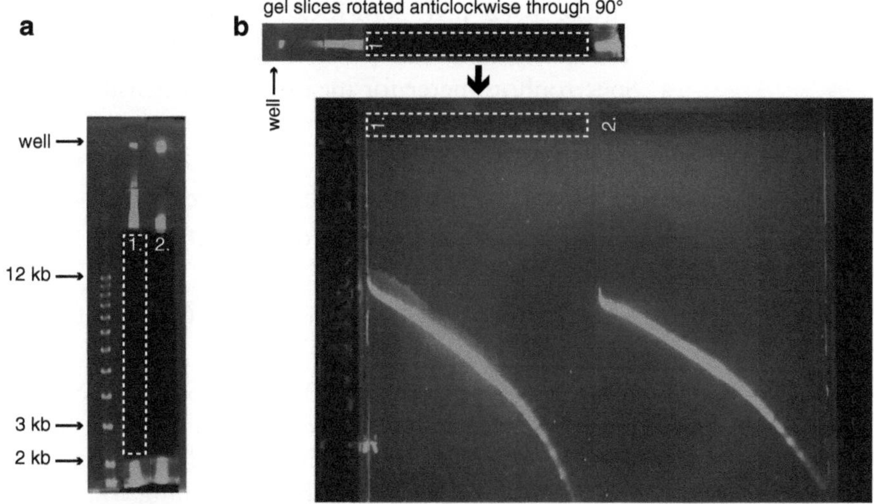

Fig. 1 Ethidium bromide stained images after the first and second dimensions of a typical native–native agarose 2D gel. In this example, two *S. solfataricus* genomic DNA plugs have been run. (**a**) The first dimension gel is stained AFTER the run is complete (0.4 % agarose, run at 0.6 V/cm for 48 h at 4 °C, in 1× TBE WITHOUT ethidium bromide in the gel or buffer). Note that the gel slices are excised PRIOR to viewing the remainder of the gel using UV light, as this treatment would otherwise damage the branched intermediates. (**b**) The excised gel slices from the first dimension are rotated through 90° and placed at the top of the second dimension gel, which is poured around the slices (1.0 % agarose, run at 5 V/cm for 7 h at 4 °C, with ethidium bromide at 0.3 μg/mL in the gel and buffer). The arcs visible by ethidium bromide staining represent the bulk linear genomic DNA. The branched replication intermediates are not present in sufficient quantities to detect by ethidium bromide staining. The DNA is therefore transferred to nylon membrane, and hybridized with a specific radiolabelled probe by standard Southern blotting

2 Materials

All reagents should be analytical grade, and solutions should be prepared using ultrapure water (deionized water purified to a sensitivity of 18 MΩ cm at 25 °C). Waste disposal regulations should be followed at the end of the procedure, with special consideration for the disposal of the radioisotope ^{32}P, and the potential mutagen ethidium bromide.

2.1 Cell Growth, Harvesting and Genomic DNA Plug Preparation

1. Oven or water bath with orbital shaker for culturing *S. solfataricus* cells at 75 °C.
2. 500 mL glass conical flask for cell culture, and aluminum foil lid.
3. DSM88 *Sulfolobus* growth medium: 10 mM $(NH_4)_2SO_4$, 2 mM KH_2PO_4, 1 mM $MgSO_4$, 0.5 mM $CaCl_2$, 75 μM $FeCl_3$, 9 μM $MnCl_2$, 12 μM $Na_2B_4O_7$, 0.75 μM $ZnSO_4$, 0.3 μM

$CuCl_2$, 0.12 μM Na_2MoO_4, 0.15 μM $VoSO_4$, 0.065 μM $CoSO_4$, 0.1 % yeast extract, pH 2.0 (adjusted with H_2SO_4).

4. Spectrophotometer for measuring cell density (Absorbance at 600 nm: A_{600}).
5. Centrifuge with rotor (e.g., rotor 19776-H; Sigma) suitable for 50 mL polypropylene conical tubes for harvesting cells.
6. Disposable plug mould (10×100 μL strips, BIORAD; Cat. 170-3713).
7. TEN solution: 50 mM Tris–HCl, 50 mM EDTA, 100 mM NaCl, pH 8.0.
8. 0.8 % low-melting point (LMP) agarose in TEN solution. Add 0.4 g LMP agarose (Biogene; Bioresolve Hi-Strength LMP agarose; Cat. 300-800) to 50 mL TEN solution and weigh the flask. Dissolve the LMP agarose by boiling and allow to cool to 37 °C. Weigh the flask again. Add warmed ultrapure water to make up to the original weight. Maintain the temperature of the agarose at 37 °C until needed.
9. NDS solution (pH 8.0): 160 mM EDTA, 1.9 mM Tris base, 138 mM NaOH, 9 mM lauroyl sarcosine. Dissolve 93 g EDTA, 0.6 g Tris base, 11 g NaOH in 350 mL in ultrapure water. Then dissolve 5 g lauroyl sarcosine in 50 mL ultrapure water and add to the solution. Adjust the pH to 8.0 with concentrated NaOH, and make up to 500 mL with ultrapure water.
10. NDS solution (pH 9.0): Prepare NDS solution as indicated above. Adjust the pH to 9.0 with concentrated NaOH, and make up to 500 mL with ultrapure water.
11. 15 mL polypropylene conical tubes.
12. Proteinase K (Sigma; Cat. P6556): prepare a 10 mg/mL stock in ultrapure water and store aliquots at −20 °C.
13. 1.5 or 2 mL microcentrifuge tubes.
14. Restriction enzymes with appropriate reaction buffers (New England Biolabs).

2.2 Native–Native Agarose 2D Gel Electrophoresis

1. 0.4 % first dimension agarose gel: add 1 g agarose (Invitrogen; electrophoresis grade; Cat. 15510-027) to 250 mL 1× TBE and weigh the flask. Dissolve the agarose by boiling. Add hot ultrapure water to make up to the original weight after boiling.
2. 1.0 % second dimension agarose gel: add 5 g agarose (Invitrogen; electrophoresis grade; Cat. 15510-027) to 500 mL 1× TBE and weigh the flask. Dissolve the agarose by boiling. Add warmed ultrapure water to make up to the original weight. Add ethidium bromide at 0.3 μg/mL final concentration.

3. Gel running buffer: 1× TBE (90 mM Tris-borate, pH~8.3, 2 mM EDTA).
4. Ethidium bromide (Sigma; Cat. E1510): add only to the second dimension gel and running buffer at 0.3 μg/mL final concentration.
5. Gel tank with a gel casting tray (15 cm × 15 cm) and a comb (for attaching genomic plugs at the first dimension).
6. Graph paper, razor blades (or scalpels), and flexible plastic ruler, to aid excision of the first dimension gel slices.
7. Large gel tank with a 20 cm × 20 cm casting tray (for second dimension).

2.3 Genomic DNA Transfer to Nylon Membrane

1. Tray (e.g., tupperware box) for containing the gel after the second dimension run.
2. Laboratory rocking platform (e.g., Stuart SSL3 rocker).
3. Acid solution (to aid transfer of genomic DNA): 0.25 M HCl.
4. Denaturation solution: 0.5 M NaOH, 1.5 M NaCl.
5. Neutralization solution: 1.0 M Tris–HCl pH 8.0, 1.5 M NaCl.
6. Nylon membrane (GE Healthcare; Amersham Hybond XL; Cat. RPN303S).
7. Transfer solution (20× SSC): 3.0 M NaCl, 0.3 M trisodium citrate ($Na_3C_6H_5O_7$). Dissolve 175.3 g NaCl and 88.2 g trisodium citrate in 800 mL of H_2O. Adjust the pH to 7.0 with a few drops of concentrated HCl. Adjust the volume to 1 L with ultrapure H_2O.
8. Rinsing solution (2× SSC): Dilute 100 mL of 20× SSC with 900 mL ultrapure H_2O.
9. Small plastic container to form blotting platform (inverted within the larger tupperware box) and two glass plates (approximately 20 cm × 20 cm).
10. Whatman blotting paper, laboratory parafilm, 10 mL plastic pipettes, and C-fold paper towels.
11. UVC Crosslinker (Hoefer).

2.4 Southern Blot Hybridization Using ^{32}P-Radiolabelled Probe

1. DNA labelling procedure: NEBlot® Kit (New England Biolabs; Cat. N1500L).
2. 1.5 mL screw-cap microcentrifuge tubes.
3. Heat block (or water bath) at 37 °C.
4. Radiolabel: α-^{32}P-dATP (3,000 Ci/mmol; Perkin Elmer; Cat. NEG512H250UC).
5. Probe purification: Illustra Microspin G-25 columns (GE Healthcare; Cat. 27-5325-01).

6. Hybridization solution (Church Gilbert solution): 0.5 M phosphate buffer (0.34 M Na_2HPO_4, 0.16 M $NaH_2PO_4 \cdot 2H_2O$), 0.24 M sodium dodecyl sulfate (SDS), 1 mM EDTA. Add 48.55 g Na_2HPO_4, 24.65 g $NaH_2PO_4 \cdot 2H_2O$, 70 g SDS, and 2 mL of 0.5 M EDTA (pH 8.0) to 700 mL ultrapure water, dissolving each component individually, and warm the solution to above 60 °C. Make up to 1 L in ultrapure water. (Alternatively use Amicon Ultra-hyb solution (Amicon Ultra-Hyb; Cat. AM8670).)
7. Hybridization wash solutions: 2× SSC (dilution of 20× SSC—*see* Subheading 2.3, **item 8**), 0.1 % SDS for the initial washes, and 0. 2× SSC, 0.1 % SDS for the final washes.
8. Heat block (or water bath) at 95 °C to denature DNA.
9. Hybridization oven and hybridization tubes (e.g., Maxi 14 oven—6247; Thermo Scientific).
10. Saran wrap, or heat-sealable plastic and heat sealer.
11. X-ray film with a cassette, or phosphor storage screen and phosphoimager (e.g., Typhoon system; GE Healthcare).

3 Methods

3.1 Preparing S. solfataricus Genomic DNA Imbedded in Agarose Plugs (See Note 1)

1. Grow 200 mL of *S. solfataricus* cell culture (in a 500 mL glass conical flask covered with aluminum foil) in DSM88 media at 75 °C to $OD_{600} \approx 0.4$ (mid-log phase).
2. Cool the culture on ice and harvest the cells by centrifugation at 7,000 × *g* for 10 min at 4 °C (rotor 19776-H; Sigma) in 50 mL polypropylene conical tubes.
3. Wash cells in 15 mL ice-cold TEN solution and spin again.
4. Repeat the wash step.
5. Resuspend the cells in 100 μL TEN (final resuspension volume will be approximately 200 μL), and warm the mixture to 37 °C for 1 min.
6. Add equal volume (200 μL) of 0.8 % low-melting point agarose and mix (LMP agarose in TEN solution, equilibrated to 37 °C after melting).
7. Pipette mixture into a plug mould (10 × 100 μL strip; BIORAD) and allow agarose to set at 4 °C (producing approximately 4 × 100 μL plugs of agarose encased cells).
8. Treat plugs overnight in 10 mL NDS, pH 9.0 + 1 mg/mL proteinase K at 37 °C in a 15 mL polypropylene conical tube, without agitation.
9. Transfer plugs to 10 mL NDS, pH 8.0 + 1 mg/mL proteinase K and treat overnight at 37 °C.

10. Transfer plugs to fresh NDS, pH 8.0 (without proteinase K) and store at 4 °C until material is required for the next step.

3.2 Digestion of Genomic DNA Within Agarose Plugs

1. Elute genomic plug in 1 mL appropriate restriction digestion buffer (e.g., 1× NEB restriction digestion buffer), without agitation, for 1 h at room temperature, in a 1.5 or 2 mL microcentrifuge tube.
2. After 1 h, discard the elutant and add another 1 mL restriction digestion buffer.
3. Change the buffer 4–5 more times (1 h washes at room temperature) and leave the final elution overnight at 4 °C.
4. Digest the agarose-imbedded DNA plug in a 500 μL volume (400 μL water, 50 μL 10× buffer, and 50 μL restriction enzyme). Incubate for 4 h, without agitation, at the appropriate temperature for the restriction enzyme (e.g., *Eco*RI at 37 °C; New England Biolabs; Cat. R0101S). Discard the buffer, and add another 500 μL restriction digestion buffer and enzyme, and incubate overnight at the appropriate temperature.

3.3 Native–Native 2D Agarose Gel Electrophoresis for Resolving *S. solfataricus* Replication Intermediates (See Note 2)

1. Attach each genomic DNA plug to the comb of the gel using a few drops of molten agarose. If multiple samples are to be run, leave at least one free lane between each sample. Insert the comb into the gel casting tray.
2. Pour the first dimension gel (0.4 % standard agarose in 1× TBE) around the comb with the attached plugs. Ethidium bromide is *not* added at this stage. The gel should be the same depth as the agarose plugs (approximately 1 cm). For a 15 cm × 15 cm gel, 250 mL of agarose will submerge the plugs fully.
3. A DNA size marker ladder (e.g., 10 μL of a 1 kb plus ladder; Invitrogen) may be run alongside the samples in the first dimension.
4. Run the first dimension at 0.6 V/cm for 48 h at 4 °C in 1× TBE.
5. After 48 h run, stain the gel with 0.3 μg/mL ethidium bromide in 1× TBE for 30 min and then excise sample lanes without viewing by UV exposure (*see* **Note 3**). A flexible plastic ruler is extremely useful for manipulating the gel slices.
6. Once the gel slices have been excised, the remaining gel including the DNA size marker ladder can then be viewed with UV light to determine how far the products of the genomic digestion have migrated (Fig. 1a).
7. Discard the top and bottom parts of the slices that contain the largest DNA fragments, and the DNA fragments that are smaller than 2 kb (e.g., excise a 9 cm long slice; cut from 2 cm below the well to 11 cm below the well—*see* Fig. 1).

8. Rotate the 9 cm long gel slices through 90° and place along the top of a new gel tray (Fig. 1. 20 cm is wide enough to run two samples concurrently). Then pour the second dimension gel (1.0 % TBE agarose + 0.3 μg/mL ethidium bromide) around the first dimension slices and allow the gel to set. 500 mL of agarose is sufficient for a 20 cm × 20 cm gel.
9. Run the second dimension at 5.0 V/cm for 7 h at 4 °C in 1× TBE with 0.3 μg/mL ethidium bromide.
10. When the run is complete, view the gel using UV light, trim the gel to remove excess agarose, and transfer the DNA onto a nylon membrane by capillary transfer.

3.4 DNA Blotting by Capillary Transfer

1. After the second dimension electrophoresis, place the gel in a container/tray and treat with 0.25 M HCl for 15 min at room temperature, with gentle agitation on a laboratory rocking platform (*see* **Note 4**).
2. Rinse the gel in distilled water and then denature the DNA strands by immersing the gel in 500 mL of denaturation solution, with gentle agitation for 30 min at room temperature.
3. Rinse the gel in distilled water and then incubate in 500 mL of neutralization solution, with gentle agitation for 30 min at room temperature.
4. Transfer the DNA to nylon membrane by capillary transfer [8].
   (a) Create a blotting platform by placing a small upturned plastic container within a larger plastic tray, and place a glass plate on top of the small upturned container.
   (b) Fill the larger tray with 20× SSC.
   (c) Prepare a "wick"—a piece of Whatman paper long enough to cover the glass platform and to extend on both sides into the reservoir containing the 20× SSC. Wet the wick with 20× SSC, and smooth any creases in the blotting paper using a 10 mL plastic pipette as a rolling pin.
   (d) Measure the length and width of the gel and prepare six pieces of Whatman paper, cut to the same size as the gel. Wet the first two pieces in 20× SSC and place them on top of the wick, and smooth out any creases with the pipette.
   (e) Place the gel on top of the two pieces of blotting paper.
   (f) Cut a piece of nylon membrane to the same dimensions as the gel. Wet the membrane in ultrapure water, and then rinse in 20× SSC. Place the membrane on top of the gel and complete the "sandwich" by laying the 4 remaining strips of Whatman paper on top of the filter, one at a time, wetting each strip in 20× SSC and smoothing out any creases (or air bubbles) with the 10 mL pipette.

(g) Place strips of parafilm around the edges of the gel on the wick (to prevent "short-circuiting" during transfer—*see* **Note 5**).

(h) Place a stack of paper towels (C-fold) on top of the "sandwich." To aid transfer place another glass plate on top of the towel stack and add a small weight (e.g., a lab bottle with 100–200 mL liquid), to maintain slight pressure during the transfer process.

5. Blot for 24–48 h.
6. Disassemble the transfer setup. Peel off the membrane from the gel and place it on a piece of blotting paper, with the side that was previously in contact with the gel facing up (this side is bound by DNA).
7. UV crosslink the DNA to the membrane (e.g., 1,200 J/m^2; Hoefer UVC Crosslinker).
8. Rinse the membrane in 2× SSC and store at 4 °C until the hybridization procedure.

3.5 Hybridization Procedures

For probe labelling, we routinely use New England Biolabs NEBlot kit and the following protocol:

3.5.1 Probe Labelling for Southern Hybridization

1. Dissolve 25 ng of template DNA in nuclease free H_2O to a final volume of 33 μL, in a 1.5 mL screw-cap microcentrifuge tube.
2. Denature the DNA at 99 °C for 5 min.
3. Quickly place the tube on ice for 5 min.
4. Centrifuge briefly.
5. Add the following components in the following order:

   5 μL 10× NEBlot labelling Buffer (which includes Random Octadeoxyribonucleotides).

   6 μL dNTP mixture (2 μL of dCTP, dTTP, and dGTP). 5 μL α-^{32}P-dATP (3,000 Ci/mmol, 50 μCi, 1.85 MBq).

   1 μL DNA polymerase I—Klenow Fragment (3′–5′ exo$^-$) (5 U.)

   Mix well.
6. Incubate at 37 °C for 1 h.
7. Terminate reaction by adding 5 μL of 0.2 M EDTA (pH 8.0).
8. Purify the probe from unincorporated nucleotides using a buffer exchange column (e.g., GE Healthcare Illustra Microspin G-25 columns).

3.5.2 Southern Hybridization

1. Pre-hybridize blotting membranes from the Subheading 3.4, **step 8** with either Church-Gilbert or Ultra-hyb (Amicon Ultra-Hyb; Cat. AM8670) solution in hybridization tubes in a

hybridization oven (e.g., Maxi 14 oven; Thermo Scientific) at 50 °C for a minimum of 1 h.

2. Denature probe in a heat block at 99 °C for 5 min.
3. Quickly place the probe on ice for 5 min.
4. Add probe directly to 20 mL of fresh pre-warmed hybridization solution in a hybridization tube. Ensure lid is secured and return tube to hybridization oven.
5. Hybridize overnight at 50 °C.
6. Discard the hybridization buffer with the probe and wash blots twice with 2× SSC, 0.1 % SDS and then twice in 0.2× SSC, 0.1 % SDS; perform all four wash steps at 50 °C for 30 min.
7. Discard the last wash, take the membrane out of the hybridization tube and enclose in heat-sealed plastic, or wrap in saran wrap.
8. Expose the membrane to X-ray film, or phosphor-storage screen for 1–7 days.
9. Develop the X-ray film, or scan the phosphor-storage screen using phosphoimager to visualize the DNA of interest.

3.6 Interpreting 2D Native–Native Agarose Gels

The native–native agarose 2D gel procedure permits the visualization of DNA replication and recombination intermediates. Different branched species, such as Y-shaped replication forks, or X-shaped joint molecules, follow defined migration patterns because the heterogeneous DNA population is differentially retarded through the second dimension. It is possible to infer a wealth of information from the 2D data by interpreting the signatures associated with different branched species. The notes below describe how these branched configurations are derived during the 2D procedure.

3.6.1 The 1n Spot (Unbranched DNA)

Figure 2a illustrates a typical 2D gel result generated from asynchronous *S. solfataricus* cells, using the protocol outlined

Fig. 2 (continued) run at 0.6 V/cm for 48 h at 4 °C, in 1× TBE without ethidium bromide in the gel or buffer. Second dimension conditions: 1.0 % agarose, run at 5 V/cm for 7 h at 4 °C, in 1× TBE with ethidium bromide at 0.3 μg/mL in the gel and buffer. (**b**) A schematic demonstrating the key features of the 2D gel, namely, the 1n spot, the X-spike, and the Y-arc, with a stalled fork spot and reversed fork "cone" signal. (**c**) The Y-shaped arc (highlighted in *grey*) is derived from replication forks moving through the restriction fragment. (**d**) The spot on the Y-arc (highlighted in *grey*) is indicative of replication forks stalling at this site. The "cone" signal above this region represents regression of these stalled forks to Holliday junction-like "chicken-foot" structures. (**e**) The X-spike (highlighted in *grey*) consists of joint molecules, such as recombination intermediates. The positioning of the crossover between the molecules dictates the degree of branching. (**f**) A schematic depicting the "bubble arc" (highlighted in *grey*), which is observed if a bidirectional origin of replication is located within the central third of the restriction fragment

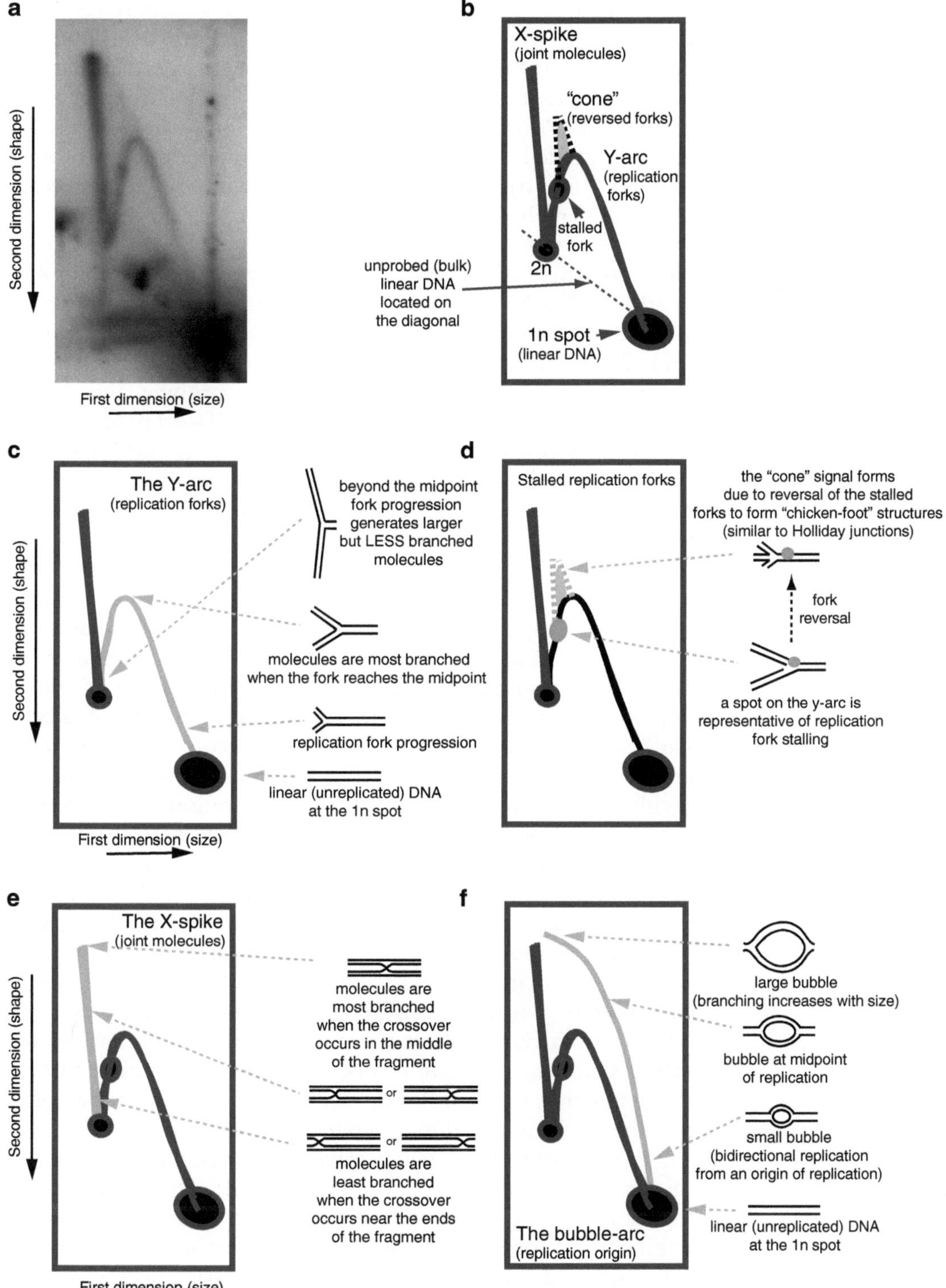

Fig. 2 A typical native–native agarose 2D gel image after Southern blotting and hybridization with a specific probe. (**a**) Branched *S. solfataricus* intermediates from an approximately 4 kb *Pvu*II restriction fragment, visualized by phosphoimaging after hybridization with a ^{32}P radiolabelled probe. First dimension conditions: 0.4 % agarose,

above. In this example, an approximately 4 kb *Pvu*II restriction fragment has been studied. The schematic in Fig. 1b illustrates the key features of this 2D result, namely, the 1n spot (unbranched material), the Y-arc (replication forks), and the X-spike (joint molecules). Prior to the Southern blot, only the unbranched, double-stranded linear DNA is visible by ethidium bromide staining (Fig. 1b). The branched intermediates, which are present at a considerably lower concentration, can only be detected after the DNA has been transferred to a nylon membrane by Southern blot, and hybridized to a specific radiolabelled probe. After hybridization, the 1n spot marks the position of the unreplicated, linear double-stranded DNA from a specific restriction fragment (Fig. 2a, b). If we briefly consider a standard one-dimensional Southern blot, the band observed after probing this blot is equivalent to the 1n spot of a 2D gel. In the 2D-gel procedure, the 1n signal is considerably stronger than in a standard Southern, because generally more genomic DNA must be processed to allow for the visualization of the much less abundant branched intermediates.

3.6.2 The "Y-Arc" Representing Replication Forks

The schematic in Fig. 2c highlights a classic "Y-arc" in grey, named after the Y-shaped branched replication fork intermediates that make up this distinctive feature. The Y-arc initially rises from the 1n spot. As the replication fork passes through the restriction fragment, the molecule becomes larger, because the DNA is being synthesized. At first, the molecules also become more branched during this process. However, Fig. 2c demonstrates that the most highly branched species occur when the replication fork has reached the midpoint of the restriction fragment. Beyond this point, the replication intermediates continue to increase in size but actually become progressively less branched (Fig. 2c). The Y-arc therefore dips, after the midpoint maximum, towards the 2n spot, as the replication of the fragment nears completion. At the 2n spot, the DNA is approximately twice the size of the original molecule, composed of two almost fully replicated unbranched copies, linked only by an extremely short stretch of the remaining unreplicated DNA. Once the replication fork has finished migrating through the fragment, the two linear copies will no longer be joined together and will therefore migrate at the 1n position.

In the specimen 2D gel displayed here, a spot can be observed on the descending arm of the Y-arc (Fig. 2d). This is indicative of replication fork pausing at this specific site within the restriction fragment. A "cone" signal can also be seen in the region above the pause site and the apex of the Y-arc (Fig. 2d). This region most likely represents stalled forks that have undergone reversal to form "chicken-foot" structures, similar to Holliday junctions, in response to the replication fork block (Fig. 2d). The replication pause site seen in this example does not completely block the movement of the replication forks passing through the fragment. In cases where an almost total replication block is introduced, such as replication termination sites in bacteria, the descending arm of the Y-arc will not continue beyond the pause site.

3.6.3 The "X-Spike" Consisting of Joint Molecules

Joint molecules formed between replicated regions, such as those involved in homologous recombination and sister chromatid cohesion, can also be observed by the 2D gel procedure. These four-way intermediates form a characteristic pattern known as the "X-spike" (Fig. 2e). This feature extends upward from the 2n spot and is representative of two fully replicated DNA fragments linked together in an X-shaped configuration. Molecules at the bottom of the X-spike are the least branched, where the cross-over point occurs at one end of the molecules. In contrast, the highest degree of branching is seen if linkage occurs in the middle of the two molecules; these intermediates are observed at the top of the spike (Fig. 2e). These X-shaped four-way branched intermediates may be linked by various topological constraints but are generally thought to be representative of Holliday junctions or hemicatenanes [1, 4, 10, 11].

The joint molecule intermediates that constitute the X-spike are prone to branch migration. This potential instability should be considered during the preparation of the genomic DNA. Crosslinking agents, such as psoralen, may be used to prevent the loss of the recombination intermediates [12]. However, in some instances it may be informative to examine the branch migration of these joint molecules. If the gel slice excised after the first dimension is heated in a branch migration buffer (e.g., 50 mM NaCl, 0.1 mM EDTA, 10 mM Tris, pH 8.0) at 50–60 °C, then the X-spike will be resolved to a discrete spot in alignment with, but below, the 2n spot. This is because these molecules migrate as joint molecules in the first dimension (with 2n content), but after branch migration they migrate separately (with 1n content) in the second dimension. It is possible to determine if the X-spike is representative of Holliday junctions by performing the branch migration in both the presence and absence of 10 mM $MgCl_2$. In the presence of magnesium ions, the arms of Holliday junctions arrange into a stacked conformation, which is refractory to branch migration. Therefore the prevention of branch migration in the presence of magnesium is indicative of Holliday junction structures [4, 13].

3.6.4 Identification of Replication Origins Using the 2D Gel Technique

The 2D gel technique can also be used to identify putative replication origins. At bidirectional replication initiation sites, the two replication forks diverge to create a replication bubble. This structure increases in size as the replication forks move apart. If an origin of replication is located within the central third of the restriction fragment, then these replication bubbles can be detected by the 2D technique. The resulting distinctive bubble arc is a hallmark of replication initiation in the region. Bubble structures are more branched and less mobile than their Y-shaped counterparts, and therefore the bubble arc lies above the Y-arc on a 2D gel (Fig. 2f). As the bubble structures increase in size, they become progressively less mobile. Therefore the bubble arc rises from the 1n spot and reaches a maximum at almost 2n content, when the arc ends at a position close to the top of the X-spike (Fig. 2f). In the specimen gel illustrated in Fig. 2a, a bubble arc is not visible. However, the schematic in Fig. 2f illustrates where the bubble arc would appear if an active origin of replication was located within the centre of this restriction fragment.

3.6.5 Fork Direction Gels

By digesting the genomic DNA in the gel slice (after running the first dimension) with an additional restriction enzyme, it is possible to determine the direction of replication fork movement within a fragment [9]. Figure 3 illustrates how the direction of fork movement can be established using this modification to the standard procedure. If the new restriction enzyme site is asymmetrically placed within the original fragment, then the direction of movement of the replication forks through the fragment will influence the shape of the modified Y-arcs. If the direction of the replication fork is such that the new restriction site removes the branched end of the forks as they enter the original restriction fragment, then these molecules will be converted to linear structures (as illustrated in Fig. 3b by the part of the grey arc highlighted within the dashed circled). In this case, the resultant Y-arc will therefore not rise immediately from the 1n spot. The arc will only rise when the molecules appear branched again; this occurs when the fork has entered the new smaller restriction fragment (Fig. 3b). In contrast, if the forks move through the fragment in the opposite direction, then the molecule will remain branched until the fork reaches the new restriction site, and consequently the Y-arc will rise immediately from the 1n spot (Fig. 3a). However, in this example, the largest intermediates will be converted to linear molecules when the fork passes the new restriction site (represented by the part of the grey arc highlighted within the dashed circled in Fig. 3a). Note that in both cases the intermediates migrate further in the second dimension, as the new restriction site has reduced the overall size of the fragment, when compared with the original restriction fragment in the unmodified 2D gel. However, the fragments remain the same size in the first dimension, as the additional cut site is not introduced until after the first dimension has been run. Native–native agarose 2D fork direction gels are discussed in further detail in Chapter 5 of this issue.

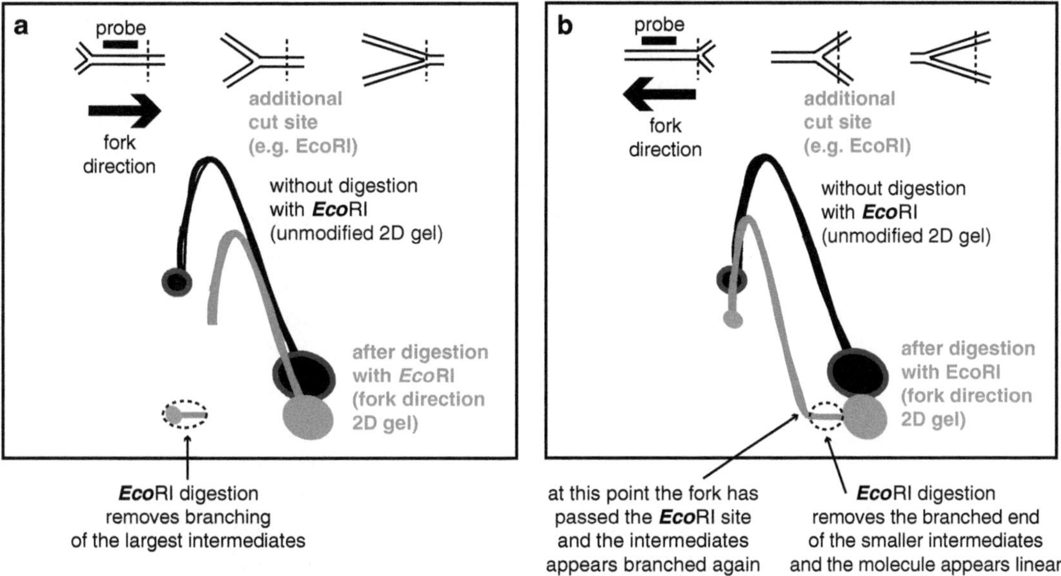

Fig. 3 2D native–native agarose fork directional gels. It is possible to determine the direction of the replication fork movement through a restriction fragment by cutting the intermediates after the first dimension with an additional restriction enzyme (*dashed line*). The *black arcs* represent the standard 2D result without the treatment between the first and second dimensions, while the *grey arcs* illustrate the fork direction arcs, when the intermediates have been digested between the dimensions with the additional restriction enzyme. Note that in this example the radiolabelled probe is specific to a region to the *left* of the additional cut site. (**a**) In this case, if the replication fork movement is from *left* to *right*, and the additional restriction site lies towards the right-hand side of the fragment, then only the larger intermediates will be converted to unbranched molecules as a result of the new restriction site (represented by the part of the *grey arc* highlighted within the *dashed circled*). (**b**) In contrast, if the replication forks move from right to left then the smallest intermediates will be cleaved into linear molecules (represented by the part of the *grey arc* highlighted within the *dashed circled*). In this example, the intermediates will only appear as branched molecules when the replication forks have passed into the smaller restriction fragment, beyond the new restriction site. *See* Subheading 3.6.5 for a more detailed explanation

4 Notes

1. The procedure outlined in this chapter has been optimized for visualizing DNA intermediates from asynchronous *S. solfataricus* cells. It may be necessary to modify the protocol when examining genomic DNA from other organisms. Branched molecules occur at very low abundance in total genomic DNA preparations, with the linear DNA making up the bulk of the signal. It is this linear DNA that can be observed with ethidium bromide staining after the second dimension (Fig. 1b). After Southern transfer and hybridization with a specific probe, the linear DNA is seen as the intense 1n spot (Fig. 2a). In principle, higher concentrations of total genomic DNA will increase the strength of the weak intermediate signals, but in practice the upper limit of the genomic material that can be utilized is

dictated by the amount of DNA that can be efficiently digested in agarose plugs by the restriction enzymes. If too much genomic DNA is used, then partial restriction digest will ensue, and this will have a detrimental effect on the end result. In cases where the abundance of the branched intermediates in total genomic DNA is too low to be detected using the standard procedure, it will be necessary to enrich for these intermediates. Cell synchronization techniques are frequently used to increase the proportion of replication intermediates [14–18]. Alternatively, if cell synchronization is not possible, the branched molecules may be enriched from the bulk linear DNA by benzoylated naphthoylated DEAE (BND) cellulose chromatography [19].

2. The protocol described above is optimal for separating branched replication fragments in the size range of 3–6 kb. This procedure will need to be modified if it is necessary to examine branched intermediates greater than 6 kb in size. For large restriction fragments, it is advisable to lower the percentage of the agarose in both the first and second dimensions, while reducing the voltage of the run and increasing the total migration time [20–22].
3. UV light will damage the branched intermediates and should be avoided until after the second dimension run. By placing a piece of graph paper underneath the gel tray after the first dimension run and using this grid as a guide, it is possible to excise the gel slice with straight edges without viewing by UV light.
4. This HCl treatment step depurinates the genomic DNA and results in the nicking of the DNA during alkali treatment at the next step, thereby facilitating the transfer of larger fragments.
5. It is important to ensure that the flow of buffer across the capillary transfer occurs through the gel, and not around the edges of the blot. This "short-circuiting" can be prevented by placing strips of laboratory parafilm around the sides of the gel on the wick, preventing contact between the wick and the blotting towels.

Acknowledgments

N.P.R. would like to thank Johanna Syrjanen (Society for General Microbiology Vacation student, 2010) for performing the 2D gels used as examples in Figs. 1 and 2.

References

1. Bell L, Byers B (1983) Separation of branched from linear DNA by two-dimensional gel electrophoresis. Anal Biochem 130:527–535
2. Brewer BJ, Fangman WL (1987) The localization of replication origins on ARS plasmids in *S. cerevisiae*. Cell 51:463–471
3. Robinson NP, Dionne I, Lundgren M, Marsh VL, Bernander R, Bell SD (2004) Identification of two origins of replication in the single chromosome of the archaeon *Sulfolobus solfataricus*. Cell 116:25–38
4. Robinson NP, Blood KA, McCallum SA, Edwards PA, Bell SD (2007) Sister chromatid junctions in the hyperthermophilic archaeon *Sulfolobus solfataricus*. EMBO J 26: 816–824
5. Robinson NP, Bell SD (2007) Extrachromosomal element capture and the evolution of multiple replication origins in archaeal chromosomes. Proc Natl Acad Sci U S A 104:5806–5811
6. Matsunaga F, Forterre P, Ishino Y, Myllykallio H (2001) In vivo interactions of archaeal Cdc6/Orc1 and minichromosome maintenance proteins with the replication origin. Proc Natl Acad Sci U S A 98:11152–11157
7. Duggin IG, Bell SD (2009) Termination structures in the *Escherichia coli* chromosome replication fork trap. J Mol Biol 387: 532–539
8. Sambrook J, Russell DW (2001) Molecular cloning. A laboratory manual, 3rd ed, Cold Spring Harbor Laboratory Press, New York.
9. Friedman KL, Brewer BJ (1995) Analysis of replication intermediates by two-dimensional agarose gel electrophoresis. Methods Enzymol 262:613–627
10. Lopes M, Cotta-Ramusino C, Liberi G, Foiani M (2003) Branch migrating sister chromatid junctions form at replication origins through Rad51/Rad52-independent mechanisms. Mol Cell 12:1499–1510
11. Lucas I, Hyrien O (2000) Hemicatenanes form upon inhibition of DNA replication. Nucleic Acids Res 28:2187–2193
12. Oh SD, Jessop L, Lao JP, Allers T, Lichten M, Hunter N (2009) Stabilization and electrophoretic analysis of meiotic recombination intermediates in *Saccharomyces cerevisiae*. Methods Mol Biol 557:209–234
13. Liberi G, Maffioletti G, Lucca C, Chiolo I, Baryshnikova A, Cotta-Ramusino C, Lopes M, Pellicioli A, Haber JE, Foiani M (2005) Rad51-dependent DNA structures accumulate at damaged replication forks in sgs1 mutants defective in the yeast ortholog of BLM RecQ helicase. Genes Dev 19:339–350
14. Duggin IG, McCallum SA, Bell SD (2008) Chromosome replication dynamics in the archaeon *Sulfolobus acidocaldarius*. Proc Natl Acad Sci U S A 105:16737–16742
15. Banfalvi G (2011) Overview of cell synchronization. Methods Mol Biol 761:1–23
16. Bates D, Epstein J, Boye E, Fahrner K, Berg H, Kleckner N (2005) The *Escherichia coli* baby cell column: a novel cell synchronization method provides new insight into the bacterial cell cycle. Mol Microbiol 57:380–391
17. Ferullo DJ, Cooper DL, Moore HR, Lovett ST (2009) Cell cycle synchronization of *Escherichia coli* using the stringent response, with fluorescence labeling assays for DNA content and replication. Methods 48:8–13
18. Helmstetter CE, Thornton M, Romero A, Eward KL (2003) Synchrony in human, mouse and bacterial cell cultures – a comparison. Cell Cycle 2:42–45
19. Mesner LD, Dijkwel PA, Hamlin JL (2009) Purification of restriction fragments containing replication intermediates from complex genomes for 2-D gel analysis. Methods Mol Biol 521:121–137
20. Krysan PJ, Calos MP (1991) Replication initiates at multiple locations on an autonomously replicating plasmid in human cells. Mol Cell Biol 11:1464–1472
21. Yasukawa T, Reyes A, Cluett TJ, Yang MY, Bowmaker M, Jacobs HT, Holt IJ (2006) Replication of vertebrate mitochondrial DNA entails transient ribonucleotide incorporation throughout the lagging strand. EMBO J 25:5358–5371
22. Hyrien O, Mechali M (1992) Plasmid replication in Xenopus eggs and egg extracts: a 2D gel electrophoretic analysis. Nucleic Acids Res 20:1463–1469

Chapter 4

Analysis of DNA Structures from Eukaryotic Cells by Two-Dimensional Native–Native DNA Agarose Gel Electrophoresis

Andreas S. Ivessa

Abstract

The neutral–neutral two-dimensional agarose gel technique is mainly used to determine the chromosomal positions where DNA replication starts, but it is also applied to visualize replication fork progression and breakage as well as intermediates in DNA recombination. Here we provide a step-by-step protocol to analyze the fairly underrepresented and fragile replication intermediates in yeast chromosomal DNA. The technique can also be adapted to analyze replication intermediates in chromosomal DNA of higher eukaryotic organisms.

Key words Replication intermediates, Chromosomal DNA, DNA replication, DNA recombination, Yeast

1 Introduction

When using the neutral–neutral two-dimensional agarose gel technique, it is possible to determine not only whether replication initiates in a particular DNA region, whether replication forks migrate passively and unimpeded, whether they slow down or come to a complete stop, but also whether two DNA molecules undergo recombination events. Originally, the technique was established by Bell and Byers and was then further developed by the Brewer–Fangman laboratory [1–3].

In the first-dimension gel DNA molecules are mainly separated by molecular weight, whereas shape of the DNA molecules has mainly no influence on its electrophoretic mobility. In contrast, in the second dimension the gel running conditions are adjusted so that branched DNA structures exhibit a different mobility in agarose gels compared to linear DNA molecules of the same molecular weight. This difference is influenced by agarose concentration, voltage applied for the DNA separation, and the presence

Svetlana Makovets (ed.), *DNA Electrophoresis: Methods and Protocols*, Methods in Molecular Biology, vol. 1054,
DOI 10.1007/978-1-62703-565-1_4,

of ethidium bromide, which is known to intercalate DNA molecules and to make them more rigid. The Huberman laboratory has developed a different 2D gel technique which uses an alkaline second dimension which allows denaturing the DNA strands in the second dimension and therefore determining the amount of nascent DNA strands [4].

Here, we will discuss only the various steps to carry out the neutral–neutral two-dimensional agarose gel technique to analyze DNA replication intermediates in yeast. Interpretations of the DNA replication patterns of this technique are briefly discussed below but further information on the method and its limitations can be found in previously published literature [3]. Although this technique is mainly used to determine DNA replication intermediates in yeast and bacteria it is also applicable to higher eukaryotic organisms to study their chromosomal and mitochondrial DNA replication [5–9].

2 Materials

All solutions are prepared with ultrapure water and analytical grade reagents. Adhere to your institute waste disposal regulations when disposing waste material.

2.1 Growing and Harvesting Yeast Cells

The following amount of solutions and reagents are required to prepare genomic DNA starting with a 500 mL yeast culture (*see* **Note 1**):

1. Deionized H_2O (dH_2O), 50 mL per strain. Autoclave dH_2O and store at 4 °C.
2. Yeast growth medium, 500 mL per strain. Usually we prepare 450 mL of the growth medium without the carbon source and 50 mL of 10× solution of the carbon source (e.g. 20 % (w/v) glucose), which is added right before use.
3. 0.1 g/mL sodium azide. 5 mL per 500 mL culture will be needed.For example, for 1 sample, 0.5 g sodium azide are dissolved in 5 mL dH_2O and stored at 4 °C.
4. NIB (Nuclear Isolation Buffer), 40 mL per sample: 50 mM MOPS free acid; 150 mM KOAc; 2 mM $MgCl_2$; 150 μM Spermine; 500 μM Spermidine; 17 % (v/v) Glycerol; pH 7.2. 0.42 g of MOPS free acid (MW: 209.26); 0.59 g of potassium acetate (MW: 98.14); 0.08 mL of 1 M $MgCl_2$; 2.1 mg of Spermine (Sigma; stored at −20 °C; MW: 348.19); 20 μL of 1 M Spermidine (Sigma; stock solution stored at −20 °C; MW: 145.25); 30 mL dH_2O is added, then 6.8 mL of 100 % (v/v) glycerol, the pH is adjusted to 7.2 with KOH, then dH_2O is added to 40 mL. Do not autoclave NIB. Store at 4 °C.

5. 2-L Erlenmeyer flasks.
6. Spectrophotometer (to measure optical density—OD_{600}).
7. Floor-top centrifuge with a GS3 rotor or equivalent.
8. 500 mL Nalgene tubes to use with the GS3 rotor.
9. Refrigerated tabletop centrifuge (with a Swinging Bucket Rotor for 50 mL conical tubes).
10. 50 mL conical tubes.

2.2 Genomic DNA Purification

1. NIB (remaining from the previous step, see above).
2. TEN (Tris–EDTA–NaCl) buffer, 10 mL per sample: 50 mM Tris–HCl, pH 8.0; 50 mM EDTA; 100 mM NaCl; 0.5 mL of 1 M Tris–HCl, pH 8.0; 1 mL of 0.5 M EDTA; 0.2 mL of 5 M NaCl; 7.5 mL of dH_2O. dH_2O is added up to 10 mL. Autoclave, store at 4 °C.
3. *N*-lauroylsarcosine sodium salt. 7.5 % (w/v) solution to be prepared later. Weigh 150 mg per sample and pour into a 15 mL conical tube.
4. Proteinase K (Sigma). 20 mg/mL to be prepared during the experiment. Weigh 2.5 mg per sample, pour into a microcentrifuge tube, and store at 4 °C.
5. 1× TE buffer, 1 mL per sample: 10 mM Tris–HCl, 1 mM EDTA pH 8.0, autoclaved, stored at 4 °C.
6. Glass beads (450–600 μm; acid washed; Sigma), 6 mL per sample.
7. CsCl (optical grade, Invitrogen), 8.3 g per sample.
8. Hoechst Dye #33258 (5 mg/mL; Sigma), 250 μL per sample.
9. 5:1 Isopropanol:dH_2O, 10 mL per sample.
10. 3 M potassium acetate, pH 5.5.
11. 100 % (v/v) Isopropanol.
12. 70 % (v/v) Ethanol.
13. Floor-top centrifuge with SA-600 rotor.
14. 30 mL oak-ridge tubes for the SA-600 rotor, one tube per strain.
15. Floor-top ultracentrifuge.
16. Vertical rotor VTi65.2 (Beckman) or equivalent with accessories (a tube sealing device and torque, if needed).
17. Ultracentrifuge tubes to use with the rotor of choice, two per strain. We use Optiseal Beckman ultracentrifuge tubes (13 mm × 48 mm; capacity 4.9 mL) with VTi65.2.
18. 50 mL conical tubes.

19. 15 mL conical tubes.
20. 1 CC syringes.
21. 16.5 G needles.
22. 18 G needles.
23. Ring stand.
24. Handheld long-wavelength UV lamp.
25. Light microscope with 40× objective.
26. 37 °C water bath.

2.3 DNA Digestion

1. Agarose (Usually we use UltraPure™ agarose from Invitrogen).
2. 1× TBE buffer (recipe for 1 L: 0.91 g Na_4 EDTA; 5.5 g boric acid; 10.3 g Tris base).
3. Gel tank with a 25 cm × 15 cm casting tray and a comb for agarose DNA electrophoresis.
4. Gel-loading dye (0.02 % (w/v) Bromophenol blue; 60 % (w/v) glycerol; 1× TBE).
5. Ethidium Bromide (10 mg EtBr resuspended in 1 mL H_2O).
6. Restriction enzymes (usually we use restriction enzymes from NEB).
7. 3 M potassium acetate, pH 5.5.
8. 100 % (v/v) Isopropanol.
9. 37 °C water bath.

2.4 Two-Dimensional Gel Electrophoresis

1. About 5 L 1× TBE buffer for first- and second-dimension gels.
2. Agarose (Usually we use UltraPure™ agarose from Invitrogen for both dimensions).
3. 1 kb DNA ladder.
4. Gel-loading dye (0.02 % (w/v) Bromophenol blue; 60 % (v/v) glycerol; 1× TBE).
5. Gel tank with a 25 cm × 15 cm casting tray and a comb for agarose DNA electrophoresis.
6. Ethidium Bromide (10 mg EtBr resuspended in 1 mL H_2O).
7. Gel tank with a self-recirculating buffer system (Fisher) and a 25 cm × 20 cm gel casting tray.
8. Glass baking dish (larger than 25 cm × 20 cm).
9. Razor blades or scalpels.
10. Ruler.
11. Handheld UV lamp (Long wave, 366 nm).

2.5 Southern Blotting

1. Stratalinker.
2. Glass baking dish or a tray.

3. Denaturation solution (about 1 L per gel): 1.5 M NaCl, 0.5 N NaOH.
4. Blotting solution (about 0.5 L per gel): 1.5 M NaCl, 0.25 N NaOH.
5. Blotting membrane (Usually we use Hybond N+ from GE Healthcare).
6. 3MM Whatman paper.
7. Saran wrap or foil.
8. Paper towels.
9. 1 L of 2× SSC for washing one membrane after the Southern transfer. Recipe for 1 L: 17.5 g NaCl (final concentration: 0.3 M); 8.82 g Sodium Citrate-$2H_2O$ (final concentration: 0.03 M).
10. Church's Buffer for hybridization (Recipe for 1 L): 2 mL of 0.5 M EDTA pH 8.0 (final concentration 1 mM); 2 mL of 85 % Phosphoric Acid (H_3PO_4); 67 g of anhydrous sodium phosphate dibasic (Na_2HPO_4) (final concentration 0.5 M); 70 g of SDS (laurel sulfate) (final concentration 7 % (w/v)); dH_2O to 1 L. Gently heat on a stir plate while mixing to get everything into solution (can take 2–3 h). Store at room temperature. Heat to 65 °C before using.
11. Carrier DNA (e.g. Calf Thymus DNA, ~10 mg/mL).
12. Hybridization oven.
13. Hybridization tubes.
14. Readiprime II Kit (GE Healthcare).
15. Radioactive labeled nucleotides (e.g. alpha-^{32}P-dCTP).
16. Geiger counter.
17. G-50 mini spin columns (GE Healthcare).
18. Heat block set at 95 °C.
19. Blot wash I: 1× SSC, 0.1 % (w/v) SDS.
20. Blot wash II: 0.1× SSC, 0.1 % (w/v) SDS.
21. X-ray film and cassettes or phosphorimager storage screens.

3 Methods

3.1 Growing and Harvesting Yeast Cells (Days 1–3)

1. On day 1, pick a single yeast colony or part of a single colony, inoculate into 5 mL cultures (or 10 mL if 1 L cultures are grown) and grow for 24 h (30 °C strains) or 48 h (ts strains, 37 °C or 23 °C) in the appropriate medium at the appropriate temperature.
2. On day 2, measure the optical density (OD_{600}).

3. Dilute the cultures in large volumes of growth medium (500 mL growth medium in 2-L Erlenmeyer flasks) so that they will grow overnight and reach the OD_{600} of ~0.5–1.0 ($1.2–2.4 \times 10^7$ cells/mL, i.e. logarithmic growth phase) when they are to be harvested. The following equation is used to dilute the samples:(Desired OD_{600}/Actual OD_{600}) × Volume of new culture = Volume to be added) (*see* **Note 2**).The progress of growth is followed by monitoring OD_{600} over time.
4. On day 3, add 5 mL of 0.1 g/mL sodium azide to 500 mL cultures to stop growth; swirl well, and place flasks on ice for 5–10 min (*see* **Note 3**).
5. Pour cultures into 500 mL Nalgene tubes and spin for 7 min at 2,700 rcf (4,000 rpm in GS3 rotor) at 4 °C. Discard supernatant.
6. Add 35 mL of prechilled (4 °C) sterile dH_2O. Resuspend cells and transfer to a 50 mL conical tube on ice. Rinse the tube with another 10–15 mL of ice-cold dH_2O and fill the conical tube up to 50 mL.
7. Spin down 5 min at 2,300 rcf (2,800 rpm in a Swinging bucket rotor) and 4 °C using a refrigerated tabletop centrifuge. Discard supernatant.
8. Wash the cell pellets with 5 mL of ice-cold NIB. Spin again 5 min at 2,300 rcf (2,800 rpm) and 4 °C in the Swinging bucket centrifuge, and discard supernatant carefully leaving almost no liquid behind.
9. Resuspend the cell pellets thoroughly in 5 mL of ice-cold NIB. This can be done by vortexing the tube.
10. Place samples into the −80 °C freezer (*see* **Note 4**) if doing the DNA purification another day or proceed to **step 2** of the Genomic DNA purification protocol (*see* Subheading 3.2 below).

3.2 Genomic DNA Purification (Days 4–7)

1. Thaw samples from −80 °C freezer on ice (frozen in NIB in 50 mL conical tubes with a volume of 5–6 mL).
2. Add 6 mL of glass beads to each sample. Vortex samples in a cold room using maximum vortexing power. Take the most efficient vortex machines in the lab. Vortex samples 8–10 times for 30 s each time, with at least 30 s breaks between the vortexing sessions. Check for ~80–90 % cell lysis under a light microscope. Vortex until at least 80 % cell lysis is observed.
3. Add 5 mL of ice-cold NIB and mix with the beads by inverting the tubes several times. Put tubes back on ice and wait until glass beads settled. Transfer the top (no glass beads) to oak-ridge tubes (SA-600 rotor). Add 10 mL NIB to the glass beads, mix, and transfer again to the same oak-ridge tube. Repeat with 10 mL NIB.

4. Spin combined samples for 20 min at 4 °C and 6,370 rcf (7,300 rpm in SA-600 rotor).
5. While the samples are spinning, prepare 20 mg/mL Proteinase K resuspended in sterile dH_2O (125 μL per sample).
6. While the samples are spinning, also make 7.5 % (w/v) solution of *N*-lauroylsarcosine (2 mL per sample). Resuspend by gentle rocking to avoid too many bubbles.
7. Carefully discard supernatant from the **step 4** centrifugation and place tubes on ice. Resuspend each pellet in 1 mL of ice-cold TEN buffer using a p1000 pipetman tip to "mush" the pellet into solution. No clumps should be visible (*see* **Note 5**).
8. Add 5 mL of ice-cold TEN buffer.
9. Add 1.6 mL of 7.5 % (w/v) *N*-lauroylsarcosine and mix gently.
10. Add 120 μL of 20 mg/mL Proteinase K (final concentration of 300 μg/mL) and mix gently.
11. Incubate 1 h in a 37 °C water bath, occasionally swirling the tubes gently.
12. Transfer the precipitated solution using cut-off (orifice) p1000 tips (the solution can be very viscous) in 1.3 mL aliquots (2 × 650 μL) into six labeled microcentrifuge tubes. Pellet the undigested cell debris by spinning 10 min at full speed and in a microcentrifuge centrifuge at 4 °C.
13. Using orifice p1000 pipette tips slowly transfer the supernatant to 15 mL conical tubes. Remove all air bubbles and bring the volume up to ~7.9 mL with ice-cold TEN buffer (meniscus should be right below the 8 mL marker).
14. Slowly pour solution into 50 mL conical tubes containing 8.3 g optical grade CsCl (1.05 g/mL of solution). Save the 15 mL conical tubes because you will transfer back into those tubes.
15. Rock tubes very slowly and gently at room temperature until all the CsCl has gone into solution. This can take up to 1 h. When fully resuspended, slowly pour the solution back into the original 15 mL tubes.
16. Add 0.025 volumes of 5 mg/mL Hoechst Dye #33258 (~250 μL) and mix gently on a rocker for approximately 10–15 min (*see* **Note 6**).
17. Carefully transfer solution into two optiseal Beckman ultracentrifuge tubes using orifice p1000 tips. This involves pipetting a 980 μL volume five times per tube (4.9 mL total per tube). Make sure there is no liquid layer at the neck of the tube.
18. Balance both tubes to within 0.025 g. Seal the tubes using a tube sealing device.

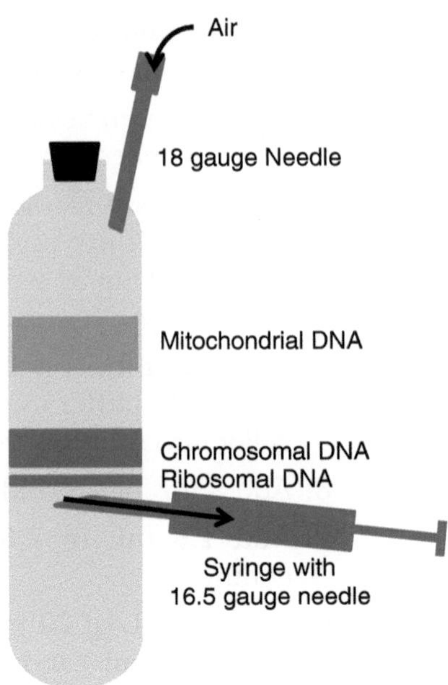

Fig. 1 Expected bands after completion of the CsCl-gradient centrifugation. When the tube is exposed to long-wave UV light from a handheld UV lamp you should see a prominent, intense band in the middle of the tube. This band contains yeast chromosomal DNA, the GC-rich ribosomal DNA (rDNA) should form a fairly light band just below. You will also observe a hazy band above the chromosomal DNA, this is the AT-rich mitochondrial DNA (mtDNA). Mark the tube (with a Sharpie pen) ~5–7 mm below the genomic DNA and the rDNA band (or below the mtDNA band, if you plan to analyze replication patterns in mtDNA). An 18 G needle is inserted at the top of the gradient tube, the chromosomal DNA is removed very slowly with a syringe and a 16.5 G needle

19. Place tubes into a VTi65.2 rotor and tighten caps to a torque of ~120 lbs/in.^{2}.
20. Spin tubes 16–20 h at 275,444 rcf (55,000 rpm) (usually we perform that at 17 °C).
21. Stop the ultracentrifuge spin. Carefully carry rotor into the dark room, try to keep the rotor leveled and do not tilt it. Remove the caps slowly.
22. Transfer a tube to a ring stand and secure it tightly.
23. Poke a hole in the top of the tube by carefully inserting an 18 G needle (leave needle in the tube; *see* Fig. 1).
24. Shine handheld UV lamp on the tube. You should see a prominent, intense band in the middle of the tube. This band contains yeast chromosomal DNA, the GC-rich ribosomal DNA (rDNA) should form a fairly light band just below (*see* Fig. 1). You will also observe a hazy band above the chromosomal

DNA band, this is the AT-rich mitochondrial DNA (mtDNA). Mark the tube with a Sharpie pencil ~5–7 mm below the genomic DNA and rDNA band (or below the mtDNA band, if you plan to analyze replication patterns in mtDNA) (*see* **Note 7**).

25. Attach a 16.5 G needle to a 1 CC syringe. "Loosen" the seal on the syringe by pulling back and forth. Insert the needle into the tube at the site of the mark with the opening of the needle facing up. Be careful not to go all the way through to the other side of the tube. Very slowly and methodically draw the DNA into the syringe. Typically we pull out about 0.7 mL per tube (stopper on syringe reads ~0.75–0.8). When done, remove the needle on top and dispose it properly. Pull the needle out of the syringe (dispose of tube + needle) and slowly transfer the DNA into a labeled 15 mL tube.
26. Extract the DNA from the remaining gradients.
27. Extract Hoechst Dye using a 5:1 Isopropanol:H_2O mix. Add 1 mL to each sample, rock gently 10–15 min in a horizontal position, let stand upright ~2 min, and discard the top layer into waste. Repeat the extractions for a total of five times. You may check the transfer of the Hoechst Dye to the top organic phase using the handheld UV lamp—but expose the tube only for a short time.
28. Using orifice p1000 tips, gently aliquot 350 μL of DNA into 1.5 mL microcentrifuge tubes (should have two microcentrifuge tubes per ultracentrifuge tube). Precipitate DNA by adding 1.1 mL of ice-cold 70 % (v/v) ethanol. Mix gently by inverting several times; do not vortex (*see* **Note 8**). Spin 20 min at top speed in a microcentrifuge at 4 °C.
29. Slowly pour off supernatants. Wash the pellets two times with 1 mL of 70 % (v/v) ethanol. Spin 5 min after each ethanol addition and discard supernatant carefully. At the final wash, remove 980 μL of the supernatant, spin down the sample in a microcentrifuge and remove the rest of the supernatant (without touching the pellet) using a p200 pipetman.
30. Air-dry the pellet on the bench for ~5 min.
31. Gently resuspend the pellets originated from the same 500 mL culture in a total volume of 500 μL 1× TE buffer (make sure that the pH is 8.0) (two tubes = 250 μL/tube) (*see* **Note 9**).
32. Gently bump sample down and store at 4 °C for 48 h. At least twice during this time gently flick the tubes to promote resuspension of the DNA.
33. After 48 h pool all the aliquots for each sample into a single tube using orifice p1000 tips and store at 4 °C for an additional 12+ h. The DNA may appear viscous.

3.3 Digestion of the Genomic DNA (Day 8)

1. We usually check the concentration of the samples in an agarose gel. Prepare a 0.6 % (w/v) agarose gel in 1× TBE buffer system.
2. Dilute the samples in a volume of 50 μL: 2 μL genomic DNA, 10 μL gel dye, 38 μL dH_2O.
3. Load 2.5, 5.0, and 10 μL of the diluted DNA samples onto the gel and run it for about 1 h at about 4 V/cm.
4. Compare different samples and aim for equal ethidium bromide (EtBr) staining intensity of genomic DNA. You may want to repeat the gels and attempt to load equal amounts. The purpose is to obtain the information on how much volume of the undigested genomic DNA you will need for the following restriction enzyme digest so that equal amounts of DNA are digested in each sample.
5. Digest the total genomic DNA with the appropriate restriction enzymes:

   Examples

   For ribosomal DNA (rDNA): Digest the DNA (100–150 μL) in the final volume of 600 μL (be sure not to vortex the DNA; pipette slowly with orifice tips and mix by inversion) with 300 U *Bgl*II or with 300 U *Stu*I for 6 h at 37 °C.

   For a DNA sequence in a single copy per genome: Digest DNA (up to 500 μL) in the final volume of 700 μL with about 330 U of a restriction enzyme for 6 h at 37 °C.
6. Confirm digest completion by running an agarose gel of the digested DNA right alongside undigested. *Bgl*II digestion yields a readily visible 4.5 kbp rDNA fragment. *Stu*I digestion yields a readily visible 5 kbp rDNA fragment (plus other sizes).
7. Precipitate digested DNA overnight at 4 °C:

   For rDNA: Add 60 μL of 3 M KOAc (pH 5.5) followed by 700 μL of 100 % isopropanol.

   For single copy: Add 70 μL of 3 M KOAc (pH 5.5) and 770 μL of 100 % isopropanol.

   Mix gently by slowly inverting the tube several times.

3.4 Two-Dimensional Gel Electrophoresis (Days 9–12)

1. Make a 0.35 % (w/v) agarose gel mix: 270 mL 1× TBE+ 0.95 g agarose in a 500 mL glass bottle (*see* **Note 10**). Heat in the microwave until all agarose has gone into solution. Check for complete resuspension by swirling the bottle and looking at the bottom for agarose that is not fully in solution. Store at 55 °C for about 30 min to cool to a good pouring temperature.

2. Pour the gel in the cold room. We usually use gel trays with the size of 15 cm × 25 cm (e.g. fits within the BioRad Sub-cell GT gel box). The gel tray should be leveled within the gel box. One gel will allow for up to eight samples. Use a 20-well comb (*see* **Note 11**). Once the gel is solidified, it is moved to room temperature.
3. Spin down the precipitated DNA (**step** 7 of Subheading 3.3) for 30 min at full-speed at 4 °C in a microcentrifuge. Pour off the supernatant and wash the pellet two times with 70 % (v/v) ethanol (5 min spins each time as above). Remove as much of the second wash as possible with a p200 pipetman. Dry on the bench for 3–5 min, resuspend in 12 μL of TE pH 8.0. Do not vortex the samples. Resuspend by flicking the tube very gently and leaving on ice for about 1 h. During that hour make about 100 μL of 1× Loading dye in 1× TE pH 8.0. After 1 h, add 4 μL of 5× dye, mix gently, and bump down by centrifugation.
4. Get ready for running the first dimension by adding 1.6–1.8 L of 1× TBE (without any EtBr) needed to fill the gel box. Remove the comb very carefully (the gel is fragile!).
5. Load 1 kb ladder in the first lane. Skip two lanes and then load the 16 μL DNA samples in every other lane. After loading the first sample, rinse each tube with 5 μL of 1× dye/TE and load that into the appropriate lanes.
6. Run the gel for 45–48 h at 22 V (about 0.73 V/cm) at room temperature.
7. For running the second dimension, prechill 2 L of 1× TBE buffer in the cold room (for one large gel box). Note: Add 60 μL of 10 mg/mL EtBr to every 2 L of the prechilled 1× TBE (0.3 μg/mL EtBr final concentration) either now or right before using.
8. On the day of setting up the second-dimension gel, make a 0.9 % (w/v) agarose second-dimension gel mix: 500 mL of 1× TBE + 4.5 g of agarose + 15 μL of 10 mg/mL EtBr (0.3 μg/mL final concentration). Mix in a 1 L glass bottle. Microwave on high until all agarose has gone into solution. Check by swirling the bottle and looking at the bottom. There should be no dense agarose visible. Store at 55 °C until ready to pour the second-dimension gel (at least 1 h).
9. At the appropriate time stop the first dimension. Carefully soak the gel in a glass baking dish with 500 mL of 1× TBE containing 0.3 μg/mL EtBr for staining the DNA. This should take 30–40 min. Since 0.31–0.35 % gels are very fragile, extra care should be taken, when the gel is transferred from the gel box to the baking dish.

10. The 1 kb ladder lane is excised and photographed with a gel imaging system (e.g. eagle-eye from Stratagene). Photograph a ruler alongside the ladder and overlay the two. Determine which 10 cm slice of the gel to run in the second dimension and excise the gel slice by cutting between the lanes and of the right length (*see* **Note 12**).
11. Transfer the slices to the 2D gel apparatus tray (four samples per apparatus; *see* Fig. 2a) (e.g. Fischer self-circulating gel box) (*see* **Note 13**). Place the DNA so that the higher molecular weights are to the left. Remove any liquid around the gel slices carefully with a paper towel (not touching the gel slices). Seal the edges of the gel and seal the slices in place with agarose (55 °C). Wait until the agarose has solidified (about 10 min).
12. Set up the gel apparatus in the cold room and level it. Pour the second-dimension gel and make sure the gel slices do not move. Remove any air-bubbles on the surface and especially between the gel slices and the second-dimension agarose using a tip. Allow to solidify for approximately 45 min to 1 h. Remove the gel dams and add the prechilled 2 L of 1× TBE/EtBr and run the gel for 18–24 h at 130–140 V (about 3.8–4.1 V/cm) (*see* **Note 14**).
13. You can monitor the DNA migration in the second dimension with a handheld UV lamp. You should run the gel so that the smallest fragments on the arc of linear fragments are just reaching the bottom of the gel (lower right hand corner of the gel, *see* Fig. 2b).

3.5 Southern Blotting and Hybridization (Days 12–14)

1. Stop the electrophoresis. Leave the gel on the tray. Photograph the gels on a gel imaging system (e.g. eagle-eye from Stratagene).
2. Nick DNA by placing the entire gel (in a tray is recommended) in the Stratalinker (*see* **Note 15**). Use the autocrosslink function (*see* **Note 16**).
3. Cut the two gels apart in the middle (horizontally), leaving you with a set of two gels. Place each gel in a glass baking dish and be sure to label the glass baking dishes with what is on each gel.
4. Incubate each gel 2 × 30 min in 500 mL of denaturation solution.
5. Incubate each gel 30 min in 500 mL of blotting solution.
6. Assemble Southern transfer in a semidry mode: A saran wrap (foil) is placed on the bench. The gel is placed in a flipped orientation (upside down) on the wrap. Hybond N+ cnylon membrane (pre-wet with blotting solution) is placed on top of

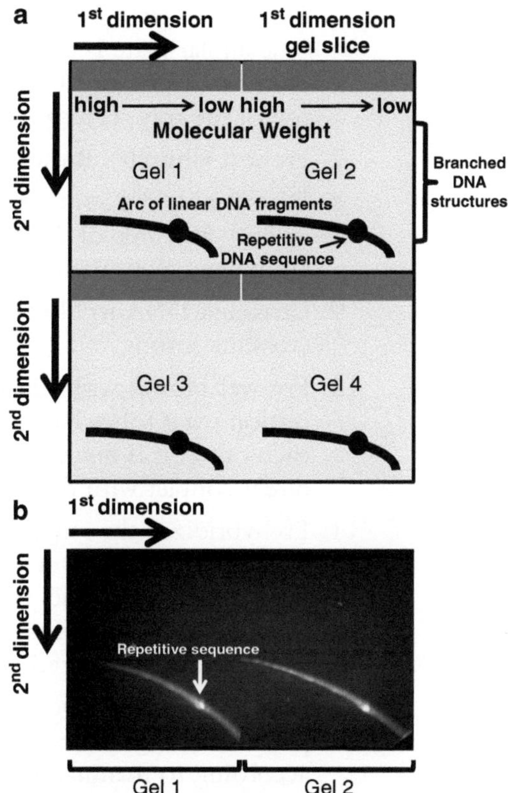

Fig. 2 A schematic of a second-dimension setup in neutral–neutral 2D gel. (**a**) The schematic displays the assembly of the first-dimension gel slices in the gel tray and the locations of the arcs of linear DNA fragments when the gels are exposed to UV-light. Any blobs of DNA within the arc of linear DNA reflect repeated DNA elements with the same sequences within the genome. For example, if yeast chromosomal DNA is digested with the restriction enzyme *Bgl*II there will be a blob of DNA at 4.5 kbp, because the ribosomal DNA array on chromosome XII consists of 100–150 repeats of an exact 9.1 kbp unit, which will be cut in half by digesting chromosomal DNA with *Bgl*II. The arc of the digested DNA fragments appears as a curve, because the separation gels in the first and second dimensions have different agarose percentages. (**b**) An image of a 2D gel (EtBr-stained gel) after completion of the second dimension is shown which is viewed by an UV-imaging system

the gel. A pipette is rolled over the membrane several times to remove any trapped bubbles between the gel and the membrane. Two wet (blotting solution) 3MM Whatman chromatography papers are placed on the membrane. The pipette is rolled over again several times to remove any trapped bubbles. Residual blotting solution on the side on the saran wrap is removed with paper towels. Two dry Whatman papers are placed on that followed by a stack of dry paper towels (about

5 cm in height). The saran wrap is wrapped up and a gel tray or a similar plate is placed on top. A small weight (e.g. aluminum heat block or a 0.5 L bottle filled with water) is placed on top of the tray. Transfer the gels 12+ h.

7. Preheat Church's Buffer to 65 °C (15 mL per large tube).
8. Disassemble the Southern transfer, gently rinse blots two times in 250–500 mL of 2× SSC for about 5 min, put blots on a saran wrap (foil) with DNA facing up.
9. Crosslink DNA to blots using the Stratalinker using the auto-crosslink setting.
10. Pre-wet blot(s) with dH_2O and place them into large hybridization oven tubes back to back (DNA sides facing away from each other; it is fine to have the DNA from one of the blots in direct contact with the glass).
11. Prehybridize the membranes with 15 mL of preheated Church's buffer for at least 5 min in a 65 °C hybridization oven with a rotisserie. Add carrier DNA (e.g. denatured calf thymus DNA; to a final concentration of 100 μg/mL) and continue prehybridization for 1–2 h.
12. To make a probe, label a specific DNA fragment with a random prime labeling kit (e.g. Readiprime II Kit (GE Healthcare)) according to manufacture's protocol (*see* **Note 17**).
13. Purify the probe by removing unincorporated radioactive nucleotides using a G-50 size exclusion column according to manufacture's recommendations.
14. Denature the probe by heating it for 5 min at 95 °C, then snap cool on ice.
15. Pour off pre-hybridization solution. Add the probe and denatured carrier DNA (e.g. calf thymus DNA; final concentration of 100 μg/mL) to fresh Church's buffer (15 mL) at 65 °C placed in a 50 mL conical tube. Pour the 15 mL probe solution into the hybridization tube (*see* **Note 18**).
16. Place the tubes back into the oven and hybridize overnight at 65 °C.
17. If you wish to save your probe for future hybridizations, pour the probe back into the 50 mL conical tube and store at 4 °C (It will turn into a solid at this temperature—this is okay). Heat the frozen probe to 65 °C for several hours before the next hybridization.
18. Wash the blots in blot wash I in the tubes at 65 °C. Do three washes of at least 5–10 min each. At least the first wash should go into the radioactive liquid waste container; the other washes may go down the drain (this has to be in accordance with your local radioactive waste disposal regulations).

19. Next, wash blots two times for 10 min each in 100 mL of 65 °C blot wash II in the tubes.
20. Take the blots out of the tube and transfer them to baking dishes filled with 65 °C blot wash II. Incubate with gentle agitation for another 10 min. Two blots, back-to back, can go in the same baking dish as long as they are not stuck together during washing.
21. Wrap each blot twice with saran wrap, make sure that no liquid is dripping out of the wrap.
22. Place blots into a film cassette and expose to X-ray film (*see* **Notes 19** and **20**).

3.6 Imaging and Image Interpretation

Interpretations of 2D gels might turn out sometimes very complicated. In Fig. 3a some of the rather simple DNA structures which can be resolved on a 2D gel, are shown as schematics. If replication forks are moving at constant speed through an examined DNA fragment just a simple arc is visible (Fig. 3a, panel #1). If replication forks are slowing down in a particular region within the examined DNA fragment, then the arc will be darker in this area (Fig. 3a, panel #2, pause). The density of the pausing signal reflects the degree of slowing of replications forks. The darker and more intense the pausing signal is, the higher is the number of replication forks that slow down in this particular area. Different DNA structures appear if replication forks initiate from within the analyzed fragment and migrate in opposite directions (Fig. 3a, panel #3, called bubble arc) or if replication forks converge within the examined fragment (Fig. 3a, panel #4). Structures observed in panel #3 indicate zones where DNA replication starts (also called origins of replications), whereas structures observed in panel #4 indicate zones where replication terminates (termination zones). A few examples of results of 2D gel experiments are displayed in Fig. 3b. If a single copy per genome sequence is studied, the intensity of the arc may then reflect how fast replication forks are migrating through the examined DNA fragment. For example, if the arc is rather light, replication forks may migrate fast through the examined fragment (Fig. 3b, #1), thus statistically there are only a few replication intermediates present in the examined fragment at the moment the cells were harvested for the analysis. In contrast, if replication forks migrate slowly through an examined fragment either a particular region of the arc is darker (Fig. 3b, #2) or the entire arc is dark (Fig. 3b, #3). More complex structures appear in 2D gels if several DNA replication processes occur in the same examined fragment (Fig. 3b, #4). In this example, DNA replication initiates from the middle and continues in both directions. Whereas the replication forks migrating to the right will leave the fragment, the replication forks migrating to the left encounter a replication fork barrier. To finish replication

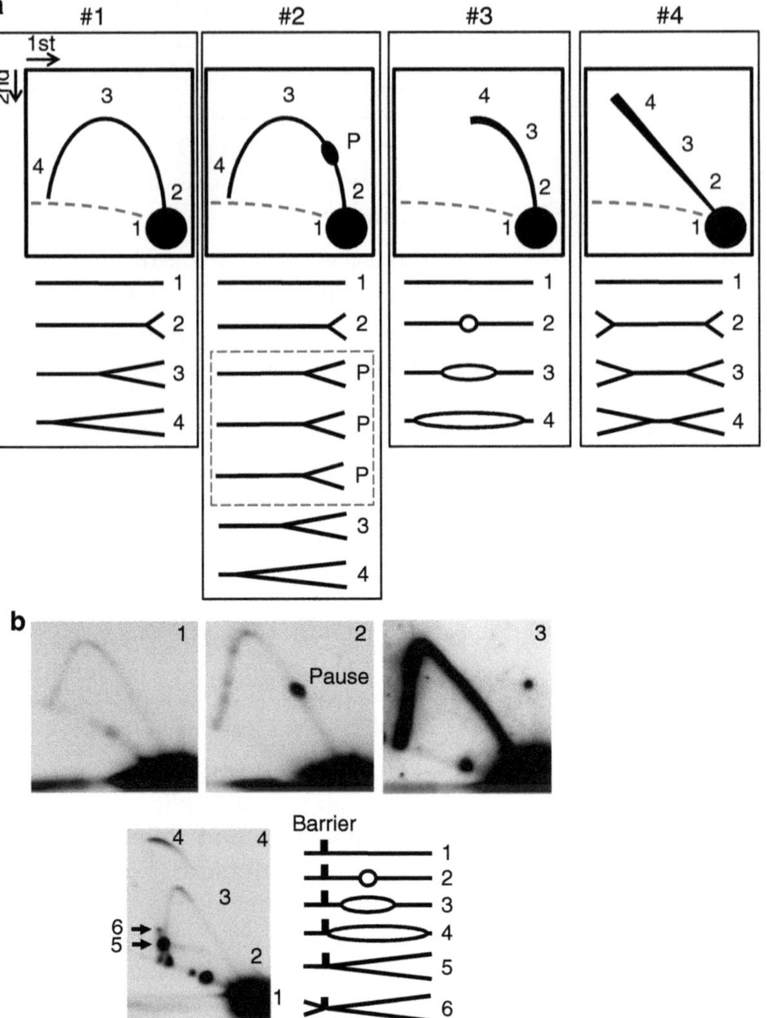

Fig. 3 Examples and interpretations of 2D gels. (**a**) Schematics of various examples of DNA structures, which can be resolved by the 2D gel technique, are displayed. These are: (*1*) Passively migrating replication forks. (*2*) Passively migrating replication forks, which pause in a specific region. (*3*) Replication, which initiates within the examined fragment (a *bubble arc*). (*4*) Replication, which terminates in the studied area (a *termination spike*). (**b**). Examples of actual 2D gels are shown (*1*) Unimpeded replication, (*2*) Replication slowed down at a certain location (*pause*), (*3*) Replication slowed down over the entire area of the examined region, (*4*) Various DNA structures within one 2D gel (replication initiation (*structures* #2–#4), replication arrest at a replication fork barrier (*structure* #5), termination of replication (*structure* #6))

successfully, replication forks coming from the other side of the barrier converge with the stalled forks. Further information about DNA structures resolved in 2D gels can be found in previous publications [3].

4 Notes

1. 500 mL are good for three gels to visualize DNA replication intermediates in ribosomal DNA, or one gel for single copy sequences.
2. Usually, we inoculate WT strains to an OD_{600} of 0.002–0.004 to reach an OD_{600} of 0.5–1 after about 15 h of growth, if YEPD medium is used.
3. There is also the option of adding EDTA to the final concentration of 30 mM to prevent DNA degradation. However, we observed no significant difference in the DNA replication patterns whether EDTA was added or not.
4. To facilitate fast thawing, we usually place the 50 mL conical tubes in a horizontal position in the −80 °C freezer. Make sure the tubes are tightly closed.
5. We avoid vortexing and pipetting up and down when resuspending the pellets in TEN buffer as doing so may sheer the DNA.
6. There might be a few yellow particles deriving from the Hoechst Dye, but the particles should not have any effect on the quality of the gradient centrifugation.
7. We usually turn off the UV lamp while extracting the DNA to avoid DNA nicking.
8. If there is a lot of DNA in the sample you should observe first a "slimy," rather translucent cloud, which after multiple tube-inversions turns into a grayish ball-like matter.
9. Do not vortex and do not pipette up and down the genomic DNA. Simply flick the tube gently.
10. A 0.35 % is good for resolving DNA fragments in the range of 3–6 kbp. The percentage of the agarose should decrease (e.g. down to 0.31 %) if larger DNA fragments (e.g. up to 10 kbp or longer—*see* ref. 10) are analyzed, and increase (more than 0.35 %) for the analysis of smaller DNA fragments (smaller than 3 kbp).This is the amount of agarose to be resuspended for a gel tray with the size of 15 cm×25 cm (fits within the BioRad Sub-cell GT gel box).
11. Though we usually use a 20-well comb that is 1.5 mm thick, a comb with 0.75 mm thickness gives even slightly better resolved replication intermediates.
12. We leave the proposed 1 N spot (where the majority of the probe binds) about 15–20 mm from the bottom end of the gel slice. It is the easiest to cut out a 10 cm block of the gel by cutting 15–20 mm below the 1 N spot and 80–85 mm above and then to split the block between the lanes while the gel is on the

UV light box. To facilitate the cutting procedure we place wet paper towels around the gel to prevent the gel from moving. Take a new razor blade for cutting the lanes. The cutting surface of the gel slices should be sharp and straight. Try to cut with the UV setting on preparative (long wave) to avoid DNA nicking.

13. Take a used, wet (water) X-ray film to transfer the gel slices. The gel slices are very fragile, avoid bending them too much. We try to straighten the gel slices in the gel tray using the wet X-ray film.
14. If the DNA fragments have a length of 6–10 kbp (or even longer) we tend to use lower voltages (i.e. 100–130 V; about 2.9–3.8 V/cm) and to run it over 2 days to prevent distortion of the arcs of DNA replication intermediates [10].
15. Be careful not to touch/hit the electrical probe in the back of the Stratalinker.
16. Alternatively, the gel can be treated with 0.25 M HCl for 30 min.
17. We prepare DNA fragments for making DNA probes by PCR using total yeast genomic DNA (usually purified by CsCl gradient centrifugation). The purity and specificity of the DNA fragment is sufficiently high if the PCR is performed consecutively two times (using genomic DNA for the first PCR, and then using 1:10,000-diluted DNA from the first PCR amplification for the second PCR).
18. Stringency can be increased or decreased by altering the temperature of hybridization (e.g. for the AT-rich mtDNA we lower the hybridization temperature to 55 °C) and/or washes as well as by adding formamide.
19. Avoid having any water or other liquid between the x-ray film and the saran wrap in which the blots are wrapped, because that may create background signals.
20. Exposure times may vary. For example, blots with highly repetitive ribosomal DNA might be quite intense in radioactive signal. If you measure between 50,000 to several 100,000 cpm with a Geiger Counter, an exposure of 30 min to a few hours to X-ray film might be sufficient [11]. Single copy per genome sequences may however need longer exposure (despite high signal in the 1 N spot), usually from overnight to a couple of days are sufficient. If the replication intermediates appear very faint, one might consider arresting cells in G1 phase (with alpha factor) and then releasing them into S-phase and harvesting samples every minute. This procedure may allow obtaining a culture enriched for cells in S-phase, i.e. cells undergoing DNA replication. You may also expose the membrane to a

phosphor storage screen and then analyze with a phosphor imaging system. This method is more sensitive and also allows determining the quantitative intensities in each replication intermediate.

References

1. Bell L, Byers B (1983) Separation of branched from linear DNA by two-dimensional gel electrophoresis. Anal Biochem 130:527–535
2. Brewer BJ, Sena EP, Fangman WL (1988) Analysis of replication intermediates by two-dimensional agarose gel electrophoresis. Cancer Cells 6:229–234
3. Friedman KL, Brewer BJ (1995) Analysis of replication intermediates by two-dimensional agarose gel electrophoresis. Methods Enzymol 262:613–627
4. Huberman JA, Spotila LD, Nawotka KA, El-Assouli SM, Davis LR (1987) The in vivo replication origin of the yeast 2 μm plasmid. Cell 51:473–481
5. Holt IJ, Lorimer HE, Jacobs HT (2000) Coupled leading- and lagging-strand synthesis of mammalian mitochondrial DNA. Cell 100:515–524
6. Yasukawa T, Yang MY, Jacobs HT, Holt IJ (2005) A bidirectional origin of replication maps to the major noncoding region of human mitochondrial DNA. Mol Cell 18:651–662
7. Le Breton C, Hennion M, Arimondo PB, Hyrien O (2011) Replication-fork stalling and processing at a single psoralen interstrand crosslink in Xenopus egg extracts. PLoS One 6:e18554
8. Gerber JK, Gogel E, Berger C, Wallisch M, Muller F, Grummt I, Grummt F (1997) Termination of mammalian rDNA replication: polar arrest of replication fork movement by transcription termination factor TTF-I. Cell 90:559–567
9. Lopez-Estrano C, Schvartzman JB, Krimer DB, Hernandez P (1999) Characterization of the pea rDNA replication fork barrier: putative cis-acting and trans-acting factors. Plant Mol Biol 40:99–110
10. Ivessa AS, Lenzmeier BA, Bessler JB, Goudsouzian LK, Schnakenberg SL, Zakian VA (2003) The *Saccharomyces cerevisiae* helicase Rrm3p facilitates replication past nonhistone protein–DNA complexes. Mol Cell 12:1525–1536
11. Ivessa AS, Zhou J-Q, Zakian VA (2000) The *Saccharomyces* Pif1p DNA helicase and the highly related Rrm3p have opposite effects on replication fork progression in ribosomal DNA. Cell 100:479–489

Chapter 5

Directionality of Replication Fork Movement Determined by Two-Dimensional Native–Native DNA Agarose Gel Electrophoresis

Andreas S. Ivessa

Abstract

The analysis of replication intermediates by the neutral–neutral two-dimensional agarose gel technique allows determining the chromosomal positions where DNA replication initiates, whether replication forks pause or stall at specific sites, or whether two DNA molecules undergo DNA recombination events. This technique does not, however, immediately tell in which direction replication forks migrate through the DNA region under investigation. Here, we describe the procedure to determine the direction of replication fork progression by carrying out a restriction enzyme digest of DNA imbedded in agarose after the completion of the first dimension of a 2D gel.

Key words Replication intermediates, Chromosomal DNA, DNA replication, Direction of replication fork movement, *In-gelo*, Yeast

1 Introduction

When using the neutral–neutral two-dimensional agarose gel technique, it is possible to determine not only whether replication initiates in a particular DNA region, whether replication forks migrate passively and unimpeded, whether they slow down or come to a complete stop, but also whether two DNA molecules undergo recombination events. However, the information about the direction of replication fork movement within the DNA region cannot be immediately gained from these experiments, unless replication forks arrest, or significantly slow down within the analyzed DNA fragment, or form a so called termination zone [1]. Here we describe a method which allows determining the direction in which replication forks move through a DNA region of interest: DNA fragments separated in the first-dimension gel are digested with another restriction enzyme directly in the gel (called *in-gelo* digest) and then subjected to another electrophoresis run. This method

Svetlana Makovets (ed.), *DNA Electrophoresis: Methods and Protocols*, Methods in Molecular Biology, vol. 1054, DOI 10.1007/978-1-62703-565-1_5, © Springer Science+Business Media New York 2013

has been previously established by the Brewer/Fangman laboratory [1, 2]. To facilitate the analysis of such *in-gelo* restriction enzyme digests, DNA regions through which replication forks are known to migrate passively are preferred for analysis.

The key steps of this procedure are summarized below:

- Genomic DNA is digested with the first restriction enzyme(s). These enzymes need to be chosen in such a way that the second enzyme (*see* below) will cleave off one quarter to one third of the analyzed DNA fragment (*see* Fig. 1a).
- The DNA fragments are separated in the first-dimension gel in the absence of ethidium bromide (EtBr) according to their molecular weights.
- A gel slice with the separated DNA fragments is cut.
- The DNA in the gel is digested with a second restriction enzyme (*in-gelo* restriction enzyme digest). This digest removes one quarter to one third from the analyzed DNA fragment (*see* Fig. 1a).
- DNA molecules are separated in the second-dimension gel in the presence of EtBr according to their shapes (degree of branching).
- DNA is transferred to a membrane and hybridized to a DNA probe recognizing the larger DNA fragment (*see* Fig. 1a).
- Depending on the direction of fork movement, two outcomes are possible (*see* Fig. 1a).

For further interpretations of *in-gelo* restrictions enzyme digest in 2D gels please *see* Subheading 3.6 below. Methods papers by the Brewer and Fangman laboratory are also recommended for further reading [1, 2].

2 Materials

All solutions are prepared with ultrapure water and analytical grade reagents. Adhere to your institute waste disposal regulations when disposing waste material.

2.1 Growing and Harvesting Yeast Cells

The following amount of solutions and reagents are required to prepare genomic DNA starting with a 500 mL yeast culture (*see* **Note 1**):

1. Deionized H_2O (dH_2O), 50 mL per strain. Autoclave dH_2O and store at 4 °C.
2. Yeast growth medium, 500 mL per strain. Usually we prepare 450 mL of the growth medium without the carbon source and

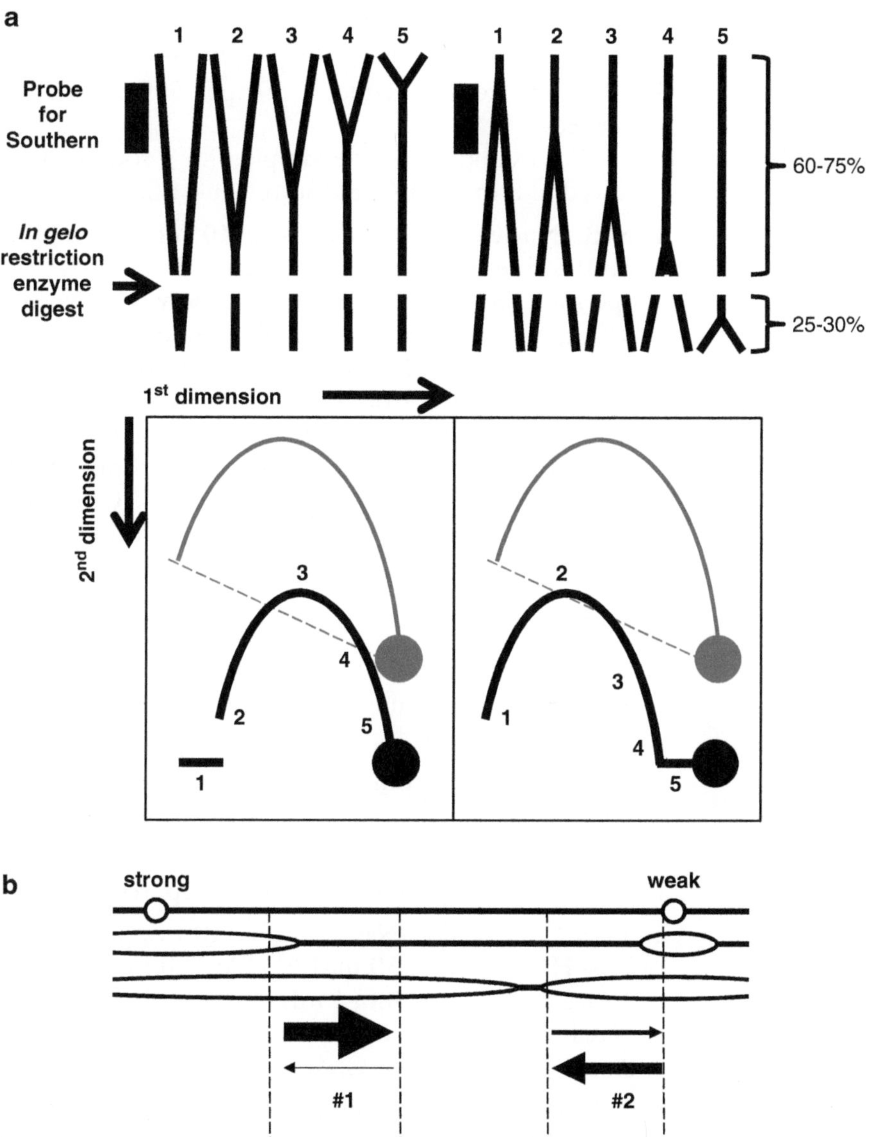

Fig. 1 Possible outcomes of an *in-gelo* restriction enzyme digest. (**a**) A schematic for experimental design to address the directionality of replication by using *in-gel* digest as part of 2D gel electrophoresis. DNA is first digested with a restriction enzyme that will generate a DNA fragment to be examined. The second restriction enzyme is used for the *in-gelo* digest. It should then cleave off one quarter to one third from the original DNA fragment. The probe is recognizing the larger DNA fragment after the *in-gelo* digest. Depending on whether the replication forks are migrating in one or the other direction through the analyzed fragment (*top panel*, DNA replication intermediates sets 1–5 on the *left* and *right*), two possible outcomes are expected (*bottom panel*, schematics on the *right* and on the *left* correspond to the replication intermediates sets on the *top*). Since the restriction enzyme in the second dimension will cut off a quarter to one third from the large DNA fragment, replication intermediates in the second dimension will have a lower molecular weight. Positioning of the replication intermediates 1–5 (see *top panels*) in the gel are indicated. (**b**) Depending on the location of the examined DNA fragment within chromosomal DNA, *in-gelo* digests may yield one of the two possible outcomes. In the case of the fragment #1, the replication forks are almost exclusively migrating in one direction. If the examined DNA fragment is close to a weak origin of replication (fragment #2), both possible outcomes may take place

50 mL of 10× solution of the carbon source (e.g. 20 % (w/v) glucose), which is added right before use.

3. 0.1 g/mL sodium azide. 5 mL per 500 mL culture will be needed.

   For example, for one sample, 0.5 g sodium azide are dissolved in 5 mL dH_2O and stored at 4 °C.

4. NIB (Nuclear Isolation Buffer), 40 mL per sample: 50 mM MOPS free acid; 150 mM KOAc; 2 mM $MgCl_2$; 150 μM Spermine; 500 μM Spermidine; 17 % (v/v) Glycerol; pH 7.2. 0.42 g of MOPS free acid (MW: 209.26); 0.59 g of potassium acetate (MW: 98.14); 0.08 mL of 1 M $MgCl_2$; 2.1 mg of Spermine (Sigma; stored at −20 °C; MW: 348.19); 20 μL of 1 M Spermidine (Sigma; stock solution stored at −20 °C; MW: 145.25); 30 mL dH_2O is added, then 6.8 mL of 100 % (v/v) glycerol, the pH is adjusted to 7.2 with KOH, then dH_2O is added to 40 mL. Do not autoclave NIB. Store at 4 °C.

5. 2-L Erlenmeyer flasks.

6. Spectrophotometer (to measure optical density—OD_{600}).

7. Floor-top centrifuge with a GS3 rotor or equivalent.

8. 500 mL Nalgene tubes to use with the GS3 rotor.

9. Refrigerated tabletop centrifuge (with a Swinging Bucket Rotor for 50 mL conical tubes).

10. 50 mL conical tubes.

2.2 Genomic DNA Purification

1. NIB (remaining from the previous step, see above).

2. TEN (Tris–EDTA–NaCl) buffer, 10 mL per sample: 50 mM Tris–HCl, pH 8.0; 50 mM EDTA; 100 mM NaCl; 0.5 mL of 1 M Tris–HCl, pH 8.0; 1 mL of 0.5 M EDTA; 0.2 mL of 5 M NaCl; 7.5 mL of dH_2O. dH_2O is added up to 10 mL. Autoclave, store at 4 °C.

3. *N*-lauroylsarcosine sodium salt. 7.5 % (w/v) solution to be prepared later. Weigh 150 mg per sample and pour into a 15 mL conical tube.

4. Proteinase K (Sigma). 20 mg/mL to be prepared during the experiment. Weigh 2.5 mg per sample, pour into a microcentrifuge tube, and store at 4 °C.

5. 1× TE buffer, 500 mL per sample: 10 mM Tris–HCl, 1 mM EDTA pH 8.0, autoclaved, stored at 4 °C.

6. Glass beads (450–600 μm; acid washed; Sigma), 6 mL per sample.

7. CsCl (optical grade, Invitrogen), 8.3 g per sample.

8. Hoechst Dye #33258 (5 mg/mL; Sigma), 250 μL per sample.

9. 5:1 Isopropanol: dH_2O, 10 mL per sample.

10. 3 M potassium acetate, pH 5.5.
11. 100 % (v/v) Isopropanol.
12. 70 % (v/v) Ethanol.
13. Floor-top centrifuge with SA-600 rotor.
14. 30 mL oak-ridge tubes for the SA-600 rotor, one tube per strain.
15. Floor-top ultracentrifuge.
16. Vertical rotor VTi65.2 (Beckman) or equivalent with accessories (a tube sealing device and torque, if needed).
17. Ultracentrifuge tubes to use with the rotor of choice, two per strain. We use Optiseal Beckman ultracentrifuge tubes (13 mm × 48 mm; capacity 4.9 mL) with VTi65.2.
18. 50 mL conical tubes.
19. 15 mL conical tubes.
20. 1 CC syringes.
21. 16.5 G needles.
22. 18 G needles.
23. Ring stand.
24. Handheld long-wavelength UV lamp.
25. Light microscope with 40× objective.
26. 37 °C water bath.

2.3 DNA Digestion (Before the First- and Second-Dimension Gels)

1. Agarose (usually we use UltraPure™ agarose from Invitrogen).
2. 1× TBE buffer (Recipe for 1 L: 0.91 g Na_4EDTA; 5.5 g boric acid; 10.3 g Tris base).
3. Gel tank with a 25 cm × 15 cm casting tray and a comb for agarose DNA electrophoresis.
4. Gel-loading dye: 0.02 % (w/v) Bromophenol blue; 60 % (w/v) glycerol; 1× TBE.
5. Ethidium Bromide (10 mg/mL in dH_2O).
6. Restriction enzymes (usually we use restriction enzymes from NEB).
7. 3 M potassium acetate, pH 5.5.
8. 100 % (v/v) Isopropanol.
9. 37 °C water bath.

2.4 Two-Dimensional Gel Electrophoresis and In-Gelo Restriction Enzyme Digestion

1. About 5 L 1× TBE buffer for first and second-dimension gels.
2. Agarose (Usually we use SeaKem® GTG® (Lonza) or UltraPure™ (Invitrogen) agarose for the first dimension (and *in-gelo* restriction enzyme digestion), and UltraPure™ agarose from Invitrogen for the second dimension).
3. 1 kb DNA ladder.

4. Gel-loading dye: 0.02 % (w/v) Bromophenol blue; 60 % (v/v) glycerol; 1× TBE.
5. Gel tank with a 25 cm × 15 cm casting tray and a comb for agarose DNA electrophoresis.
6. Small container for *in-gelo* restriction enzyme digestion (e.g. plastic cover of tip box as long they measure a minimum length of 12 cm and a height of 2 cm. These boxes usually fit well the gel slices of 10 cm length).
7. 1× TE (*see* Subheading 2.2 above).
8. 500 mL 1× restriction enzyme digestion buffer (*see* **Note 2**).
9. Ethidium Bromide (10 mg/mL in dH_2O).
10. Gel tank with a self-recirculating buffer system (Fisher) and a 25 cm × 20 cm gel casting tray.
11. Glass baking dish (larger than 25 cm × 20 cm).
12. Razor blades or scalpels.
13. Ruler.
14. Handheld UV lamp (Long wave, 366 nm).

2.5 Southern Blotting

1. Stratalinker.
2. Glass baking dish or a tray.
3. Denaturation solution (about 1 L per gel): 1.5 M NaCl, 0.5 N NaOH.
4. Blotting solution (about 0.5 L per gel): 1.5 M NaCl, 0.25 N NaOH.
5. Blotting membrane (Usually we use Hybond N+ from GE Healthcare).
6. 3MM Whatman paper.
7. Saran wrap or foil.
8. Paper towels.
9. 1 L of 2× SSC for washing one membrane after the Southern transfer. Recipe for 1 L: 17.5 g NaCl (final concentration: 0.3 M); 8.82 g Sodium Citrate-$2H_2O$ (final concentration: 0.03 M).
10. Church's Buffer for hybridization (Recipe for 1 L): 2 mL of 0.5 M EDTA pH 8.0 (final concentration 1 mM); 2 mL of 85 % Phosphoric Acid (H_3PO_4); 67 g of anhydrous sodium phosphate dibasic (Na_2HPO_4) (final concentration 0.5 M); 70 g of SDS (laurel sulfate) (final concentration 7 % (w/v)); dH_2O to 1 L. Gently heat on stir plate while mixing to get everything into solution (can take 2–3 h). Store at room temperature. Heat to 65 °C before using.
11. Carrier DNA (e.g. Calf Thymus DNA) ~10 mg/mL.
12. Blot wash I: 1× SSC, 0.1 % (w/v) SDS.

13. Blot wash II: 0.1× SSC, 0.1 % (w/v) SDS.
14. Hybridization oven.
15. Hybridization tubes.
16. Readiprime II Kit (GE Healthcare).
17. Radioactive labeled nucleotides (e.g. alpha-^{32}P-dCTP).
18. Geiger counter.
19. G-50 mini spin columns (GE Healthcare).
20. Heat block set at 95 °C.
21. X-ray film and cassettes or phosphorimager storage screens.

3 Methods

In this chapter we will particularly focus on how to carry out a restriction enzyme digestion of DNA in an agarose gel (*in-gelo* restriction enzyme digestion). This kind of digestion is an important step in the determination of the direction in which replication forks migrate through DNA.

Although most parts of the protocol are the same as used for the analysis of yeast replication intermediates by the neutral–neutral two-dimensional agarose gel technique, which is described in Chapter 4, for clarity we have also incorporated these sections in this protocol. Further differences are also mentioned in the Subheading 4 below.

3.1 Growing and Harvesting Yeast Cells (Days 1–3)

1. On day 1, pick a single yeast colony or part of a single colony, inoculate into 5 mL cultures (or 10 mL if 1 L cultures are grown) and grow for 24 h (30 °C strains) or 48 h (ts strains, 37 °C or 23 °C) in the appropriate medium at the appropriate temperature.
2. On day 2, measure the optical density (OD_{600}).
3. Dilute the cultures in large volumes of growth medium (500 mL growth medium in 2-L Erlenmeyer flasks) so that they will grow overnight and reach the OD_{600} of ~0.5–1.0 ($1.2–2.4 \times 10^7$ cells/mL, i.e. logarithmic growth phase) when they are to be harvested. The following equation is used to dilute the samples:

   (Desired OD_{600}/Actual OD_{600}) × Volume to be added (*see* **Note 3**).

   The progress of growth is followed by monitoring OD_{600} over time.
4. On day 3, add 5 mL of 0.1 g/mL sodium azide to 500 mL cultures to stop growth; swirl well, and place flasks on ice for 5–10 min (*see* **Note 4**).

5. Pour cultures into 500 mL Nalgene tubes and spin for 7 min at 2,700 RCF (4,000 RPM in GS3 rotor) at 4 °C. Discard supernatant.
6. Add 35 mL of prechilled (4 °C) sterile dH_2O. Resuspend cells and transfer to a 50 mL conical tube on ice. Rinse the tube with another 10–15 mL of ice-cold dH_2O and fill the conical tube up to 50 mL.
7. Spin down 5 min at 2,300 RCF in a Swinging bucket rotor and 4 °C using a refrigerated tabletop centrifuge. Discard supernatant.
8. Wash the cell pellets with 5 mL of ice-cold NIB. Spin again 5 min at 2,300 RCF and 4 °C in the Swinging bucket centrifuge, and discard supernatant carefully leaving almost no liquid behind.
9. Resuspend the cell pellets thoroughly in 5 mL of ice-cold NIB. This can be done by vortexing the tube.
10. Place samples into the –80 °C freezer (*see* **Note 5**) if doing the DNA purification another day or proceed to **step 2** of the Genomic DNA purification protocol (*see* Subheading 3.2 below).

3.2 Genomic DNA Purification (Days 4–7)

1. Thaw samples from –80 °C freezer on ice (frozen in NIB in 50 mL conical tubes with a volume of 5–6 mL).
2. Add 6 mL of glass beads to each sample. Vortex samples in a cold room using maximum vortexing power. Take the most efficient vortex machines in the lab. Vortex samples 8–10 times for 30 s each time, with at least 30 s breaks between the vortexing sessions. Check for ~80–90 % cell lysis under a light microscope. Vortex until at least 80 % cell lysis is observed.
3. Add 5 mL of ice-cold NIB and mix with the beads by inverting the tubes several times. Put tubes back on ice and wait until glass beads settled. Transfer the top (no glass beads) to oak-ridge tubes (SA-600 rotor). Add 10 mL NIB to the glass beads, mix, and transfer again to the same oak-ridge tube. Repeat with 10 mL NIB.
4. Spin combined samples for 20 min at 4 °C and 6,370 RCF (7,300 RPM in SA-600 rotor).
5. While the samples are spinning, prepare 20 mg/mL Proteinase K resuspended in sterile dH_2O (125 μL per sample).
6. While the samples are spinning, also make 7.5 % (w/v) solution of *N*-lauroylsarcosine (2 mL per sample). Resuspend by gentle rocking to avoid too many bubbles.
7. Carefully discard supernatant from the **step 4** centrifugation and place tubes on ice. Resuspend each pellet in 1 mL of ice-cold

TEN buffer using a p1000 pipetman tip to "mush" the pellet into solution. No clumps should be visible (*see* **Note 6**).

8. Add 5 mL of ice-cold TEN buffer.
9. Add 1.6 mL of 7.5 % (w/v) *N*-lauroylsarcosine and mix gently.
10. Add 120 μL of 20 mg/mL Proteinase K (final concentration of 300 μg/mL) and mix gently.
11. Incubate 1 h in a 37 °C water bath, occasionally swirling the tubes gently.
12. Transfer the precipitated solution using cut-off (orifice) p1000 tips (the solution can be very viscous) in 1.3 mL aliquots (2×650 μL) into 6 labeled microcentrifuge tubes. Pellet the undigested cell debris by spinning 10 min at full speed in a microcentrifuge centrifuge at 4 °C.
13. Using orifice p1000 pipette tips, slowly transfer supernatant to 15 mL conical tubes. Remove all air bubbles and bring the volume up to ~7.9 mL with ice-cold TEN buffer (meniscus should be right below the 8 mL marker).
14. Slowly pour solution into 50 mL conical tubes containing 8.3 g optical grade CsCl (1.05 g/mL of solution). Save the 15 mL conical tubes because you will transfer back into those tubes.
15. Rock tubes very slowly and gently at room temperature until all the CsCl has gone into solution. This can take up to 1 h. When fully resuspended, slowly pour the solution back into the original 15 mL tubes.
16. Add 0.025 volumes of 5 mg/mL Hoechst Dye #33258 (~250 μL) and mix gently on a rocker for approximately 10–15 min (*see* **Note 7**).
17. Carefully transfer solution into two optiseal Beckman ultracentrifuge tubes using orifice p1000 tips. This involves pipetting a 980 μL volume five times per tube (4.9 mL total per tube). Make sure there is no liquid layer at the neck of the tube.
18. Balance both tubes to within 0.025 g (25 mg). Seal the tubes using a tube sealing device.
19. Place tubes into a VTi65.2 rotor and tighten caps to a torque of ~120 lbs/inch^2.
20. Spin tubes 16–20 h at 275,444 RCF (55,000 RPM) (usually we perform that at 17 °C).
21. Stop the ultracentrifuge spin. Carefully carry rotor into the dark room, try to keep the rotor leveled and do not tilt it. Remove the caps slowly.
22. Transfer a tube to a ring stand and secure it tightly.

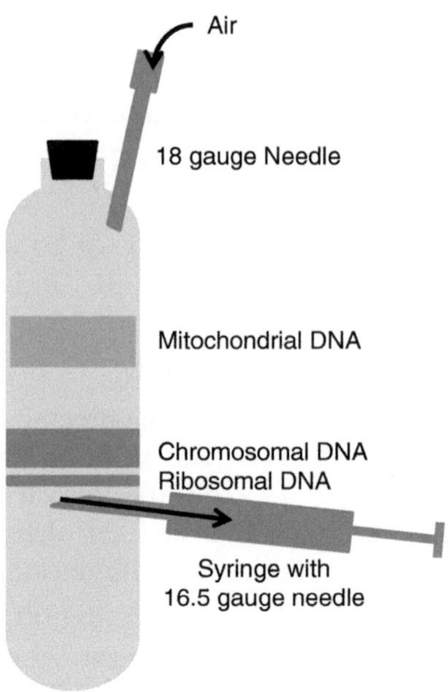

Fig. 2 Expected bands after completion of the CsCl-gradient centrifugation. When the tube is exposed to long-wave UV light from a handheld UV lamp you should see a prominent, intense band in the middle of the tube. This band contains yeast chromosomal DNA, the GC-rich ribosomal DNA (rDNA) should form a fairly light band just below. You will also observe a hazy band above the chromosomal DNA, this is the AT-rich mitochondrial DNA (mtDNA). Mark the tube (with a Sharpie pen) ~5–7 mm below the genomic DNA and the rDNA band (or below the mtDNA band, if you plan to analyze replication patterns in mtDNA). An 18 G needle is inserted at the top of the gradient tube, the chromosomal DNA is removed very slowly with a syringe and a 16.5 G needle

23. Poke a hole in the top of the tube by carefully inserting an 18 G needle (leave needle in the tube; *see* Fig. 2).
24. Shine handheld UV lamp on the tube. You should see a prominent, intense band in the middle of the tube. This band contains yeast chromosomal DNA, the GC-rich ribosomal DNA (rDNA) should form a fairly light band just below (*see* Fig. 2). You will also observe a hazy band above the chromosomal DNA band, this is the AT-rich mitochondrial DNA (mtDNA). Mark the tube with a Sharpie pen ~5–7 mm below the genomic DNA and rDNA band (or below the mtDNA band, if you plan to analyze replication patterns in mtDNA) (*see* **Note 8**).
25. Attach a 16.5 G needle to a 1 CC syringe. "Loosen" the seal on the syringe by pulling back and forth. Insert the needle into the tube at the site of the mark with the opening of the needle facing up. Be careful not to go all the way through to the other side of the tube. Very slowly and methodically draw the DNA into the syringe. Typically we pull out about 0.7 mL per tube

(stopper on syringe reads ~0.75–0.8). When done, remove the needle on top and dispose it properly. Pull the needle out of the syringe (dispose of tube + needle) and slowly transfer the DNA into a labeled 15 mL tube.

26. Extract the DNA from the remaining gradients.
27. Extract Hoechst Dye using a 5:1 Isopropanol:H_2O mix. Add 1 mL to each sample, rock gently 10–15 min in a horizontal position, let stand upright ~2 min, and discard the top layer into waste. Repeat the extractions for a total of five times. You may check the transfer of the Hoechst Dye to the top organic phase using the handheld UV lamp—but expose the tube only for a short time.
28. Using orifice p1000 tips, gently aliquot 350 μL of DNA into 1.5 mL microcentrifuge tubes (should have two microcentrifuge tubes per ultracentrifuge tube). Precipitate DNA by adding 1.1 mL of ice-cold 70 % (v/v) ethanol. Mix gently by inverting several times; do not vortex (*see* **Note 9**). Spin 20 min at top speed in a microcentrifuge at 4 °C.
29. Slowly pour off supernatants. Wash the pellets two times with 1 mL of 70 % (v/v) ethanol. Spin 5 min after each ethanol addition and discard supernatant carefully. At the final wash, remove 980 μL of the supernatant, spin down the sample in a microcentrifuge and remove the rest of the supernatant (without touching the pellet) using a p200 pipetman.
30. Air-dry the pellet on the bench for ~5 min.
31. Gently resuspend the pellets originated from the same 500 mL culture in a total volume of 500 μL 1× TE buffer (make sure that the pH is 8.0); (two tubes = 250 μL/tube) (*see* **Note 10**).
32. Gently bump sample down and store at 4 °C for 48 h. At least twice during this time gently flick the tubes to promote resuspension of the DNA.
33. After 48 h, pool all the aliquots for each sample into a single tube using orifice p1000 tips and store at 4 °C for an additional 12+ h. The DNA may appear viscous.

3.3 Digestion of the Genomic DNA (Before the First- and Second-Dimension Gels) (Day 8)

1. We usually check the concentration of the samples in an agarose gel. Prepare a 0.6 % (w/v) agarose gel in 1× TBE buffer system.
2. Dilute the samples in a volume of 50 μL: 2 μL genomic DNA, 10 μL gel dye, 38 μL dH_2O.
3. Load 2.5, 5.0, and 10 μL of the diluted DNA samples onto the gel and run it for about 1 h at about 4 V/cm.
4. Compare different samples and aim for equal ethidium bromide (EtBr) staining intensity of genomic DNA. You may want to repeat the gels and attempt to load equal amounts.

The purpose is to obtain the information on how much volume of the undigested genomic DNA you will need for the following restriction enzyme digest so that equal amounts of DNA are digested in each sample.

5. Digest the total genomic DNA with the appropriate restriction enzymes (*see* **Note 11**):

   Examples

   – For ribosomal DNA (rDNA): Digest the DNA (100–150 μL) in the final volume of 600 μL (be sure not to vortex the DNA; pipette slowly with orifice tips and mix by inversion) with 300 U *Bgl*II or with 300 U *Stu*I. Digest for 6 h at 37 °C.

   – For a DNA sequence in a single copy per genome: Digest DNA (up to 500 μL) in the final volume of 700 μL with about 330 U of a restriction enzyme for 6 h at 37 °C.

6. Confirm digest completion by running an agarose gel of the digested DNA right alongside undigested. *Bgl*II digestion yields a readily visible 4.5 kbp rDNA fragment. *Stu*I digestion yields a readily visible 5 kbp rDNA fragment (plus other sizes).

7. Precipitate digested DNA overnight at 4 °C:

   For rDNA: Add 60 μL of 3 M KOAc (pH 5.5) followed by 700 μL of 100 % isopropanol.

   For single copy: Add 70 μL of 3 M KOAc (pH 5.5) and 770 μL of 100 % isopropanol.

   Mix gently by slowly inverting the tube several times.

3.4 Two-Dimensional Gel Electrophoresis and In-Gelo DNA Digest (Days 9–12)

1. Make a 0.35 % (w/v) agarose gel mix: 270 mL 1× TBE + 0.95 g agarose in a 500 mL glass bottle (*see* **Note 12**). Heat in the microwave until all agarose has gone into solution. Check for complete resuspension by swirling the bottle and looking at the bottom for agarose that is not fully in solution. Store at 55 °C for about 30 min to cool to a good pouring temperature.

2. Pour the gel in the cold room. We usually use gel trays with the size of 15 cm × 25 cm (e.g. fits within the BioRad Sub-cell GT gel box). The gel tray should be leveled within the gel box. One gel will allow for up to eight samples. Use a 20-well comb (*see* **Note 13**). Once the gel is solidified, it is moved to room temperature.

3. Spin down the precipitated DNA (**step** 7 of Subheading 3.4) for 30 min at full-speed at 4 °C in a microcentrifuge. Pour off the supernatant and wash the pellet two times with 70 % (w/v) ethanol (5 min spins each time as above). Remove as much of the second wash as possible with a p200 pipetman. Dry on the

bench for 3–5 min. Resuspend in 12 μL of TE, pH 8.0. Do not vortex the samples, resuspend by flicking the tube very gently and leaving on ice for about 1 h. During that hour make about 100 μL of 1× Loading dye in 1× TE pH 8.0. After 1 h, add 4 μL of 5× dye, mix gently, and bump down by centrifugation.

4. Get ready for running the first dimension by adding 1.6–1.8 L of 1× TBE (without any EtBr) needed to fill the gel box. Remove the comb very carefully (the gel is fragile!).
5. Load 1 kb ladder in the first lane. Skip two lanes and then load the 16 μL DNA samples in every other lane. After loading the first sample, rinse each tube with 5 μL of 1× dye/TE and load that into the appropriate lanes.
6. Run the gel for 45–48 h at 22 V (about 0.73 V/cm) at room temperature.
7. For running the second dimension, prechill 2 L of 1× TBE buffer in the cold room (for one large gel box). Note: Add 60 μL of 10 mg/mL EtBr to every 2 L of the prechilled 1× TBE (0.3 μg/mL EtBr final concentration) either now or right before using.
8. On the day of the *in-gelo* digest and at the appropriate time stop the first-dimension run. Carefully soak the gel in a glass baking dish with 500 mL of 1× TBE containing 0.3 μg/mL EtBr for staining the DNA. This should take 30–40 min. Since 0.31–0.35 % gels are very fragile, extra care should be taken when the gel is transferred from the gel box to the baking dish.
9. The 1 kb ladder lane is excised and photographed with a gel imaging system (e.g. eagle-eye from Stratagene). Photograph a ruler alongside the ladder and overlay the 2. Determine which 10 cm slice of the gel to run in the second dimension and excise the gel slice by cutting between the lanes and of the right length (*see* **Note 14**).
10. Using a used wet X-ray film, transfer the gel slice to a small plastic container (e.g. cover of a tip box) (*see* **Note 15**).
11. Incubate the gel slice in 1× TE buffer 2 × 15 min at room temperature. Fill up the container so that the slice is covered with the liquid. Put on a shaker with very slow movement. Suck off the liquid carefully with vacuum without getting close to the gel slice.
12. Incubate the gel slice multiple times (e.g. 4 × 15 min) in 1× restriction enzyme buffer (including BSA, if recommended from the restriction enzyme supplier) which is used for the *in-gelo* restriction enzyme digestion (*see* **Note 2**). Each time fill up the container so that the slice is covered with the restriction enzyme buffer. Put on a shaker with very slow movement.

Remove the liquid carefully with vacuum without getting close to the gel slice.

13. After the incubation, remove all liquid carefully (with vacuum) and dry the surface of the gel slice carefully with a kimwipe. Remember that the first-dimension gel is quite fragile.
14. Pipette 10–20 μL of the restriction enzyme directly on top of the gel slice, distributing it evenly along the slice (*see* **Note 16**).
15. Seal the container carefully and incubate at the recommended temperature (usually 37 °C) in a moist chamber for 2 h (*see* **Note 17**).
16. After 2 h, add 10–20 μL restriction enzyme on top of the gel slice. We found that this will increase the efficiency of the digest. Continue the incubations at the recommended temperature (usually 37 °C) for up to 8 h in total.
17. During the incubation time, make a 0.9 % (w/v) agarose second-dimension gel mixes for the second-dimension gel: 500 mL of 1× TBE + 4.5 g of agarose + 15 μL of 10 mg/mL EtBr (0.3 μg/mL final concentration). Mix in a 1 L glass bottle. Microwave on high until all agarose has gone into solution. Check by swirling the bottle and looking at the bottom. There should be no dense agarose visible. Store at 55 °C until ready to pour the second-dimension gel (at least 1 h).
18. After the incubation, cover the gel slice with 1× TE buffer and place for about 30 min on a shaker with slow movements.
19. Stain the gel slice in 1× TBE containing 0.3 μg/mL EtBr for about 10–15 min.
20. Transfer the gel slice to the 2D gel apparatus tray (up to four samples per apparatus; *see* Fig. 3) (e.g. Fischer self-circulating gel box) (*see* **Note 15**). Orient so that the higher molecular weight DNA fragments are to the left. Carefully remove any liquid around the gel slices with a paper towel (not touching the gel slices). Seal the edges of the gel and seal the slices in place with agarose (55 °C). Wait until the agarose has solidified (about 10 min).
21. Set up the gel apparatus in a cold room and level it. Pour the second-dimension gel and make sure the gel slices do not move. Using a tip, remove any air-bubbles on the surface and especially between the gel slices and the second-dimension agarose. Allow to solidify for 45 min to 1 h. Remove the gel dams and add the prechilled 2 L of 1× TBE/EtBr.
22. Run the gel for 18–24 h at 130–140 V (3.8–4.1 V/cm). You can monitor the DNA migration in the second dimension with a handheld UV lamp (*see* **Note 18**).

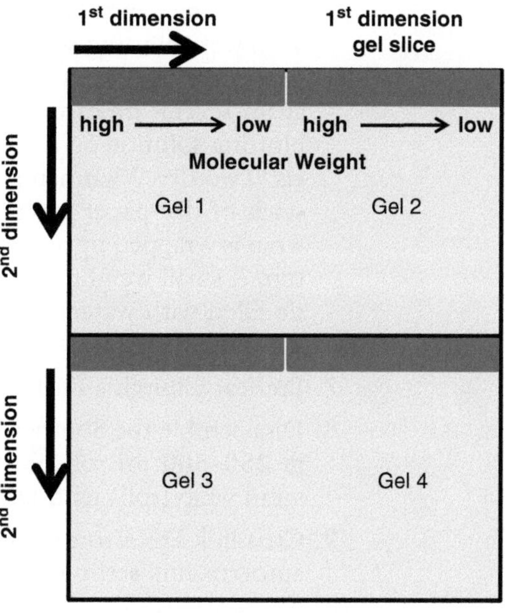

Fig. 3 A schematic of a second-dimension setup in neutral–neutral 2D gel. The first-dimension gel slices are transferred to the 2D gel apparatus tray (four samples per apparatus). The gel slices are oriented so that the higher molecular weight DNA fragments are to the left. Any liquid is carefully removed around the gel slices with a paper towel (not touching the gel slices). The edges of the gel are sealed and the slices are put in place with agarose (55 °C). The second-dimension gel is poured in the cold room

3.5 Southern Blotting and Hybridization (Days 12–14)

1. Stop the gel runs. Leave the gel on the tray. Photograph the gels on a gel imaging system (e.g. Eagle-eye from Stratagene).
2. Nick the DNA by placing the entire gel (in a tray is recommended) in the Stratalinker (*see* **Note 19**). Use the autocross-link function (*see* **Note 20**).
3. Cut the two gels apart in the middle (horizontally), leaving you with a set of two gels. Place each gel in a glass baking dish and make sure to label the glass baking dishes with sample names.
4. Incubate each gel 2 × 30 min in 500 mL of denaturation solution.
5. Incubate each gel 30 min in 500 mL of blotting solution.
6. Assemble Southern transfer in a semidry mode: A saran wrap (foil) is placed on the bench. The gel is placed in a flipped orientation (upside down) on the wrap. Hybond N+ nylon membrane (pre-wet with blotting solution) is placed on top of the gel. A pipette is rolled over the membrane several times to

remove any trapped bubbles between the gel and the membrane. Two wet (blotting solution) 3MM Whatman chromatography papers are placed on the membrane. The pipette is rolled over again several times to remove any trapped bubbles. Residual blotting solution on the saran wrap is removed with paper towels. Two dry Whatman papers are placed on top followed by a stack of dry paper towels (about 5 cm in height). The saran wrap is wrapped up and a gel tray or a similar plate is placed on top. A small weight (e.g. aluminum heat block or a 0.5 L bottle filled with water) is placed on top of the tray. Transfer the gels for over 12 h.

7. Preheat Church's Buffer to 65 °C (15 mL per large tube).
8. Disassemble the Southern transfer, gently rinse blots two times in 250–500 mL of 2× SSC for about 5 min, put blots on a saran wrap (foil) with DNA facing up.
9. Crosslink DNA to the membranes using the Stratalinker at the autocrosslink setting.
10. Pre-wet blot(s) with dH_2O and place them into large hybridization oven tubes back to back (DNA sides facing away from each other; it is fine to have the DNA from one of the blots in direct contact with the glass).
11. Prehybridize the membranes with 15 mL of preheated Church's buffer for at least 5 min in a 65 °C hybridization oven with a rotisserie. Add carrier DNA (e.g. denatured calf thymus DNA; to a final concentration of 100 μg/mL) and continue prehybridization for 1–2 h.
12. To make a probe, label a specific DNA fragment with a random prime labeling kit (e.g. Readiprime II Kit (GE Healthcare)) according to the manufacture's protocol (*see* **Note 21**).
13. Purify the probe by removing unincorporated radioactive nucleotides using a G-50 size exclusion column according to the manufacture's recommendations.
14. Denature the probe by heating it for 5 min at 95 °C, then snap cool on ice.
15. Pour off pre-hybridization solution. Add the probe and denatured carrier DNA (e.g. calf thymus DNA; final concentration of 100 μg/mL) to fresh Church's buffer (15 mL) at 65 °C placed in a 50 mL conical tube. Pour the 15 mL probe solution into the hybridization tube (*see* **Note 22**).
16. Place the tubes back into the oven and hybridize overnight at 65 °C.
17. If you wish to save your probe for future hybridizations, pour the probe back into the 50 mL conical tube and store at 4 °C (It will turn into a solid at this temperature—this is okay).

Heat the frozen probe to 65 °C for several hours before the next hybridization.

18. Wash the blots in blot wash I in the tubes at 65 °C. Do three washes of at least 5–10 min each. At least the first wash should go into the radioactive liquid waste container, the other washes may go down the drain (this has to be in accordance with your local radioactive waste disposal regulations).
19. Next, wash blots two times for 10 min each in 100 mL of 65 °C blot wash II in the tubes.
20. Take the blots out of the tube and transfer them to baking dishes filled with 65 °C blot wash II. Incubate with gentle agitation for another 10 min. Two blots, back-to back, can go in the same baking dish as long as they are not stuck together during washing.
21. Wrap each blot twice with saran wrap, make sure that no liquid is dripping out of the wrap (*see* **Note 23**).
22. Place blots into a film cassette and expose to X-ray film (*see* **Note 24**).

3.6 Imaging and Interpretation

As displayed in Fig. 1a, in the most ideal situation either one of two possible results can be obtained. In this scenario, replication forks are almost entirely migrating in just one direction through the examined DNA fragment. This information tells that there is a strong and active origin of replication placed close to one side of the examined fragment, whereas on the opposite side of the examined fragment the next active replication origin is either quite far away or exhibits only a weak activity (Fig. 1b, fragment #1). Thus, replication forks will migrate in fragment #1 primarily from left to right, but there will be almost no forks moving in the opposite direction in this particular fragment. In contrast, in examined fragment #2 the majority of the replication forks will not only migrate in the opposite direction compared to fragment #1, but there will also be considerably more forks migrating towards the weak promoter. In this scenario one should observe a strong pattern for forks migrating from right to left, but there will also be a considerably significant signal indicating forks migrating in the opposite direction (thus, coming in from the strong origin). Another problem in interpretation of such gels is that the *in-gelo* restriction enzyme digest might not occur to the completion, thus in addition to two arcs of *in-gelo* cut fragment, a larger arc of original, undigested fragments might be present (shown in gray on Fig. 1a schematics). An experimental example of an incomplete digestion is presented in Fig. 4. Further information about *in-gelo* digests can be found in previously published reviews and methodological articles [1, 2].

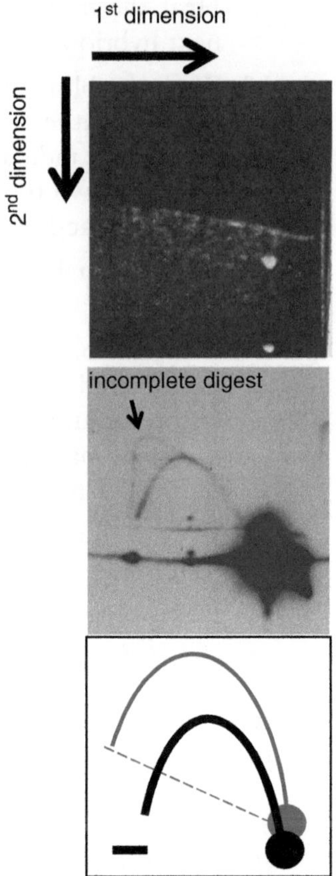

Fig. 4 Second-dimension gel after the DNA was subjected to an *in-gelo* restriction digest prior to the run. *Top panel*: an image of an ethidium bromide stained gel after running the second dimension. The spotty appearance of the DNA is characteristic of *in-gelo* digested DNA. *Center panel*: Southern blot hybridization of the gel from the *top panel*. The *arrow points* to the arc of larger molecular weight fragments indicating the incompleteness of the *in-gelo* digest. *Bottom panel*: a schematic corresponding to the Southern blot in the *middle panel*. *In-gelo* digested DNA is shown in *black*, residual *in-gelo* undigested DNA is in *gray*

4 Notes

1. 500 mL are good for three gels to visualize DNA replication intermediates in ribosomal DNA, or one gel for single copy sequences.
2. Many manufactures of restriction enzymes, such as New England Biolabs, provide recipes for their restriction enzyme buffers in the catalogues.
3. Usually, we inoculate WT strains to an OD_{600} of 0.002–0.004 to reach an OD_{600} of 0.5–1 after about 15 h of growth, if YEPD medium is used.

4. There is also the option of adding EDTA to the final concentration of 30 mM to prevent DNA degradation. However, we observed no significant difference in the DNA replication patterns whether EDTA was added or not.
5. To facilitate fast thawing, we usually place the 50 mL conical tubes in a horizontal position in the −80 °C freezer. Make sure the tubes are tightly closed.
6. We avoid vortexing and pipetting up and down when resuspending the pellets in TEN buffer as doing so may sheer the DNA.
7. There might be a few yellow particles deriving from the Hoechst Dye, but the particles should not have any effect on the quality of the gradient centrifugation.
8. We usually turn off the UV lamp while extracting the DNA to avoid DNA nicking.
9. If there is a lot of DNA in the sample you should observe first a "slimy," rather translucent cloud, which after multiple tube-inversions turns into a grayish ball-like matter.
10. Do not vortex and do not pipette up and down the genomic DNA. Simply flick the tube gently.
11. The restriction enzyme(s) for the generation of the large DNA fragment need to be chosen carefully. Usually the sites of this digest are chosen in such a way that the second digest will remove about a quarter to one third from the large DNA fragment (*see* Fig. 1a).
12. The type of agarose used for the first dimension may have an effect on the efficiency of the *in-gelo* restriction enzyme digestion. We have good experience with SeaKem® GTG® (Lonza) and UltraPure™ (Invitrogen) agarose to use them for *in-gelo* restriction enzyme digestions. The New England Biolabs Catalogue is also a good source of information about the activity of various restriction enzymes in various types of agarose. I would recommended to test the restriction enzyme digest on a defined DNA molecule first (e.g. plasmid or lambda DNA).A 0.35 % is good for resolving DNA fragments in the range of 3–6 kbp. The percentage of the agarose should decrease (e.g. down to 0.31 %) if larger DNA fragments (e.g. up to 10 kbp or longer—*see* ref. 3) are analyzed, and increase (more than 0.35 %) for the analysis of smaller DNA fragments (smaller than 3 kbp).This is the amount of agarose to be resuspended for a gel tray with the size of 15 cm × 25 cm (fits within the BioRad Sub-cell GT gel box).
13. Though we usually use a 20-well comb that is 1.5 mm thick, a comb with 0.75 mm thickness gives even slightly better resolved replication intermediates.

14. We leave the proposed 1 N spot (where the majority of the probe binds) about 15–20 mm from of the bottom of the gel slice. It is the easiest to cut out a 10 cm block of the gel by cutting 15–20 mm below the 1 N spot and 80–85 mm above and then to split the block into slices by cutting between the lanes while the gel is on the UV light box. To facilitate the cutting procedure we place wet paper towels around the gel to prevent the gel from moving. Take a new razor blade for cutting the lanes. The cutting surface of the gel slices should be sharp and straight. Try to cut with the UV setting on preparative (long wave) to avoid DNA nicking.
15. Take a used, wet (water) X-ray film to transfer the gel slices. The gel slices are very fragile, avoid bending them too much.
16. You may also dilute the restriction enzyme in some 1× restriction enzyme buffer (containing BSA) before pipetting it onto the gel slice. For example, dilute 10–20 μL of the restriction enzyme with about 20 μL of the 1× restriction enzyme buffer, then pipette it on the gel slice.
17. Place a wet kimwipe next to the gel slice to keep a moist environment, but the kimwipe should not be dripping-wet.
18. There are some differences in the DNA pattern after the *in-gelo* digest second dimension in comparison to a conventional 2D gel. The most obvious difference is that the DNA in the second dimension often does not form a round-shaped arc of linear fragments but rather comes in a spotty and hazy appearance (*see* Fig. 4, top panel). Repeated sequences (i.e. 2 μ and ribosomal DNA) may allow determining whether the *in-gelo* restriction enzyme digest was complete. It is important to note that many fragments are smaller than they would have been without the *in-gelo* restriction digest and that possibly a higher concentration of agarose is needed and/or the second-dimension gel needs to be run shorter. A second arc in the gel (Fig. 4, X-ray film, middle panel) might indicate that the *in-gelo* restriction enzyme digest may be incomplete.
19. Be careful not to touch/hit the electrical probe in the back of the Strata-linker.
20. Alternatively, the gel can be treated with 0.25 M HCl for 30 min.
21. We prepare DNA fragments for making DNA probes by PCR using total yeast genomic DNA (usually purified by CsCl gradient centrifugation). The purity and specificity of the DNA fragment is sufficiently high if the PCR is performed consecutively two times (using genomic DNA for the first PCR, and then using 1:10,000-diluted DNA from the first PCR amplification for the second PCR).

22. Stringency can be increased or decreased by altering the temperature of hybridization (e.g. for the AT-rich mtDNA we lower the hybridization temperature to 55 °C) and/or washes as well as by adding formamide.
23. Avoid having any water or other liquid between the X-ray film and the saran foil in which the blots are wrapped, because that may create background signals.
24. Exposure times may vary. For example, blots with highly repetitive ribosomal DNA might be quite intense in radioactive signal. If you measure between 50,000 to several 100,000 cpm with a Geiger Counter, an exposure of 30 min to a few hours to X-ray film might be sufficient. Single copy genes may, however, need longer exposure (despite high radioactive signal in the 1 N spot), usually from overnight to a couple of days is sufficient. You may also expose the membrane to a phosphor storage screen and then analyze with a phosphor imaging system. This method is more sensitive and also allows determining the quantitative intensities in each replication intermediate.

References

1. Friedman KL, Brewer BJ (1995) Analysis of replication intermediates by two-dimensional agarose gel electrophoresis. Methods Enzymol 262:613–627
2. Brewer BJ, Lockshon D, Fangman WL (1992) The arrest of replication forks in the rDNA of yeast occurs independently of transcription. Cell 71:267–276
3. Ivessa AS, Lenzmeier BA, Bessler JB, Goudsouzian LK, Schnakenberg SL, Zakian VA (2003) The *Saccharomyces cerevisiae* helicase Rrm3p facilitates replication past nonhistone protein–DNA complexes. Mol Cell 12:1525–1536

22. Stringency can be increased or decreased by altering the temperature of hybridization (e.g., for the 41-mer probe, we [illegible] lower the hybridization temperature to 35 °C, [illegible] as well as by adding formamide).

23. Avoid having any water or other liquids between the X-ray film and the saran foil in which the blots are wrapped, as these may create background signals.

24. Exposure times may vary. For example, if [illegible] have ribosomal DNA, might be quite [illegible] only if your measures between 50,000 [illegible] with a Geiger counter, an exposure of [illegible] to X-ray film might be sufficient. Since only [illegible] even need longer exposure time [illegible] the 2-N spots, usually [illegible] sufficient. [illegible]

Chapter 6

Native/Denaturing Two-Dimensional DNA Electrophoresis and Its Application to the Analysis of Recombination Intermediates

Jessica P. Lao, Shangming Tang, and Neil Hunter

Abstract

Two-dimensional (2D) gel electrophoresis employs distinct electrophoretic conditions to better resolve complex mixtures of molecules. In combination with Southern analysis, 2D agarose gel electrophoresis is routinely employed to detect and analyze DNA intermediates that arise during the replication and repair of chromosomes. By separating intermediates into their component single-strands, native/denaturing 2D gels can reveal structure that is not apparent under native conditions alone. Here, we describe a general method for native/denaturing two-dimensional gel electrophoresis and its application to understanding the DNA strand-composition of recombination intermediates formed during meiosis.

Key words Two-dimensional agarose gel electrophoresis, 2D gel, Native/denaturing, Neutral/alkaline, Component strand analysis, Southern hybridization, Homologous recombination, Double-strand break

1 Introduction

Two-dimensional (2D) agarose gel electrophoresis of DNA has been used extensively to detect and analyze DNA intermediates that arise during DNA replication and repair [1–5]. Two distinct types of 2D gel electrophoresis are routinely employed: native–native and native-denaturing.

Native-native (a.k.a. neutral/neutral) 2D electrophoresis maintains the native duplex structure of DNA throughout, but applies different conditions in the two dimensions to separate branched DNA molecules from linear DNA [1, 4]. The first-dimension gel separates DNA fragments based on molecular weight. The second-dimension gel resolves in the orthogonal plane and employs conditions that retard the migration of branched molecules, such as replication forks and recombination intermediates, relative to linear DNA molecules of identical molecular weight [1].

Svetlana Makovets (ed.), *DNA Electrophoresis: Methods and Protocols*, Methods in Molecular Biology, vol. 1054, DOI 10.1007/978-1-62703-565-1_6, © Springer Science+Business Media New York 2013

Native/denaturing (a.k.a. neutral/alkaline) 2D gels resolve duplex DNA in the first dimension, but apply denaturing conditions in the second dimension to separate DNA into its component single strands (Fig. 1b) [5, 6]. Thus, the molecular weight of a native duplex is indicated by its migration position in the first dimension and its strand composition is revealed in the second dimension. Native/denaturing 2D gel electrophoresis can reveal a variety of structures that are not apparent under native conditions. These include single-strand nicks, partial duplexes (gaps and overhangs), and hairpins/stem-loops [5, 7–10].

Seminal studies of homologous recombination in microorganisms have employed native/denaturing 2D gel electrophoresis to identify and analyze key intermediates formed during this essential chromosome repair process [11–16]. These include broken chromosome ends with long 3′-overhanging tails ("resected DNA

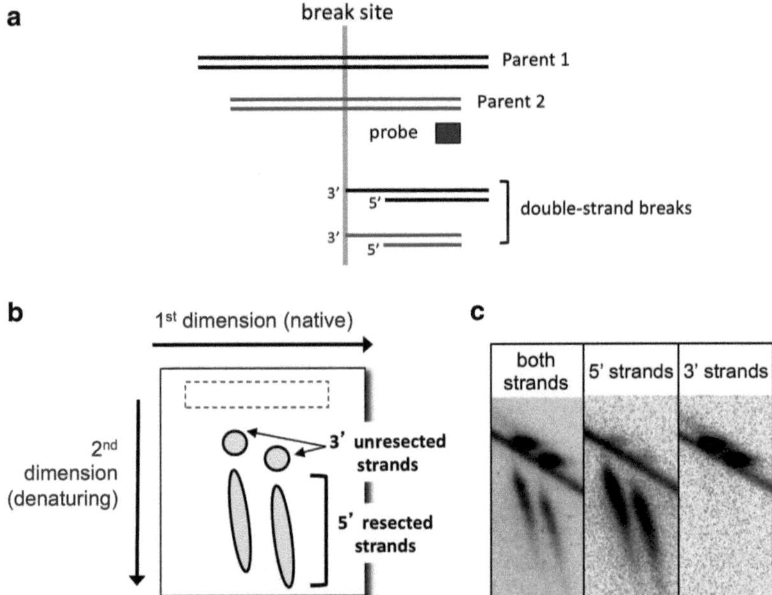

Fig. 1 Native/denaturing agarose gel analysis of break processing during meiotic recombination. (**a**) Illustration of a typical assay locus showing a DNA double-strand break "hotspot" flanked by restriction endonuclease site polymorphisms between the parental homologous chromosomes of diploid cells. Details of the well characterized *HIS4LEU2* locus can be found in ref. 7. The probe detects broken DNA fragments on one side of the break site. (**b**) Schematic of a native-denaturing 2D gel showing the migration patterns of DNA double-strand breaks formed at the assay locus shown in *panel A*. Discrete 3′-strands signals indicate that only the 5-strands undergo detectable degradation. The 5′-strand signals are smeared due to the heterogeneity of the resection process. (**c**) Representative Southern images of a native-denaturing 2D gel, successively hybridized with probes that recognize both DNA strands (double stranded probe), or just one of the strands (strand-specific, single-stranded probes)

double-strand-breaks"; Fig. 1a), hybrid or "heteroduplex" DNA formed by strand exchange between homologous chromosomes and a variety of joint-molecule strand-exchange intermediates.

Here we describe a method for native/denaturing 2D agarose gel electrophoresis and showcase its specific application to measure and quantify the processing of DNA double-strand breaks formed during the process of meiotic recombination (Fig. 1) [7]. This method is generally applicable to the quantitative analysis of a variety of DNA structures formed during chromosome metabolism.

2 Materials

Materials suppliers are listed only when deemed important. In general, any high-quality source is acceptable. All solutions should be prepared with sterile deionized water at room temperature unless otherwise stated.

2.1 DNA Extraction

1. 50 mL disposable plastic tubes.
2. 10 % sodium azide: 10 % w/v sodium azide.
3. Spheroplasting Buffer: 1 M Sorbitol, 50 mM KPO_4 buffer pH 7.0, 10 mM EDTA pH 7.5. Filter sterilize and store at 4 °C.
4. β-mercaptoethanol.
5. Zymolyase100T.
6. Guanidine Lysis Solution pH 8.0: 4.5 M Guanidine–HCl, 0.1 M EDTA pH 8.0, 0.15 M NaCl, 0.05 % sodium lauryl sarcosinate (Sarkosyl). Adjust to pH 8.0 with 50 % NaOH.
7. 200 proof ethanol.
8. RNase Stock Solution: 10× TE pH 8.0 (100 mM Tris–HCl, 10 mM EDTA pH 8.0), 50 mg/mL RNase (DNase-free). Store at −20 °C.
9. Proteinase-K Solution: 20 mg/mL proteinase-K in 20 mM $CaCl_2$, 10 mM Tris–HCl pH 7.5, 50 % glycerol. Store at −20 °C. Mix well before use.
10. Phenol pH 8.0/chloroform/isoamyl alcohol mixed in a 25:24:1 ratio.
11. 3 M NaOAc pH 5.2: 3 M sodium acetate, adjust pH to 5.2 with glacial acetic acid. Sterilize by autoclaving.
12. 10× TE pH 8.0: 100 mM Tris–HCl, 10 mM EDTA.
13. 37 °C waterbath; 65 °C waterbath.
14. 2.0 mL microcentrifuge tubes.
15. Wide-bore 1 mL micropipette tips.

2.2 Native-Denaturing Gel Electrophoresis

1. Large gel box with two large gel trays (20 cm wide × 25 cm long) and combs. We use the Buffer Puffer™ gel box, Model A5 (Owl Separation Systems) with a 36-well comb.
2. SeaKem LE Agarose (Lonza).
3. TBE: 90 mM Tris base, 90 mM boric acid, 2 mM EDTA. Prepare as 10× stock.
4. TE: 10 mM Tris–HCl pH8.0, 1 mM EDTA. Prepare as 10× stock.
5. DNA molecular weight standards covering the size range of interest.
6. Large pyrex dishes (39 cm × 45 cm) for staining and equilibrating gels.
7. Platform shaker.
8. 10 mg/mL ethidium bromide in dH_2O. Store at room temperature wrapped in aluminum foil.
9. Long-wave (350 nm) UV transilluminator box.
10. Clean razorblades.
11. Flexible plastic ruler.
12. 6× Gel Loading Dye: 0.25 % (w/v) bromophenol blue, 0.25 % (w/v) xylene cyanol FF, 15 % (w/v) Ficoll (Type 400, Pharmacia) in dH_2O. Store at room temperature.
13. Native Gel Loading Buffer: 100 μL 6× loading dye + 60 μL 10× NEB3 restriction enzyme buffer (New England Biolabs; *see* **Note 1**).
14. 5× Alkaline Running Buffer: 250 mM NaOH, 5 mM EDTA in dH_2O. Make fresh for each use.
15. 1× Alkaline Running Buffer: 50 mM NaOH, 1 mM EDTA in dH_2O. Make fresh for each use.
16. 6× Alkaline Gel Loading Buffer: 300 mM NaOH, 6 mM EDTA, 18 % (w/v) Ficoll (Type 400, Pharmacia), 0.15 % (w/v) bromocrescol green, 0.25 % (w/v) xylene cyanol in dH_2O. Store in plastic tubes at room temperature.

2.3 Southern Blotting

Make sure all containers are clean. Rinse well with dH_2O prior to use. Dry with lint-free Kimwipes.

1. Large pyrex dishes (39 cm × 45 cm) for staining and equilibrating gels, and to use as a reservoir for the Southern transfer.
2. Platform shaker.
3. 0.25 M HCl made fresh.
4. 0.4 M NaOH made fresh.
5. Glass plate (22.5 cm × 43 cm) to span large pyrex dish that will be used as a reservoir for the transfer.

6. Whatman 3MM paper (46 cm × 57 cm sheets).
7. ZetaProbe nylon membrane (BioRad).
8. Single-fold paper towels.
9. Plastic/plexi-glass plate (22 cm × 28 cm).

2.4 Southern Hybridization

1. Radioactive workstation with appropriate protective screening, Geiger counter, microcentrifuge, 95 °C heating block and waste containers for dry and liquid waste.
2. Hybridization oven and bottles (Autoblot mini hybridization oven, BELLCO glass, Inc. or equivalent).
3. 1 M sodium-phosphate pH 7.2.
4. Large pyrex dish (39 cm × 45 cm) to accommodate the Southern blot membrane.
5. Hybridization Buffer: 0.25 M sodium-phosphate pH 7.2, 0.25 M NaCl, 1 mM EDTA, 7 % SDS, 5 % Dextran Sulfate. For 500 mL: 125 mL 1 M Na-phosphate pH 7.2, 25 mL 5 M NaCl, 1 mL 0.5 M EDTA, 35 g SDS, 25 g Dextran Sulfate. Add liquids to a 1 L beaker. Make up to ~300 mL with sterile deionized water. Slowly add SDS and Dextran Sulfate while stirring. Heat to 65 °C and stir until fully dissolved. Aliquot into 50 mL plastic tubes. Store at room temperature. SDS will come out of solution so heat to 65 °C and mix well before use.
6. Sheared Salmon sperm DNA (10 mg/mL, Boehringer) in 0.2 mL aliquots. Store at −20 °C.
7. Prime-It® RmT Random Primer Labeling Kit.
8. Radiolabeled nucleotide: P32 α dCTP (Amersham PB10-205 or equivalent). Buy only as needed and store at −20 °C.
9. ProbeQuant G-50 Micro Columns (GE Healthcare).
10. 20× SCC stock solution.
11. 20 % SDS solution.
12. Low Stringency Wash: 2× SSC, 0.1 % SDS. For 1 L, place 100 mL 20× SSC in a 1 L measuring cylinder and make up to 990 mL. Add 5 mL 20 % SDS.
13. High Stringency Wash: 0.1× SSC, 0.1 % SDS. For 1 L, place 5 mL 20× SSC in a 1 L measuring cylinder and make up to 990 mL. Add 5 mL 20 % SDS.
14. Large microwave oven.
15. Large shaking water bath at 65 °C.
16. Large plastic container (Tupperware or similar) with a sealing lid to accommodate the Southern blot membrane and 0.5 L of wash buffer.

17. Long round-nosed forceps to remove membrane from hybridization bottle.
18. Whatman paper.
19. Plastic wrap.

2.5 Imaging and Analysis

1. Phosphorimager (Typhoon™, Storm™, or PhosphorImager™; GE Healthcare).
2. Storage Phosphor screens and cassettes (GE Healthcare).
3. ImageQuant™ (GE Healthcare) and Excel™ (Microsoft) software packages. Open-source software such as ImageJ (http://rsb.info.nih.gov/ij/) can also be used for image analysis.

3 Methods

3.1 DNA Extraction and Purification

DNA samples can be prepared using many standard protocols or kits, but great care should be taken to minimize extraction-induced shearing and nicking. To this end, methods that employ harsh extraction conditions (high heat or extremes of pH), vortexing and excessive pipetting must be avoided. We have published several detailed protocols for purifying high molecular weight genomic DNA from yeast cells [17]. Here, we describe our standard method.

1. Collect cell samples containing ~3×10^8 cells (~20 mL of culture with an OD_{600} of ~1) in 50 mL disposable tubes. Add 10 % Sodium Azide to cell cultures for a final concentration of 0.1 %. Harvest the cells for 4 min at $1{,}900 \times g$ in a benchtop centrifuge. Drain, spin again briefly, and remove any remaining supernatant with a pipette. Store cell pellet overnight at −20 °C, or continue.
2. Prepare sufficient spheroplasting mixture on ice; for each sample: 0.5 mL spheroplasting buffer, 5 μL β-mercaptoethanol and 0.25 mg 100-T zymolyase. Quickly and thoroughly resuspend the cell pellet in 0.5 mL spheroplasting mixture and incubate at 37 °C for 15 min. Gently mix twice during this incubation (*see* **Note 2**).
3. Harvest the spheroplasted cells for 4 min at $1{,}900 \times g$. Using a P1000 pipette, carefully remove as much of the supernatant as possible without disturbing the loose cell pellet (*see* **Note 3**). Respin if necessary. Remove supernatant to a hazardous waste container (*see* **Note 4**).
4. Add 1.5 mL Guanidine lysis solution and resuspend the stringy spheroplast pellet by flicking the bottom of the tube ("finger-vortexing"). Place at 65 °C for 20 min; finger vortex several times during this incubation to completely lyse the cells.

5. Cool tubes on ice. Add 1.5 mL 100 % ethanol and mix well by inversion. Store at −20 °C overnight, or incubate in a −20 °C freezer for 20 min.
6. Pellet for 15 min at 3,400×*g* in a benchtop centrifuge. Pour supernatant into a hazardous waste container (*see* **Note 4**) and drain tubes onto a paper towel. Respin briefly and remove the last traces of supernatant with a pipette.
7. Add 0.5 mL RNase solution. Break up and disperse pellet with a 1 mL pipette tip but do not vortex or pipette up and down. Incubate at 37 °C on a shaker or rollerdrum for 1 h. Finger vortex several times during this incubation to disperse the pellet (*see* **Note 5**).
8. Add 15 μL proteinase-K solution and incubate at 65 °C for 1 h. Finger vortex several times during this incubation to mix. Spin tubes briefly to collect the solution and then transfer to 2.0 mL eppendorf tubes using wide-bore pipette tips.
9. Add 0.5 mL phenol/chloroform/isoamyl alcohol: shake and invert for ~30 s to mix thoroughly, let stand for 3 min, shake again and then spin at full speed for 10 min in a microcentrifuge. *Carefully* remove the upper aqueous layer using wide-bore pipette tips into a new 2.0 mL microcentrifuge tube, avoiding the white interface.
10. Repeat the extraction with phenol/chloroform/isoamyl alcohol (*see* **Note 6**).
11. Ethanol precipitate the DNA: add 20 μL 3 M sodium acetate pH 5.2 and 1 mL 100 % ethanol. Invert well by inversion. If a visible spool of DNA forms, pulse spin and carefully pour off the supernatant. Otherwise leave to stand for 20 min, spin for 5 min at full speed and pour off the supernatant.
12. Rinse the DNA pellet in 1.5 mL 70 % ethanol. If the pellet is loose repeat the spin. Pour off the ethanol and drain the tubes. Spin briefly and remove the last traces of ethanol with a pipette tip.
13. Air-dry for at least 10 min, or until the pellet looks just dry. Add 50 μL 1× TE and allow the DNA to hydrate overnight in at 4 °C. Flick the tube well to mix and store at −20 °C (*see* **Note 7**).

3.2 Native-Alkaline Gel Electrophoresis

Electrophoresis conditions are optimized for analysis of DNA fragments in the 2–10 kb size range. Conditions should be optimized for fragments outside this range.

1. *Day 1*: To release DNA fragments of interest, digest an appropriate amount of sample DNA to completion. For analysis of DNA break processing in budding yeast (Fig. 1), we digest 4–10 μg of genomic DNA at a concentration of 1 μg per 20 μL, with a fourfold excess of restriction enzyme.

2. Using standard methods, ethanol-precipitate and then air-dry the digested DNA for 10 min or until the pellet looks just dry.
3. Add 15 μL 1× TE and allow to DNA to rehydrate for several minutes at 4 °C (*see* **Note 8**). Gently flick the tube to resuspend the DNA and collect drops by brief centrifugation.
4. Prepare a 0.6 % agarose gel (*see* **Note 9**): add 2.1 g SeaKem LE agarose to 350 mL 1× TBE (*without* ethidium bromide). Dissolve completely by heating in the microwave and boiling steadily for 30 s. Cool to 50 °C and pour into a clean, leveled gel tray. Allow gel to harden for at least 30 min and then add 1× TBE to just cover the gel.
5. Add 5 μL native gel loading buffer to the DNA sample(s). Mix gently.
6. Load the gel, leaving at least one lane space between different samples and two lane spaces between the molecular weight standards and the samples.
7. Run the first-dimension gel at 70 V (~2 V/cm) for 24 h at room temperature.
8. Chill at least 7 L of distilled water in a 4 °C cold room overnight (for preparation of the alkaline running buffers).
9. *Day 2*: Soak the gel in 1 L 5 mM EDTA for 15 min at room temperature with gentle shaking.
10. Using the prechilled dH_2O, prepare 1 L, 5× alkaline running buffer and 6 L, 1× alkaline running buffer. Store in the cold room.
11. Stain the gel in 1 L 1 mM EDTA+0.5 μg/mL ethidium bromide for 30 min at room temperature with gentle shaking.
12. While staining the gel, prepare 1.2 % agarose in *water* for the second denaturing dimension: add 4.4 g SeaKem LE agarose in 400 mL dH_2O (*without* ethidium bromide). Dissolve completely by heating in the microwave and boiling steadily for 30 s. Place in a 50 °C water bath to cool.
13. Lay a piece of SaranWrap on top of the long-wave UV transilluminator and carefully slide the gel on top. Visualize the DNA and using clean razorblades carefully and cleanly cut slices from the lanes to cover the size range of interest. Minimize any excess agarose on the sides of the slices. For analysis of breaks in budding yeast at the *HIS4LEU2* meiotic recombination hotspot [7] (Fig. 1), we typically excise 5 cm-long gel slices encompassing DNA in the 2.0–4.0 kb range. Cut off one corner of the gel slice in order to keep track of the orientation.
14. Using a flexible plastic ruler, transfer the slices to a clean gel tray and arrange in an orientation that is orthogonal to the first dimension (*see* **Note 10**).

15. Rig up a gel-comb to make wells for loading the molecular weight DNA standards (*see* **Notes 11** and **12**).
16. Promptly pour the cooled 1.2 % agarose to *just* cover the gel slices (there will be a little left over agarose). Carefully pour away from the slices to avoid disturbing them. Allow the gel to harden in the cold room for 30 min.
17. Soak the gel in 1 L, 5× alkaline running buffer for 30 min to denature the DNA.
18. Soak the gel in 1 L, 1× alkaline running buffer, twice for 15 min each to equilibrate the gel.
19. Place the gel in the electrophoresis apparatus. Add 1× alkaline running buffer to just cover the gel.
20. Load ~100–200 ng of appropriate molecular weight standards (for DSB analysis, we use 200 ng of Promega's 1 kb ladder), made up *fresh* in 1× denaturing buffer and alkaline loading buffer.
21. Lay a piece of Saran Wrap on top of the buffer to limit diffusion out of the gel.
22. Run the gel in the cold room at 55 V (~1.7 V/cm) for 30 h with a change of buffer after ~15 h (*see* **Note 13**).
23. Rinse the gel twice in 1 L dH_2O with gentle shaking before blotting the gel by alkaline transfer overnight (*see* **Note 14**).

3.3 Southern Transfer

Southern transfer is performed using routine methods [18].

1. *Day 1*: Soak the gel in 1 L dH_2O for 10 min. All subsequent steps, in 1 L volumes with gentle shaking. Prepare 0.25 M HCl and 0.4 M NaOH solutions. You will need sufficient NaOH to soak the gel and do the transfer, i.e. 2 L.
2. Flip the gel so that the flat side is facing up. To do this, slide the gel onto the bottom of a second gel tray, gently sandwich the gel with the other tray, quickly flip and slide the gel gently back into the buffer.
3. Soak gel in 0.25 M HCl for 20 min with gentle shaking.
4. Pour the HCl into a large beaker and rinse the gel briefly with 1 L dH_2O (*see* **Note 15**).
5. Soak the gel in 0.4 M NaOH for 30 min with gentle shaking.
6. Prepare the following: four pieces of Whatman blotting paper the same size as the gel, one piece of Whatman paper suitable for a wick (*see* **Note 16**), and one piece of Zeta Probe nylon membrane the same size as the gel. Use gloves and a clean dry surface when measuring and cutting the membrane (do not remove the protective paper; keep the membrane free of dust; avoid any crumpling or other mechanical stress on the membrane; all will increase background in subsequent hybridization). Use a fine-tipped sharpie or pencil to write the date and identifier on the top right hand side of the membrane.

7. Set up the blot. Fold the wick evenly across a clean glass plate (*see* **Note 16**). Place the glass plate and wick on top of a clean pyrex dish with both ends of the wick in the tray. Wet the surface of the wick with 0.4 M NaOH and flatten by rolling with a 25 mL glass pipette. Add more 0.4 M NaOH to the dish to create a reservoir. Place two gel-size pieces of Whatman in the center of the wick, wet well with NaOH and roll flat. You will use ~1 L total NaOH solution.
8. Carefully slide the gel from the tray to the wick. Gently push any trapped air bubbles from under the gel with gloved fingers. Empty the used NaOH into the beaker containing the used HCl (*see* **Note 17**). Rinse the dish and fill with dH_2O.
9. Wet the nylon membrane in the dH_2O (*see* **Note 18**). Minimize handling and hold only at the edges. Carefully align the top edges of the membrane and the gel and lower the membrane onto the gel. Do not move the membrane around once in contact with gel.
10. Soak the two additional gel-size pieces of Whatman in the dH_2O and place on top of the membrane. Take a 25 mL pipette and carefully, but firmly, roll out any bubbles working out from the center.
11. Place a large piece of plastic wrap across the whole construction and cut around the gel with a sharp razor blade. The plastic wrap should seal around the edges of the gel so that capillary action occurs only via the gel and membrane.
12. Take a pack of single-fold paper towels; unfold several towels and lay them flat on top of the Whatman. Evenly split the rest of the stack into two and place on top in a side-by-side arrangement. Place a plexi-glass plate (not glass) and a *small* weight (small bottle containing less than 50 mL of liquid) on top of the blot.
13. Blot for at least 6 h; generally overnight.
14. *Day 2*: Prepare 1 L, 50 mM sodium phosphate pH7.2 and pour into a clean pyrex dish that can accommodate the nylon membrane. Deconstruct the blot: remove the weight, plate, paper towels, and top two pieces of Whatman paper. Immediately peel off the membrane and place into the sodium phosphate solution to neutralize. Do not allow the membrane to dry out prior to neutralization as this will result in high background. Gently shake membrane for 10 min with one change of buffer.
15. Move directly to the pre-hybridization step (Subheading 3.4, **step 1**), or briefly drain the membrane on a piece of Whatman paper and store the damp membrane in plastic wrap at 4 °C (short term) or –20 °C (longer than 2 days).

3.4 Southern Hybridization

Southern hybridization with ^{32}P-radiolabeled DNA probes is performed using routine methods [18]. Probes should hybridize to the ends of the DNA fragments being analyzed. This "indirect end-labeling" approach allows unambiguous estimation of the sizes of DNA strands. To detect the molecular weight standards, we spike probe-labeling reactions with 0.05 ng of the DNA standard. If strand-specific probes are required (e.g. Fig. 1c), we amplify and radiolabel probes using linear PCR, as described [19].

1. Prewarm the Hybridization Solution to 65 °C and mix well by swirling. Turn on the hybridization oven and set to 65 °C. Rinse and the hybridization bottles, lids, and O-rings (if used) with dH_2O. Prewarm bottles in the oven.
2. Add 10 mL Hybridization Solution per small bottle or 20 mL per large bottle. Place bottles back into the oven.
3. Denature a 0.2 mL aliquot of sheared Salmon sperm DNA by heating in a 95 °C dry block for 5 min (*see* **Note 19**). Immediately quench on ice/water and then store on ice.
4. Submerge the membrane to be hybridized, with the DNA-side facing up, in 50 mM sodium phosphate pH7.2 in a clean pyrex dish. Carefully roll-up the membrane with DNA side on the inside. Place into hybridization bottle and slowly unroll by turning the bottle, keeping the opening face wet with hybridization solution. Avoid bubbles. Add the Salmon sperm DNA, promptly return to the oven and pre-hybridize for >30 min, preferably several hours.
5. Remove radiolabeled nucleotide from the freezer and allow to thaw. Sign-out the amount to be used in the logbook. Sign into the radioactive workstation and monitor the area using a Geiger counter.
6. Label 25–50 ng of probe DNA using a Prime-It® RmT Random Primer Labeling Kit (or equivalent) and 32P αdCTP according to the manufacturer's protocol (*see* **Note 20**).
7. Separate the probe from unincorporated nucleotides using a Probe-Quant G-50 Micro Column, according to the manufacturer's protocol.
8. Denature the probe in 95 °C heating block or boiling water bath for 2 min. Quench on ice for 2 min (*see* **Note 19**).
9. Add the denatured probe to the hybridization bottle and promptly return to oven. Monitor the work station with a Geiger counter, clean up any contamination immediately and sign-out.
10. Hybridize for >6 h, typically overnight.
11. Turn on a shaking water bath and prewarm to 65 °C. Sign into the radioactive workstation.

12. Make up 1 L Low Stringency Wash and add to a 1 L flask. Heat to 65 °C (for 1 L this takes ~3.5 min in a microwave on full power). Check the temperature carefully—do NOT heat above 65 °C or you will strip the probe.
13. Carefully pour off the hybridization mix into the radioactive liquid waste (record all volumes of liquid waste added). Fill the bottle ~1/2 full with preheated wash and return promptly to the oven. Incubate for 10 min.
14. Pour the wash into the radioactive liquid waste and record the volume. Repeat the low-stringency wash. Prepare 1.5 L High-Stringency wash and heat to 65 °C.
15. Pour the second low stringency wash into liquid waste and record the volume. Fill the bottle ~1/2 full with preheated High Stringency wash and return promptly to the oven. Incubate for 30 min.
16. Pour the wash into the radioactive liquid waste and record the volume. Carefully and quickly remove the membrane from the tube with long round-nosed forceps and place into a clean plastic container with 0.5 L prewarmed High Stringency wash. Cover the container and incubate in a 65 °C shaking water bath for 20 min with gentle shaking. To prevent the container from floating, you will need to add a weight (we use a piece of lead flashing).
17. Repeat the high-stringency wash.
18. Monitor the blot—the edges should be essentially free of counts (~2× background with the Geiger counter). If there are significant background counts, wash again. Distinct signals corresponding to specific hybridizing species should be detectable.
19. Drain excess wash from the blot onto a wad of paper towels and briefly dry DNA side up on a piece of Whatman paper. Carefully wrap the damp blot in plastic wrap. Using small pieces of tape, mount onto a piece of Whatman paper with the DNA side face-up.

3.5 Imaging and Analysis

1. Expose the hybridized Southern blots to storage-phosphor screens for >30 min to overnight and scan the screens on a Phosphorimager. To obtain the optimum exposure for quantification, we aim to have the strongest signals in the upper third of the dynamic range of the phosphorimager.
2. Open the Southern image in ImageQuant and draw a line trace through the lane containing the DNA molecular weight standards (Fig. 2a).
3. Create a graph using the analysis tool (Fig. 2a). The graph will show distance in millimeter versus radioactive signal counts.

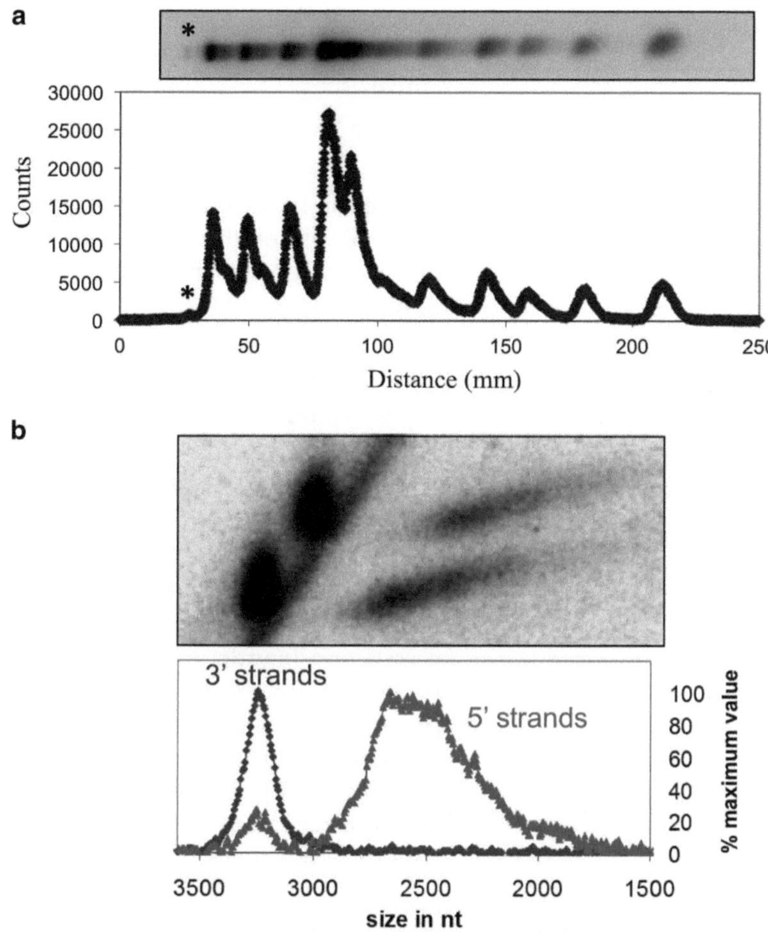

Fig. 2 Estimation of strand lengths from native/denaturing 2D gels. (**a**) *Top panel*: Southern image from a phosphorimager of a native-denaturing 2D gel showing the DNA standard (1 kb ladder). *Asterisk* indicates the location of the well. *Bottom panel*: corresponding line trace of signal intensity (counts) versus migration distance (mm). *Asterisk* indicates the small signal peak corresponding to the well. (**b**) *Top panel*: Southern image of native/denaturing 2D analysis of DNA double-strand breaks (as in Fig. 1c, but flipped 90°). *Bottom panel*: phosphorimager signal profiles (% maximum value), measured from native/denaturing 2D gels successively hybridized with strand-specific probes, plotted as a function of strand length (nucleotides, nt). Strand lengths were calculated via linear regression using DNA molecular weight standards, as described in the main text

Distinct peaks correspond to the individual fragments of the DNA standards. The position of the wells is the reference point for migration distance and is indicated by a tiny peak located above the largest signal peak for the DNA standards (Fig. 2a, the position of the well is indicated by an asterisk).

4. Import the ImageQuant graph data into Excel. Assign a size (bp) to the migration distance (mm) of each band of the DNA standards.
5. Obtain a linear regression equation by graphing migration distances (x) versus *log* (size in bp) (y).
6. For each signal of interest, calculate the migration distance (x) (Fig. 2b). Enter this value into the regression equation and solve for y. To determine fragment size in bp, take the inverse *log* of y.
7. To quantify the relative levels of hybridizing species, integrated signal intensities are measured in ImageQuant by drawing shapes around individual bands/spots and subtracting background signal from regions of equivalent size (Fig. 2b). The level of a given species is expressed as a fraction of the total hybridization signal (the sum of all analyzed bands/spots minus background).
8. For the analysis of break processing during recombination, we calculate a "resection profile," which describes the distribution of resection lengths for the 5-strands of DNA double-strand breaks (Fig. 2b); *see* [7]. To do this, the signal intensities for 5′ strands within successive length bins of 100 nt are calculated and plotted in a bar graph as percentages of the total hybridizing signal [7]. To correct for background signal, the same process is repeated for an adjacent region of the Southern blot without DSB signals.

4 Notes

1. Additional salt prevents the samples from drifting out of the wells.
2. Mix by flicking the tube.
3. Time this step carefully. Over-spheroplasted cells will not form a tight pellet and may require repeat centrifugation. Under-spheroplasted cells will form a dense pellet (resembling intact cells) and should be gently resuspended and incubated for an additional 2–5 min.
4. Follow institutional guidelines for storage and disposal of hazardous waste.
5. The pellet will not dissolve but should become a well-dispersed suspension.
6. If a distinct white interface is still visible, perform another extraction.
7. At this stage we measure the concentration of all DNA samples and adjust them to a standard concentration (typically 100–500 ng/μL).

8. DNA pellets should not be over dried. DNA can be rehydrated for several hours to overnight at 4 °C.
9. For gels of less than 0.6 % agarose, use SeaKem Gold Agarose (Lonza), which has much greater gel strength.
10. For 5 cm slices, each second-dimension gel can accommodate a maximum of two rows of three gel slices.
11. A makeshift comb for DNA standards can be made by turning a regular gel comb over in its holder (comb side facing up) and attaching well-sized plastic sticks using adhesive tape. Alternatively, teeth can be broken from an existing comb to leave wells at desired positions (typically at the two ends). When running two rows of samples, marker lanes should be staggered.
12. To accurately estimate the size of DNA strands, take special care to align the gel slices with the marker wells.
13. NaOH buffer is rapidly depleted, which will slow electrophoresis and lead to overheating. Buffer should be replaced at least once.
14. Denaturing gels are slippery and brittle. Take extra care when handling.
15. Firmly support the gel with a gloved hand when pouring off buffers.
16. Blot dimensions are as follows: pyrex dish, 39 cm × 45 cm; glass plate, 22.5 cm × 43 cm; Whatman wick: 23 cm × 28.5 cm (half of a large sheet); plexi-glass plate, 22 cm × 28 cm.
17. Check the pH of the final acid/base mixture and neutralize for disposal.
18. Float the membrane on top of the water and then tilt the tray up and down to completely submerge.
19. Use plastic lid-locks to prevent tubes from popping open. This is especially important for labeling reactions to prevent aerosol contamination.
20. To maximize specificity, DNA probes should be highly purified. Cloned probes should be gel purified from the vector backbone. PCR amplified probes should be prepared by nested PCR and gel purified.

Acknowledgements

This work was supported by National Institutes of Health National Institute of General Medical Sciences (NIH NIGMS) grant GM074223 to N.H. N.H. is an Early Career Scientist of the Howard Hughes Medical Institute.

References

1. Bell L, Byers B (1983) Separation of branched from linear DNA by two-dimensional gel electrophoresis. Anal Biochem 130:527
2. Bell LR, Byers B (1983) Homologous association of chromosomal DNA during yeast meiosis. Cold Spring Harb Symp Quant Biol 47(Pt 2):829–840
3. Brewer BJ, Fangman WL (1991) Mapping replication origins in yeast chromosomes. Bioessays 13:317
4. Dijkwel PA, Hamlin JL (1997) Mapping replication origins by neutral/neutral two-dimensional gel electrophoresis. Methods 13:235
5. Huberman JA (1997) Mapping replication origins, pause sites, and termini by neutral/alkaline two-dimensional gel electrophoresis. Methods 13:247
6. McDonell M, Simon MN, Studier FW (1977) Analysis of restriction fragments of T7 DNA and determination of molecular weights by electrophoresis in neutral and alkaline gels. J Mol Biol 110:119
7. Zakharyevich K et al (2010) Temporally and biochemically distinct activities of Exo1 during meiosis: double-strand break resection and resolution of double Holliday junctions. Mol Cell 40:1001
8. Wahls WP, DeWall KM, Davidson MK (2005) Mapping of ssDNA nicks within dsDNA genomes by two-dimensional gel electrophoresis. J Arkansas Acad Sci 59:178
9. Schwacha A, Kleckner N (1994) Identification of joint molecules that form frequently between homologs but rarely between sister chromatids during yeast meiosis. Cell 76:51
10. Lao JP, Oh SD, Shinohara M, Shinohara A, Hunter N (2008) Rad52 promotes post-invasion steps of meiotic double-strand-break repair. Mol Cell 29:517
11. Allers T, Lichten M (2001) Intermediates of yeast meiotic recombination contain heteroduplex DNA. Mol Cell 8:225
12. Schwacha A, Kleckner N (1995) Identification of double holliday junctions as intermediates in meiotic recombination. Cell 83:783
13. Cromie GA et al (2006) Single holliday junctions are intermediates of meiotic recombination. Cell 127:1167
14. Bzymek M, Thayer NH, Oh SD, Kleckner N, Hunter N (2010) Double holliday junctions are intermediates of DNA break repair. Nature 464:937
15. Oh SD et al (2007) BLM ortholog, Sgs1, prevents aberrant crossing-over by suppressing formation of multichromatid joint molecules. Cell 130:259
16. Oh SD, Lao JP, Taylor AF, Smith GR, Hunter N (2008) RecQ helicase, Sgs1, and XPF family endonuclease, Mus81-Mms4, resolve aberrant joint molecules during meiotic recombination. Mol Cell 31:324
17. Oh SD et al (2009) Stabilization and electrophoretic analysis of meiotic recombination intermediates in *Saccharomyces cerevisiae*. Methods Mol Biol 557:209
18. Green MR, Sambrook J (2012) Molecular cloning: a laboratory manual (fourth edition). Cold Spring Harbor Laboratory Press, Cold Spring Harbor, NY
19. Ruven HJ, Seelen CM, Lohman PH, Mullenders LH, van Zeeland AA (1994) Efficient synthesis of ^{32}P-labeled single-stranded DNA probes using linear PCR; application of the method for analysis of strand-specific DNA repair. Mutat Res 315:189

Chapter 7

Plasmid DNA Topology Assayed by Two-Dimensional Agarose Gel Electrophoresis

Jorge B. Schvartzman, María-Luisa Martínez-Robles, Pablo Hernández, and Dora B. Krimer

Abstract

Two-dimensional (2D) agarose gel electrophoresis is nowadays one of the best methods available to analyze DNA molecules with different masses and shapes. The possibility to use nicking enzymes and intercalating agents to change the twist of DNA during only one or in both runs, improves the capacity of 2D gels to discern molecules that apparently may look alike. Here we present protocols where 2D gels are used to understand the structure of DNA molecules and its dynamics in living cells. This knowledge is essential to comprehend how DNA topology affects and is affected by all the essential functions that DNA is involved in: replication, transcription, repair and recombination.

Key words DNA topology, DNA replication, Supercoiling, Catenanes, Knots, Topoisomerases

1 Introduction

Scientists are often criticized for their reductionist arguments. There are countless articles and textbooks with theoretical models where DNA replication and segregation are explained. Surprisingly, many describe the DNA molecule in two dimensions neglecting the fact that the cell nucleus and DNA are three-dimensional structures. DNA is a dynamic double-helix where topology affects and is affected by all the basic functions DNA plays in nature: replication, transcription, repair and recombination [1]. The fact that a DNA molecule must experience constant topological changes was recognized by Watson and Crick [2] as soon as they proposed their original model for the structure of DNA [3]. John Cairns proposed that a putative enzyme, a "swivelase," could relieve the topological tension that would eventually accumulate ahead of a transcription and replicating fork [4] and a few years later James Wang discovered a protein from

Svetlana Makovets (ed.), *DNA Electrophoresis: Methods and Protocols*, Methods in Molecular Biology, vol. 1054, DOI 10.1007/978-1-62703-565-1_7, © Springer Science+Business Media New York 2013

Escherichia coli that converts highly twisted superhelical DNA into less twisted, covalently closed forms [5, 6].

Theoretical models apart, it is not an easy task to distinguish DNA molecules with different topology experimentally. Weil and Vinograd [7], as well as Dulbecco and Vogt [8] were among the first to use electron microscopy to provide direct evidence for a ring structure of polyoma virus DNA. These electron micrographs also revealed that DNA molecules could be supercoiled. The first quantitative study of polyoma virus DNA supercoiling was achieved by Thorne [9] using agarose gel electrophoresis. Electrophoresis is one of the best methods so far to separate molecules with different masses and shapes [10]. Slight modifications and adjustments of the basic protocol are used to improve the capacity of agarose gel electrophoresis to separate molecules of the same molecular weight. Intercalating agents such as chloroquine or ethidium bromide change the twist of DNA and, in the case of covalently closed molecules, affects supercoiling [11–17]. DNA can be analyzed in two consecutive runs of electrophoresis performed in the same or in different conditions. The second run occurs perpendicular to the first. This type of analysis is usually referred to as: Two-dimensional (2D) agarose gel electrophoresis [18–25]. For non-replicating molecules, 2D gels allow not only the precise calculation of supercoiling density [26–28] as illustrated in Fig. 1, but also computation of the exact number of nodes in catenated and knotted molecules [29, 30] as shown in Fig. 2. For partially replicated molecules, 2D gels allow to reveal and characterize intra (Fig. 3) and inter-chromatid knots [29, 31–36] as well as to confirm the occurrence of replication fork reversal and the extrusion of nascent-nascent duplexes both in vivo and in vitro [34, 37–45]. Finally, 2D gels can also be used as a preparative method to enrich DNA samples for specific molecules that can be later analysed by other means [29, 37, 38, 44–46].

Here we present protocols where 2D gels are used to analyse DNA topology to improve our understanding of DNA replication, transcription, repair and recombination.

2 Materials

2.1 Electrophoretic Tanks and Trays

There are many submarine horizontal gel systems on the market. For a good separation of molecules with different shapes, though, the tanks and trays must be sufficiently long. We recommend tanks with inter-electrode distances of at least 30 cm that can hold trays of up to 20 cm long (*see* **Note 1**).

2.2 Stock Solutions

Prepare all solutions using ultrapure water and analytical grade reagents. Store all reagents at room temperature unless indicated otherwise. Follow all waste disposal regulations when disposing waste materials. No sodium azide is added to the list of reagents described below.

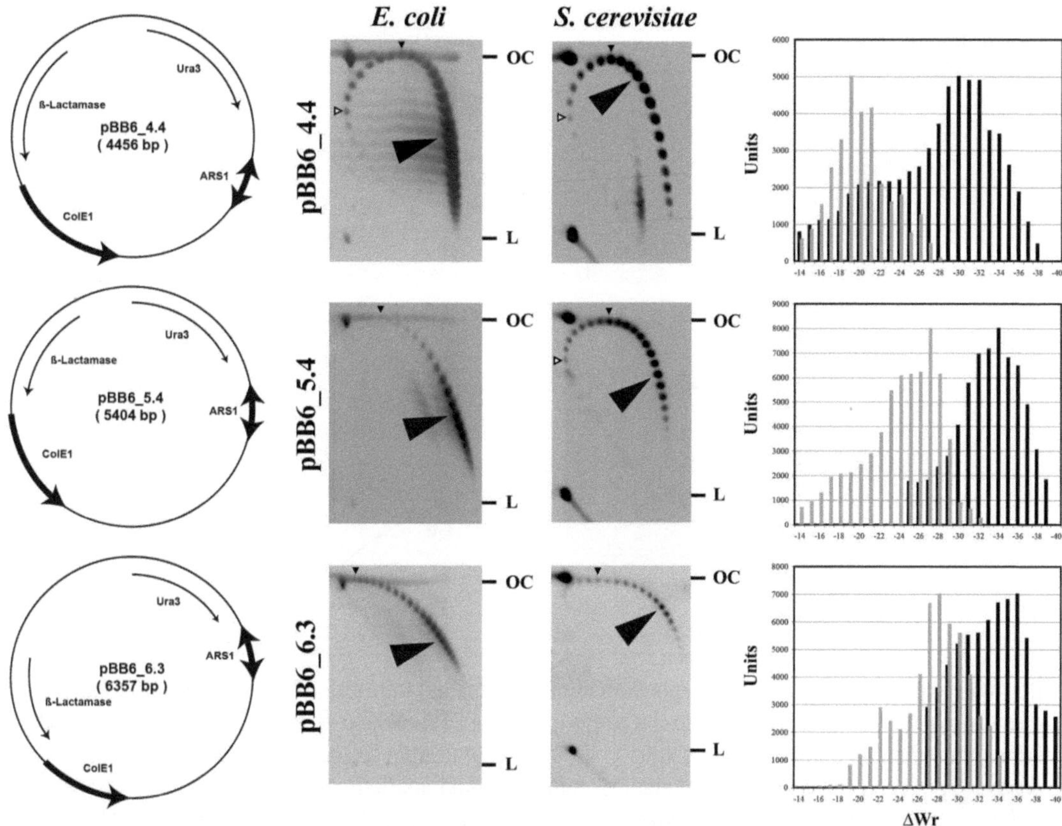

Fig. 1 The genetic maps of pBB6_4.4, pBB6_5.4 and pBB6_6.3 showing the relative position of their most relevant features: the ColE1 unidirectional origin, the gene coding for β-lactamase, the bidirectional ARS1 origin and the *URA3* gene are shown to the *left*. In the *middle*, representative autoradiograms of 2D gels isolated from *E. coli* and *S. cerevisiae* cells where the first as well as the second dimensions were run in the presence of different concentrations of chloroquine in order to resolve all the populations of topoisomers. For bacterial plasmids, the first dimension was performed in the presence of 1 μg/mL and the second dimension with 2 μg/mL CHL. For mini-chromosomes isolated from budding yeast, was run in the presence of CHL, 0.5 μg/mL the first dimension and 1 μg/mL in the second dimension. In each autoradiogram, the most abundant topoisomers are marked with *large black arrowheads*. Topoisomers that have migrated with a ΔLk = 0 during the first dimension are indicated with a *small open arrowhead* while *small black arrowhead* point to those that migrated with a ΔLk = 0 during the second dimension. To the *right* of the autoradiograms, histograms representing the corresponding densitometric figures: the values corresponding to topoisomers isolated from *E. coli* are depicted in *black* and those corresponding to *S. cerevisiae* are in *gray* (reproduced from ref. 27 with permission from WILEY-VCH Verlag GmbH & Co. KGaA, Weinheim)

2.3 Bacterial Transformation

1. LB (Luria Broth): 1 % bacto tryptone, 0.5 % bacto yeast extract, 1 % sodium chloride, pH 7.5 (*see* **Note 2**).
2. 1 M magnesium sulfate.
3. 0.1 M calcium chloride (stored at 4 °C).

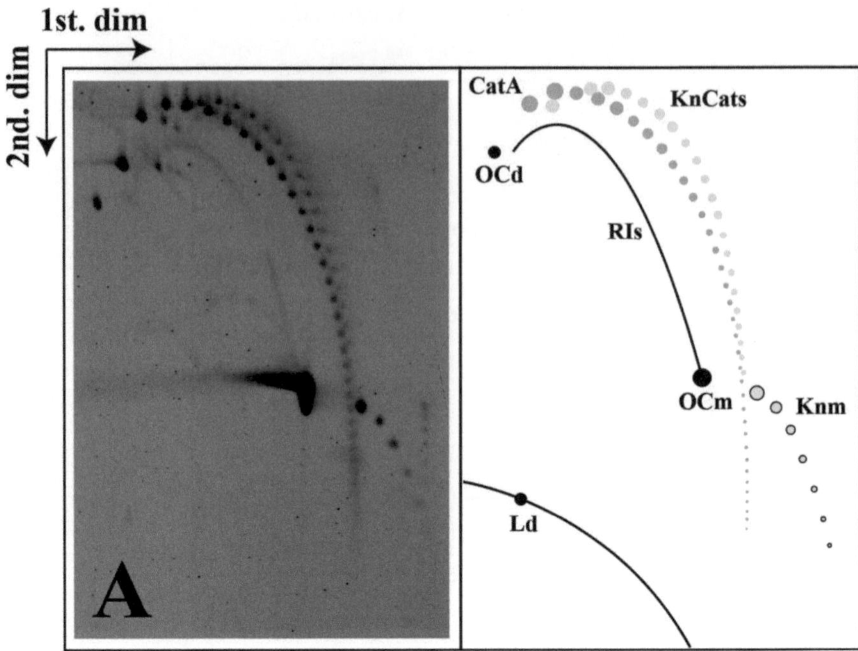

Fig. 2 Autoradiogram of a 2D gel corresponding to pBR18 isolated from DH5αF′ *E. coli* cells after exposure to norfloxacin and complete digestion with the single-stranded DNA nicking enzyme Nb-BsmI. A diagrammatic interpretation of the autoradiogram is shown to its *right* where CatAs are depicted in *blue* and knotted CatAs in *yellow*. *RIs* Nicked Replication Intermediates, *KnCats* Nicked knotted catenanes, *OCd* open circles corresponding to dimers, *OCm* open circles corresponding to monomers, *Knm* nicked knotted monomers, *Ld* linearized dimers, *Lm* linearized monomers (reproduced from ref. 30 with permission from Oxford University Press)

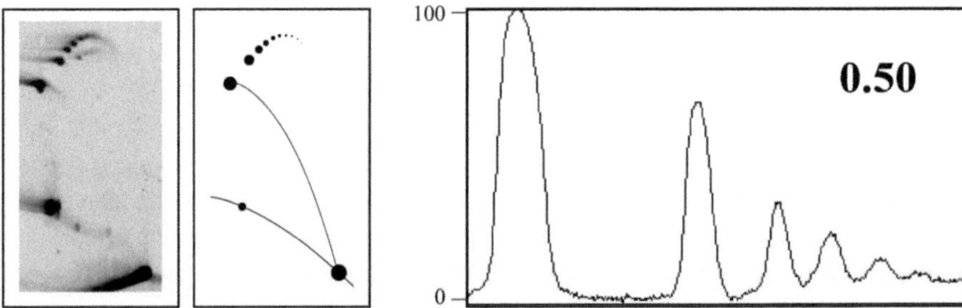

Fig. 3 The autoradiogram of a 2D gel corresponding to pBR-TerE@DraI isolated from parE10 *E. coli* cells grown at the permissive temperature (30 °C) is shown to the *left* together with an interpretative diagram to its *right*. DNA was digested with AlwNI and the portion of the autoradiogram where unknotted and knotted RIs migrated was scanned and analysed by densitometry. The corresponding densitometric profile is shown to the *right*. The number at the *top right* corner of the profile indicates the ratio of knotted/unknotted molecules (reproduced from ref. 29 with permission from Oxford University Press)

2.4 Isolation of Plasmid DNA from E. coli Cells

1. LB (Luria Broth).
2. STE (sodium chloride Tris-EDTA—ethylenediaminetetraacetic acid) buffer: 0.1 M sodium chloride, 10 mM Tris–HCl, pH 8.0 and 1 mM EDTA (stored at 4 °C).
3. 25 % sucrose in 0.25 M Tris–HCl, pH 8.0 (stored at 4 °C).
4. Lysozyme/RNase A solution: 10 mg/mL Lysozyme and 0.1 mg/mL RNase A in 0.25 M Tris–HCl, pH 8.0.
5. 0.25 M EDTA, pH 8.0 (stored at 4 °C).
6. Lysis buffer: 1 % Brij-58, 0.4 % sodium deoxycholate, 63 mM EDTA and 50 mM Tris–HCl, pH 8.0 (stored at 4 °C).
7. TE (Tris-EDTA): 10 mM Tris–HCl and 1 mM EDTA, pH 8.0.
8. 25 % polyethylene glycol 6,000 and 1.25 M sodium chloride in TE (stored at 4 °C).
9. Proteinase K buffer: 1 M sodium chloride, 10 mM Tris–HCl, pH 9, 1 mM EDTA and 0.1 % SDS.
10. 20 mg/mL proteinase K (stored at −20 °C).
11. Phenol:chloroform:isoamyl alcohol (25:24:1) equilibrated with 10 mM Tris–HCl, pH 8.0.
12. Chloroform:isoamyl alcohol (24:1).
13. 100 % ethanol.

2.5 Neutral/Neutral (N/N) Two-Dimensional (2D) Agarose Gel Electrophoresis

1. Agarose (Seakem LE; Lonza).
2. 5× TBE (Tris-Borate-EDTA) 0.445 M Tris base, 0.445 M boric acid and 0.01 M EDTA.
3. 10× gel loading buffer: 0.1 % xylene cyanol, 0.1 % bromophenol blue, 30 % glycerol and 10 mM EDTA, pH 8.0.
4. 10 mg/mL ethidium bromide (Sigma).
5. 10 mg/mL chloroquine (Sigma).

2.6 Southern Blotting

1. 0.25 M hydrochloric acid.
2. 4 N sodium hydroxide.
3. 20× SSC (saline sodium citrate): 3 M sodium chloride and 0.3 M tri-sodium citrate.
4. Zeta-Probe blotting membranes (Bio-Rad).

2.7 Non-radioactive Hybridization

1. Digoxigenin-High Prime kit (Roche).
2. 20× SSPE (saline sodium phosphate EDTA): 3.6 M sodium chloride, 0.2 M disodium phosphate, 20 mM EDTA.
3. 10 % Blotto (non-fat powdered milk).
4. 20 % SDS (sodium dodecyl sulfate).

5. 20 % dextran sulfate (stored at 4 °C).
6. 10 mg/mL sonicated and denatured salmon sperm DNA (stored at 4 °C).
7. 20× SSC.
8. Antidigoxigenin-AP conjugate antibody (Roche).
9. CDP-Star (Perkin Elmer).

3 Methods

3.1 Bacterial Transformation

Transform competent *E. coli* cells with monomeric forms of the plasmids to be studied.

1. Inoculate and grow the selected bacterial strain in LB overnight at 37 °C.
2. Make a 100× dilution in LB plus 10 mM magnesium sulfate.
3. Grow the culture at 37 °C up to an OD_{600} = 0.3–0.4.
4. Centrifuge 1 mL of the culture in an eppendorf tube at room temperature and 9,300 × *g* for 4 min.
5. Carefully resuspend the pellet in 100 μL 0.1 M cold calcium chloride.
6. Incubate the tube on ice for 20 min.
7. Add 10 ng of plasmid DNA in a 10 μL volume and incubate for another 10 min on ice.
8. Heat-shock the cells by placing the tube at 37 °C for 5 min or at 42 °C for 2 min.
9. Add 1 mL of LB and incubate it at 37 °C for 1 h with moderate agitation.
10. Spread the cells on an agar plate containing the appropriate antibiotic to select for cells with the plasmid.

3.2 Isolation of Plasmid DNAs from E. coli Cells

1. Dilute an overnight culture of plasmid containing bacteria 40-fold into 1 L of fresh LB medium with the antibiotic required.
2. Grow the cells at 37 °C to exponential phase at an OD_{600} = 0.4–0.6, chill the culture quickly on ice and centrifuge at 4,000 × *g* for 15 min at 4 °C.
3. Wash cells with 20 mL of STE buffer.
4. Harvest the cells by centrifugation and resuspend them in 5 mL of 25 % sucrose in 0.25M Tris–HCl, pH 8.0.
5. Add 1 mL of the Lysozyme/RNase A solution and incubate the suspension on ice for 5 min.
6. Add 2 mL of 0.25 M EDTA and keep the suspension on ice for another 5 min.

7. Lyse the cells by adding 8 mL of lysis buffer and inverting the tubes 3–4 times to allow lysis to happen. Incubate the samples on ice for 15 min.
8. Centrifuge samples at 26,000 × *g* for 60 min at 4 °C to pellet chromosomal DNA as well as bacterial debris.
9. Pipette the supernatant with plasmid DNA to a fresh tube and precipitate it by adding 2/3 volume of 25 % polyethylene glycol 6,000 and 1.25 M sodium chloride in TE.
10. Incubate overnight at 4 °C.
11. Pellet the precipitated DNA by centrifugation at 6,000 × *g* for 15 min at 4 °C.
12. Resuspend the pellet in 5 mL of preheated proteinase K digestion buffer with 100 μg/mL of Proteinase K at 37 °C for 30 min.
13. Extract the proteins at least twice with pH 8.0-equilibrated phenol:chloroform:isoamyl alcohol (25:24:1), and once with chloroform:isoamyl alcohol (24:1).
14. Precipitate the DNA with 2.5 volumes of absolute ethanol overnight at −20 °C and resuspend it in 1× TE.

3.3 Neutral/Neutral (N/N) Two-Dimensional (2D) Agarose Gel Electrophoresis (See Note 3)

1. Assemble a gel casting set with a 20-teeth (0.5 × 0.15 cm each) comb and pour a 0.28–0.4 % agarose gel in 1× TBE buffer. Let the gel solidify.
2. Place the gel into a gel tank and pour 1× TBE buffer to cover the gel. Carefully remove the comb.
3. Add 1× gel loading dye to the DNA samples and load the samples into the gel wells.
4. Run the first dimension at 0.4–0.9 V/cm at room temperature for 24–72 h.
5. Stop the run. Take the gel casting tray with the gel out of the tank and place it on clean surface. Cut out each lane from the first dimension and place it on top of another gel casting tray oriented 90° with respect to the first dimension and so that the well side of the first dimension slice is on the left of the second dimension.
6. Melt 0.5–1.2 % agarose gel in 1× TBE, cool it down to 50–55 °C and pour over the slices from the first dimension. Let the gel solidify.
7. Place the gel into a gel tank and pour cold 1XTBE to cover the gel. Run the second dimension at 0.9–5 V/cm at 4 °C for 8–80 h.
8. If necessary add different concentrations of chloroquine or ethidium bromide to the TBE buffer into both the agarose gel

and the running buffer, as well as during the first and/or the second dimension of the 2D gel system (*see* **Note 4**).

9. After electrophoresis, transfer the DNA from the gels to an appropriate membrane by Southern blotting.

3.4 Southern Blotting

1. To depurinate the DNA prior to transfer, submerge the gels for 10–15 min in 0.25 M hydrochloric acid, with moderate shaking at room temperature.
2. Next, set up the blot transfer as follows, avoiding the formation of air bubbles: Place three sheets of Whatman 3 MM paper that has been soaked with 0.4 N sodium hydroxide atop of a "bridge" that rests in a shallow reservoir of 0.4 N sodium hydroxide. Place the gel atop of the three soaked sheets of Whatman 3 MM paper. Roll a sterile pipette over the sandwich to remove all air bubbles that often form between the gel and the paper. Cut a piece of positively charged Zeta-probe blotting membrane to the size of the gel. Pre-soak the membrane in distilled water and place it on top of the gel. Use a pipette to eliminate air bubbles as above. Complete the blot assembly by adding three sheets of Whatman 3 MM paper soaked in 0.4 N sodium hydroxide, a stack of paper towels, a glass plate, and a 200–500 g weight (*see* **Note 5**).
3. Allow the blot transfer overnight in transfer buffer (0.4 N sodium hydroxide).
4. After the transfer, peel the membrane from the gel, rinse it briefly in 2× SSC, and air dry.

3.5 Non-radioactive Hybridization

1. Label the DNA probes with digoxigenin using the DIG-High Prime kit (Roche) according to manufacture's recommendations.
2. Meantime, pre-hybridize the membranes in a 20 mL pre-hybridization solution (2× SSPE, 0.5 % Blotto, 1 % SDS, 10 % dextran sulfate and 0.5 mg/mL sonicated and denatured salmon sperm DNA) in hybridization bottles on a rotisserie inside a hybridization oven set at 65 °C for 4–6 h.
3. Denature the labelled DNA probe by heating it at 95–100 °C for 5 min and chill it quickly in an ice bath.
4. Add the probe to the hybridization bottles, place the bottles back into the oven and hybridize for 12–16 h.
5. Wash the hybridized membranes sequentially with 2× SSC and 0.1 % SDS at room temperature for 5 min twice, and then twice with 0.1× SSC and 0.1 % SDS at 68 °C for 15 min.
6. Perform the detection with an antidigoxigenin-AP conjugate antibody (Roche) and CDP-Star (Perkin Elmer) according to the instructions provided by the manufacturer.

3.6 Densitometry

DNA linking number (Lk) or any other parameter can be calculated by quantifying the amount of every given signal of interest. Scan the autoradiograms and quantify the signals of interest by densitometry using a NIH (National Institutes of Health) Image J64 software on probed 2D gel blots.

3.7 Calculation of DNA Supercoiling Density

DNA linking number (Lk) can be calculated on an autoradiogram by quantifying the signal amount of every given topoisomer by densitometry. DNA supercoiling density (σ) is calculated according to the equation $\sigma = \Delta Lk/Lk_0$ [27, 47, 48]. Linking number differences (ΔLk) are determined using the equation $\Delta Lk = Lk - Lk_0$, in which $Lk_0 = N/10.5$, where N is the size of the molecule in bp and 10.5 is the number of base pairs per one complete turn in B-DNA, the most probable helical repeat of DNA under the relaxation conditions used [49, 50].

4 Notes

1. The electrophoretic mobility of DNA molecules in agarose gels depends on their mass and shape. DNA is negatively charged. Thus, when submerged in an appropriate medium, usually an electricity transmitting buffer, molecules migrate towards the anode (the positive pole) at a rate that is inversely proportional to the log of the molecular mass. Agarose beads in the gel behave as obstacles that DNA molecules must overcome in their route towards the anode. For this reason, small linear molecules migrate faster than large or non-linear ones. Submarine agarose gels are routinely used in Cell and Molecular Biology to analyze DNA. Agarose, dissolved in the appropriate buffer at concentrations that may vary between 0.3 and 2.0 %, are polymerized and submerged in the same buffer in a horizontal electrophoretic apparatus. The gel matrix, submerged in the buffer, lies in the middle on a kind of viaduct between the electrode compartments. When connected to a power supply, DNA molecules run through the gel matrix towards the anode and electrophoresis is interrupted when the molecules of interest have been conveniently separated.
2. Luria broth (LB), a nutritionally rich medium, is primarily used for the growth of bacteria. The formula of the LB medium was originally published in 1951. LB media formulations have been an industry standard for the cultivation of *E. coli* as far back as the 1950s. It is widely used in molecular microbiology applications for the preparation of plasmid DNA and recombinant proteins. It continues to be one of the most common media used for maintaining and cultivating laboratory recombinant strains of *E. coli*. There are several common

formulations of LB. Although somewhat different, they generally share a similar composition of ingredients used to promote bacterial growth, including the following: Peptides and casein peptones, vitamins (including B vitamins), trace elements (e.g., nitrogen, sulfur, magnesium), and minerals.

3. Circular DNA molecules are analysed in N/N 2D gels as described before [19, 51]. Electrophoresis conditions, though, vary significantly depending on the mass of the molecules to be analyzed. Finding the correct electrophoresis conditions for small or large plasmids may require some experimentation using plasmids of known sizes. In our hands, for plasmids ~2.5 kb the first dimension is usually run in a 0.5 % agarose gel at 1 V/cm for approximately 20–24 h. The second dimension is run in a 1.2 % agarose gel at 5–6 V/cm for 8–12 h at 4 °C. Plasmids larger than 6–8 kb must be run under conditions of lower agarose concentration and lower voltage in both dimensions in order to successfully separate different topoisomers. For example, for plasmids of ~10 kb the first dimension is usually run in a 0.28 % agarose gel at 0.45 V/cm for approximately ~70 h. The second dimension is run in a 0.58 % agarose gel at 0.9 V/cm for ~90 h at room temperature.
4. Chloroquine and ethidium bromide are planar molecules that intercalate between the two strands of the DNA double helix. This intercalation causes a reduction of DNA twist. As in closed topological domains, like covalently closed circular bacterial plasmids, the linking number is a constant, changes in twist result in compensatory changes in writhe [1, 30, 34]. For this reason if a plasmid is negatively supercoiled, increasing concentrations of these compounds progressively remove the negative supercoiling first and add net positive supercoiling only after all the native negative supercoiling had been removed. Therefore, by adding the appropriate concentration of chloroquine or ethidium bromide during the first and the second dimensions of a 2D gel system, a wide range of topoisomers with negative and positive supercoiling can be analysed allowing the identification of the most abundant topoisomer for each particular population.
5. Make sure there is not short-cut between the wick and the paper towels, i.e., they do not touch each other. The capillary force created by the dry towels has to go through the gel only for the DNA to migrate from the gel onto the membrane. Cling film or other wrap is often placed around the gel to prevent short-cuts.

Acknowledgements

We acknowledge current and past members of the laboratory for their continuous suggestions and support. We would like to strengthen that this work could not be accomplished without the continuous support and constructive criticism of Andrzej Stasiak. This work was sustained by grant BFU2011-22489 to J.B.S. from the Spanish Ministerio de Economía y Competitividad.

References

1. Schvartzman JB, Stasiak A (2004) A topological view of the replicon. EMBO Rep 5: 256–261
2. Watson JD, Crick FHC (1953) Genetical implications of the structure of deoxyribonucleic acids. Nature 171:964–967
3. Watson JD, Crick FHC (1953) Molecular structure of nucleic acids. Nature 161:737–738
4. Cairns J (1963) The bacterial chromosome and its manner of replication as seen by autoradiography. J Mol Biol 6:208–213
5. Wang JC (1969) Degree of superhelicity of covalently closed cyclic DNA's from *Escherichia coli*. J Mol Biol 43:263–272
6. Wang JC (1971) Interaction between DNA and an *Escherichia coli* protein omega. J Mol Biol 55:523–533
7. Weil R, Vinograd J (1963) The cyclic helix and cyclic coil forms of polyoma viral DNA. Proc Natl Acad Sci U S A 50:730–739
8. Dulbecco R, Vogt M (1963) Evidence for a ring structure of polyoma virus DNA. Proc Natl Acad Sci U S A 50:236–243
9. Thorne HV (1966) Electrophoretic separation of polyoma virus DNA from host cell DNA. Virology 29:234–239
10. Stellwagen NC (2009) Electrophoresis of DNA in agarose gels, polyacrylamide gels and in free solution. Electrophoresis 30(Suppl 1):S188–S195
11. Bauer W, Vinograd J (1968) The interaction of closed circular DNA with intercalative dyes. J Mol Biol 33:141–171
12. Bauer W, Vinograd J (1970) Interaction of closed circular DNA with intercalative dyes. II. The free energy of superhelix formation in SV40 DNA. J Mol Biol 47:419–435
13. Bauer W, Vinogradj (1970) The interaction of closed circular DNA with intercalative dyes. 3. Dependence of the buoyant density upon superhelix density and base composition. J Mol Biol 54:281–298
14. Keller W (1975) Determination of the number of superhelical turns in simian virus 40 DNA by gel electrophoresis. Proc Natl Acad Sci U S A 72:4876–4880
15. LePecq JB, Paoletti C (1967) A fluorescent complex between ethidium bromide and nucleic acids. Physical–chemical characterization. J Mol Biol 27:87–106
16. Lerman LS (1961) Structural considerations in the interaction of DNA and acridines. J Mol Biol 3:18–30
17. Pulleyblank DE, Morgan AR (1975) The sense of naturally occurring superhelices and the unwinding angle of intercalated ethidium. J Mol Biol 91:1–13
18. Bell L, Byers B (1983) Separation of branched from linear DNA by two-dimensional gel electrophoresis. Anal Biochem 130:527–535
19. Brewer BJ, Fangman WL (1987) The localization of replication origins on ARS plasmids in *S. cerevisiae*. Cell 51:463–471
20. Huberman JA, Spotila LD, Nawotka KA, El-Assouli SM, Davis LR (1987) The in vivo replication origin of the yeast 2 mm plasmid. Cell 51:473–481
21. Lee CH, Mizusawa H, Kakefuda T (1981) Unwinding of double-stranded DNA helix by dihydration. Proc Natl Acad Sci U S A 78: 2838–2842
22. Minden JS, Marians KJ (1985) Replication of pBR322 DNA in vitro with purified proteins. J Biol Chem 260:9316–9325
23. Oppenheim A (1981) Separation of closed circular DNA from linear DNA by electrophoresis in two dimensions in agarose gels. Nucleic Acids Res 9:6805–6812
24. Sundin O, Varshavsky A (1980) Terminal stages of SV40 DNA replication proceed via multiply intertwined catenated dimers. Cell 21:103–114
25. Sundin O, Varshavsky A (1981) Arrest of segregation leads to accumulation of highly intertwined catenated dimers dissection of the final stages of SV40 DNA replication. Cell 25:659–669
26. Ferrandiz MJ, Martin-Galiano AJ, Schvartzman JB, de la Campa AG (2010) The genome of *Streptococcus pneumoniae* is organized in topology-reacting gene clusters. Nucleic Acids Res 38:3570–3581

27. Mayan-Santos MD, Martinez-Robles ML, Hernandez P, Krimer D, Schvartzman JB (2007) DNA is more negatively supercoiled in bacterial plasmids than in minichromosomes isolated from budding yeast. Electrophoresis 28:3845–3853
28. Mirkin SM (2001) DNA topology: fundamentals, John Wiley & Sons, Inc. DOI: 10.1038/npg.els.0001038
29. Lopez V, Martinez-Robles ML, Hernandez P, Krimer DB, Schvartzman JB (2012) Topo IV is the topoisomerase that knots and unknots sister duplexes during DNA replication. Nucleic Acids Res 40:3563–3573
30. Martinez-Robles ML et al (2009) Interplay of DNA supercoiling and catenation during the segregation of sister duplexes. Nucleic Acids Res 37:5126–5137
31. Adams DE, Shekhtman EM, Zechiedrich EL, Schmid MB, Cozzarelli NR (1992) The role of topoisomerase-IV in partitioning bacterial replicons and the structure of catenated intermediates in DNA replication. Cell 71:277–288
32. Olavarrieta L, Hernández P, Krimer DB, Schvartzman JB (2002) DNA knotting caused by head-on collision of transcription and replication. J Mol Biol 322:1–6
33. Olavarrieta L, Martínez-Robles ML, Hernández P, Krimer DB, Schvartzman JB (2002) Knotting dynamics during DNA replication. Mol Microbiol 46:699–707
34. Olavarrieta L et al (2002) Supercoiling, knotting and replication fork reversal in partially replicated plasmids. Nucleic Acids Res 30:656–666
35. Sogo JM et al (1999) Formation of knots in partially replicated DNA molecules. J Mol Biol 286:637–643
36. Viguera E et al (1996) The ColE1 unidirectional origin acts as a polar replication fork pausing site. J Biol Chem 271:22414–22421
37. Fierro-Fernandez M, Hernandez P, Krimer DB, Schvartzman JB (2007) Replication fork reversal occurs spontaneously after digestion but is constrained in supercoiled domains. J Biol Chem 282:18190–18196
38. Fierro-Fernandez M, Hernandez P, Krimer DB, Stasiak A, Schvartzman JB (2007) Topological locking restrains replication fork reversal. Proc Natl Acad Sci U S A 104:1500–1505
39. Long DT, Kreuzer KN (2008) Regression supports two mechanisms of fork processing in phage T4. Proc Natl Acad Sci U S A 105: 6852–6857
40. Long DT, Kreuzer KN (2009) Fork regression is an active helicase-driven pathway in bacteriophage T4. EMBO Rep 10:394–399
41. Lopes M et al (2001) The DNA replication checkpoint response stabilizes stalled replication forks. Nature 412:557–561
42. Pohlhaus J, Kreuzer K (2006) Formation and processing of stalled replication forks—utility of two-dimensional agarose gels. Methods Enzymol 409:477–493
43. Postow L et al (2001) Positive torsional strain causes the formation of a four-way junction at replication forks. J Biol Chem 276:2790–2796
44. Sogo JM, Lopes M, Foiani M (2002) Fork reversal and ssDNA accumulation at stalled replication forks owing to checkpoint defects. Science 297:599–602
45. Viguera E, Hernandez P, Krimer DB, Lurz R, Schvartzman JB (2000) Visualisation of plasmid replication intermediates containing reversed forks. Nucleic Acids Res 28:498–503
46. Santamaría D, Hernández P, Martínez-Robles ML, Krimer DB, Schvartzman JB (2000) Premature termination of DNA replication in plasmids carrying two inversely oriented ColE1 origins. J Mol Biol 300:75–82
47. Vologodskii A (1998) Exploiting circular DNA. Proc Natl Acad Sci U S A 95:4092–4093
48. Wang JC, Peck LJ, Becherer K (1983) DNA supercoiling and its effects on DNA structure and function. Cold Spring Harb Symp Quant Biol 47(Pt 1):85–91
49. Horowitz DS, Wang JC (1984) Torsional rigidity of DNA and length dependence of the free energy of DNA supercoiling. J Mol Biol 173:75–91
50. Salceda J, Fernandez X, Roca J (2006) Topoisomerase II, not topoisomerase I, is the proficient relaxase of nucleosomal DNA. EMBO J 25:2575–2583
51. Friedman KL, Brewer BJ (1995) Analysis of replication intermediates by two-dimensional agarose gel electrophoresis. In: Campbell JL (ed) DNA replication, methods in enzymology, vol 262. Academic Press Inc., San Diego, CA, pp 613–627

Chapter 8

A Neutral Glyoxal Gel Electrophoresis Method for the Detection and Semi-quantitation of DNA Single-Strand Breaks

Brian Pachkowski and Jun Nakamura

Abstract

Single-strand breaks are among the most prevalent lesions found in DNA. Traditional electrophoretic methods (e.g., the Comet assay) used for investigating these lesions rely on alkaline conditions to denature DNA prior to electrophoresis. However, the presence of alkali-labile sites in DNA can result in the introduction of additional single-strand breaks upon alkali treatment during DNA sample processing. Herein, we describe a neutral glyoxal gel electrophoresis assay which is based on alkali-free DNA denaturation and is suitable for qualitative and semi-quantitative analyses of single-strand breaks in DNA isolated from different organisms.

Key words Agarose, Alkaline, DNA damage, DNA repair, Electrophoresis, Genotoxicant, Glyoxal, Neutral, Single-strand breaks

1 Introduction

Exposure to endogenous and exogenous genotoxic agents can lead to a number of DNA lesions [1]. Depending on the genotoxic agent, DNA single-strand breaks (SSBs) are prevalent lesions formed either by oxidative attack on the DNA backbone or during the course of DNA repair [2]. Methods used to assess SSB content, such as the Comet assay, have generally relied on alkaline conditions used for DNA denaturation which then allows the separation of single-stranded DNA molecules based on their molecular weight. The extent of SSB content is estimated from the fraction of the lower molecular weight DNA molecules [3]. However, the use of alkaline conditions can cause DNA breakage at alkali-labile sites (ALSs), which include apurinic/apyrimidinic sites, thereby forming additional (i.e., artifactual or inadvertent) SSBs in the analyzed DNA samples [4]. This phenomenon presents a methodological challenge in trying to assess the extent of SSB formation in vivo

Svetlana Makovets (ed.), *DNA Electrophoresis: Methods and Protocols*, Methods in Molecular Biology, vol. 1054,
DOI 10.1007/978-1-62703-565-1_8, © Springer Science+Business Media New York 2013

accurately, and some modifications of alkaline-based methods have been developed to protect ALSs [5]. A completely different approach has also been developed: it is based on coupling DNA denaturation under neutral conditions with glyoxal treatment to prevent DNA renaturation [6, 7].

Herein we describe this alternative approach, a neutral gel electrophoresis assay, to assess SSB content. Briefly, DNA is denatured under neutral conditions, during which glyoxal adduction of DNA bases prevents the reannealing of DNA strands. Denatured DNA samples are then loaded onto neutral agarose gels and subjected to electrophoresis. After sufficient separation of DNA samples, the gels are stained and gel images are obtained. The images can be used for qualitative or semi-quantitative analyses of SSBs in the DNA samples [8–10].

2 Materials

Prepare all solutions at room temperature using deionized water (dH_2O) and analytical grade reagents. Store solutions at room temperature unless otherwise indicated. Follow your institutional safety regulations and those from the manufacturers when handling reagents and waste disposal.

2.1 DNA Isolation

1. Commercially available DNA isolation kit (5 PRIME, Inc.) (*see* **Note 1**).
2. 2 M stock solution of 2,2,6,6-tetramethylpiperidine-*N*-oxyl solution (TEMPO, Acros Organics) in 100 % methanol. Store at −20 °C.
3. Ribonuclease A (~30 μg/μL) from bovine pancreas (RNase, Sigma). Store at −20 °C.
4. 70 % ethanol (v/v) in dH_2O. Store at −80 °C.
5. 100 % isopropanol. Store at −20 °C.
6. Microcentrifuge set up at 4 °C.
7. Orbital shaker set up at 4 °C.

2.2 DNA Sample Preparation

1. 200 mM sodium phosphate stock solution, pH 7.0. Add 800 mL of dH_2O to a beaker. Weigh 16.4 g of Na_2HPO_4 and transfer to the beaker. Weigh 10.2 g of NaH_2PO_4 and transfer to the beaker. Adjust pH to 7.0 with NaOH. Add water to bring the volume to 1 L (*see* **Note 2**).
2. Denaturing buffer: 10 mM sodium phosphate, pH 7.0, 1.5 M glyoxal, and 50 % DMSO (v/v) (*see* **Note 3**). Add 23.8 μL of 200 mM sodium phosphate stock solution; 83.3 μL of 40 % glyoxal (w/v); 238 μL 99.7 % dimethyl sulfoxide (DMSO)

to a 1.5 mL microcentrifuge tube, vortex to mix, and store on ice until use (*see* **Note 4**).

3. Calf thymus DNA (Sigma).
4. DNA size marker: HindIII Digested Lambda DNA with a size range from 0.125 to 23.1 kbp (New England Biolabs).

2.3 DNA Gel Electrophoresis

1. Running buffer: 10 mM sodium phosphate, pH 7.0. Add 50 mL of 200 mM sodium phosphate stock solution to a glass container and then add 950 mL of dH_2O (*see* **Note 5**).
2. Sample loading buffer: 10 mM sodium phosphate, pH 7.0, 50 % glycerol (v/v), 0.01 % bromophenol blue (w/v), and 0.01 % xylene orange (w/v). Add 50 mL of glycerol and 5 mL of 200 mM sodium phosphate stock solution to a glass beaker. Weigh 10 mg of bromophenol blue and 10 mg of xylene cyanol and transfer to the beaker. Mix well and adjust the volume to 100 mL with dH_2O (*see* **Note 6**).
3. Gel tank (Horizontal Gel electrophoresis system Model H4, 35 cm × 20.5 cm, ~3 L capacity, BRL Life Technologies Inc.).
4. Casting tray and comb (Gasketed EasyCast™ B1A-UVT Gel Tray and comb, Owl separation system).
5. Peristaltic pump (Masterflex Model 7523, Cole-Parmer).
6. Agarose (Low EEO electrophoresis grade, Fisher).
7. Aluminum foil.

2.4 Gel Staining and Imaging

1. 10 mg/mL acridine orange in water stock solution (Invitrogen). Store at −80 °C.
2. Acridine orange staining solution: 5 μL/mL acridine orange. Add 475 mL of dH_2O to 25 mL of 200 mM sodium phosphate stock solution in a glass beaker. Then add 250 μL of 10 mg/mL acridine orange stock solution. Mix and store at room temperature either in the dark or away from ambient light (*see* **Note** 7).

3 Methods

3.1 DNA Isolation

1. Thaw cell pellets at room temperature for 5 min, keeping them out of direct sunlight, and then place them on ice (*see* **Note 8**).
2. Supplement cell lysis solution with 2 M TEMPO stock solution to give a final TEMPO concentration of 20 mM. For example, add 10 μL 2 M TEMPO stock solution to 1 mL of cell lysis solution and keep it on ice until needed.
3. Add 600 μL of cell lysis/TEMPO solution to each cell pellet.

4. Resuspend cell pellets so that they are not adhered to the bottom of the microcentrifuge tubes.
5. Incubate the tubes on ice for 5 min to lyse the cells.
6. Add 200 μL of protein precipitation solution to each tube.
7. Vortex each tube for approximately 20 s.
8. Centrifuge the tubes at 16,110×*g* for 15 min at 4 °C (*see* **Note 9**).
9. After centrifugation, decant the supernatants, which contain a mix of DNA, RNA, and nuclear proteins, into new ice-cold microcentrifuge tubes.
10. Add 600 μL of isopropanol to each tube to precipitate nucleic acids.
11. Place the tubes into a tube rack and mix the content of the tubes by inverting the rack approximately 50 times (*see* **Note 10**).
12. Centrifuge the tubes at 16,110×*g* for 15 min at 4 °C to pellet DNA and RNA.
13. Remove supernatant and place tubes on ice.
14. To wash the DNA/RNA pellet, add 500 μL of 70 % ethanol to each microcentrifuge tube. Gently dislodge the DNA/RNA pellet from the bottom of the microcentrifuge tube with a pipette tip.
15. Centrifuge the tubes at 16,110×*g* for 15 min at 4 °C.
16. Decant ethanol and place tubes on ice.
17. Add 600 μL of cell lysis/TEMPO solution to each microcentrifuge tube.
18. Gently resuspend each DNA/RNA pellet by dislodging it from the bottom of the microcentrifuge tube with a pipette tip.
19. Place microcentrifuge tubes in an orbital shaker and mix for 30 min at 4 °C (*see* **Note 11**).
20. After mixing, ensure that each DNA/RNA pellet is floating in solution and not attached to the bottom of the tube.
21. Place tubes back onto the orbital shaker and mix for additional 30 min at 4 °C.
22. Add 2.7 μL of RNase to each microcentrifuge tube and vortex for 1 s.
23. Incubate the DNA/RNA/RNase mixture for 30 min on a rocking platform set at 37 °C (*see* **Note 12**).
24. After the RNase incubation, add 200 μL of protein precipitation solution to each microcentrifuge tube (*see* **Note 13**).
25. Vortex each tube for 20 s.
26. Centrifuge the tubes at 16,110×*g* for 15 min at 4 °C.
27. Decant the supernatant from each tube into a fresh ice-cold microcentrifuge tube.

28. Add 600 μL of isopropanol to each tube.
29. Place microcentrifuge tubes containing the DNA/isopropanol mixture into a tube rack and mix by inverting the rack approximately 50 times.
30. Centrifuge the tubes at 16,110 × *g* for 15 min at 4 °C.
31. Decant the supernatants.
32. To wash the DNA, add 500 μL of 70 % ethanol to each tube and gently dislodge the DNA pellet from the bottom of the microcentrifuge tube with a pipette tip.
33. Centrifuge the tubes at 16,110 × *g* for 15 min at 4 °C.
34. During centrifugation, prepare a 1 mM TEMPO solution. For example, add 2 μL 2 M TEMPO stock solution to 4 mL of dH_2O and keep it on ice until needed.
35. Decant ethanol from each microcentrifuge tube.
36. Add 30–50 μL of the 1 mM TEMPO solution to each DNA pellet ensuring that the pellet is dislodged from the bottom of the microcentrifuge tube (*see* **Note 14**).
37. Place the tubes in an orbital shaker and mix for 1 h at 4 °C.
38. Store samples at −80 °C until further analysis.

3.2 Gel Preparation

1. To prepare a 0.7 % agarose gel, 7 cm × 8.5 cm with 1.5 cm depth, mix 0.467 g of agarose, 3.3 mL of 200 mM sodium phosphate stock solution, and 63.3-mL dH_2O in a temperature resistant glass container that can be placed in a microwave oven. Heat the mix in a microwave oven until the agarose is completely melted (*see* **Note 15**).
2. Place a clean thermometer into the molten agarose and place the glass vessel into a container filled with ice water. Allow the mixture to cool to approximately 70 °C. While the molten agarose is cooling, assemble the gel-casting apparatus.
3. Slowly pour the cooled agarose solution into the gel-casting tray, taking care to avoid introducing air bubbles into the agarose. Carefully insert a comb into the agarose, avoiding introducing air bubbles (*see* **Note 16**).
4. Loosely cover the top of the gel casting tray/comb assembly with aluminum foil until the agarose gel solidifies completely (*see* **Note 17**).

3.3 DNA Sample Preparation

1. To a 500-μL microcentrifuge tube, add predetermined volumes of sample components in the following order: dH_2O, isolated DNA, and 29-μL denaturing solution, to give the final sample volume of 40 μL. In addition to the analyzed DNA, include samples of the DNA size marker and calf thymus DNA (*see* **Note 18**).

2. Vortex each sample for 1 s to mix the sample components and then spin briefly (<3 s) to collect the sample at the bottom of the tube (*see* **Note 19**).
3. Place sample tubes into a heat block at 50 °C for 1 h (*see* **Note 20**). Cover the heat block with foil to minimize sample exposure to light (*see* **Note 21**).
4. Remove each tube from the heating block and briefly centrifuge to collect any condensation formed on the top of the microcentrifuge tubes.
5. Add 5 μL of sample loading buffer to each DNA sample and keep the samples on ice until you are ready to load them into a gel (*see* **Note 22**).

3.4 Neutral Agarose Gel Electrophoresis

1. Connect a horizontal electrophoresis chamber to a pump in a 4 °C cold room and fill the chamber with precooled running buffer (*see* **Note 23**).
2. Place the gel(s) into the gel tank.
3. Slowly load all the DNA samples into separate wells (*see* **Note 24**), including the DNA size marker and the positive control DNA (i.e., calf thymus DNA) (*see* **Note 25**).
4. After all of the samples have been loaded, turn on the pump to begin circulating the electrophoresis buffer. Electrophorese at 0.86 V/cm for 16 h (*see* **Note 26**).

3.5 Gel Staining, Imaging, and Analysis

1. After electrophoresis, place the gel(s) into a plastic container with 500 mL of acridine orange staining solution. Completely cover the container with aluminum foil to protect the stain from light.
2. Place the covered container on a platform rocker and gently agitate. Incubate the gel(s) in the staining solution for 1 h at room temperature.
3. Take an image of the gel(s).
4. Destain the gel(s) by replacing the staining solution with fresh dH_2O and returning the gel(s) to the covered plastic container. Continue the gentle agitation at room temperature. Take images at regular intervals (i.e., every hour) until the desired image is captured (*see* Fig. 1 and **Note 27**).
5. Evaluate the DNA smears visually to compare DNA samples qualitatively (*see* **Note 28**). Alternatively, gel analysis software can be used to provide a more quantitative assessment of DNA SSB content by comparing intensity values for each lane (*see* **Note 29**).

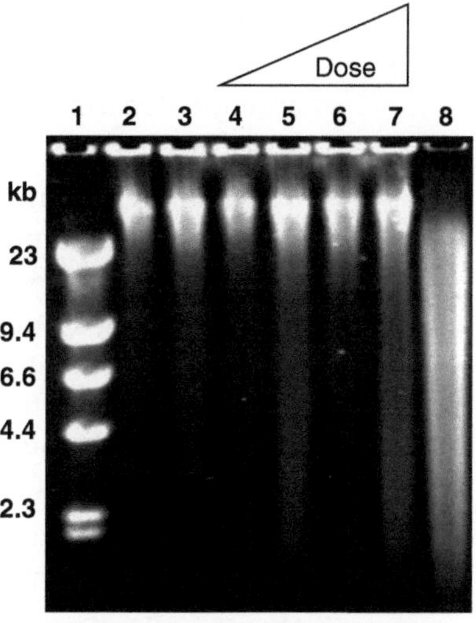

Fig. 1 SSB detection in DT40 cells grown in the presence of MMS (methyl-methane-sulphonate) for 4 h. Adapted with modifications from (Pachkowski B.F., et al. Mutat Res. 2009 671:93–99). Representative visualization of SSB content in genomic DNA after neutral glyoxal gel electrophoresis analysis. Images obtained using the Kodak Image Station 440CF system. *Lane 1*: DNA size marker; *Even numbered lanes*: samples from wild-type cells; *Odd numbered lanes*: samples from base excision repair deficient cells; *Lane 8*: commercially available calf thymus DNA (positive control for SSBs)

4 Notes

1. The DNA isolation protocol we use is a modification of the protocol coming with the commercially available kit (*see* ref. 11), which contains only a cell lysis solution and a protein precipitation solution. We use additional reagents including an antioxidant (e.g., TEMPO), RNase, isopropanol, and ethanol. To minimize the artifactual or inadvertent production of SSBs during DNA isolation, we suggest that an antioxidant be used and that all steps of isolation be conducted on ice or at 4 °C with minimal manipulation (i.e., shearing) of the DNA. Other DNA isolation methods (i.e., phenol-based isolation) should be suitable for this gel assay, if the above precautions are taken.
2. To facilitate dissolving of the solids in the solution, use a stirring bar and/or apply some heat but avoid boiling. Dissolve one salt completely before adding the other one to the beaker. Store at room temperature for up to 2 weeks. This stock solution is used for a number of applications in this assay.
3. These concentrations approximate the final concentrations once the denaturing buffer has been added to dH_2O and isolated DNA samples (*see* Subheading 3.3, **step 1**).

4. We prepare denaturing buffer fresh for each experiment. The volumes provided represent those needed in order to have enough denaturing buffer (29 μL per sample) for up to 12 samples.
5. The volume of running buffer needed for an experiment will depend on the dimensions of the electrophoresis chamber. We typically prepare 3 L of fresh running buffer for each experiment. Of this volume, 2.8 L is poured into the electrophoresis chamber at 4 °C to precool it while the DNA samples are being denatured. The remaining buffer can be kept at room temperature and used for storing newly made gels until they are needed.
6. The sample loading buffer can be stored at room temperature in the dark or in a tube covered with foil. We usually prepare fresh sample loading buffer every 3–6 months.
7. We typically prepare the acridine orange staining solution fresh for each experiment during the final 30 min of electrophoresis and discard used buffer according to health and safety regulations. The 10 mg/mL acridine orange stock solution can be prepared ahead of time and stored at −80 °C until needed.
8. Here we describe the isolation of DNA from DT40 (i.e., chicken B-lymphoma cells) cell pellets stored at −80 °C. We store cell pellets in 500-μL microcentrifuge tubes. The protocol can also be used to isolate DNA from freshly harvested cultured cell and can be applied to other eukaryotic cell types. In the case of DT40 cells, we perform our experiments at a cell density of 1×10^6 cells/mL. To harvest cells after exposure, we decant approximately 2×10^7 cells into a 50-mL conical tube. After centrifugation, $260 \times g$ for 5 min, the supernatant is removed, and the cell pellet is resuspended in ~1 mL of complete culture media. The resuspended cells are transferred to a 500-μL microcentrifuge tube, centrifugated at $13{,}600 \times g$ for 10 s and the supernatant is removed. The cell pellets are then stored at −80 °C or used for DNA isolation as described above.
9. At this point in the DNA isolation protocol, cells have been lysed open, thereby releasing the DNA into solution while cellular debris is pelleted at the bottom of the microcentrifuge tube during centrifugation.
10. After mixing the supernatant/isopropanol mixtures, precipitated nucleic acids will appear as white threads.
11. We set the orbital shaker on a low setting to minimize any physical shearing of the DNA/RNA pellet.
12. Place the tubes on their sides in order to increase the interaction of the RNase with the DNA/RNA pellet.
13. After the incubation with RNase, only DNA and residual proteins, including RNase, remain in each microcentrifuge tube.

14. In our experience, adding this volume of TEMPO will give a DNA concentration of approximately 1 μg/μL. Regardless of the DNA concentration, we use TEMPO as an antioxidant to prevent the formation of oxygen free radicals which can introduce single-strand breaks during DNA isolation and storage.
15. Heat the agarose suspension in a microwave oven in 30-s intervals until it starts boiling. At the end of each interval, swirl the molten agarose in order to dissolve any solid agarose particles found on the walls of the vessel.
16. If necessary, remove any air bubbles from the agarose by using a pipette.
17. Covering the gel helps to prevent any debris (e.g., dust) from settling on top of the cooling agarose gel, thereby improving the resulting image of the stained gel. We have found that gels typically solidify within 1 h. Once the gels are completely solid, we store them in a plastic container filled with the leftovers of running buffer (*see* Subheading 2.3, **item 1**).
18. We have found that 5–10 μg of isolated DNA provides a good smear for determining SSB content using the assay described herein. Each sample to be prepared must contain an equivalent amount of DNA. We typically dilute the purified DNA solutions to the final concentration of 1 μg/μL. Accordingly, to prepare a sample with 10 μg of DNA, we add 10 μL of a 1 μg/μL isolated DNA solution to 1 μL of dH_2O, followed by 29 μL of denaturation solution. Volumes of isolated DNA solution and dH_2O will depend on the amount of DNA isolated from the starting material (e.g., cells, tissues). We also prepare marker DNA (5 μg of HindIII digested lambda DNA) and a positive control DNA (10 μg) in parallel with the isolated DNA samples. The marker DNA indicates the extent of DNA migration for DNA fragments based on the number of bases. For positive control DNA, we use commercially available calf thymus DNA, which we have found to contain a high level of DNA damage, including SSBs. As with the isolated DNA, the DNA size marker and calf thymus DNA are initially double stranded but from sample processing will be analyzed as single-stranded DNA after gel electrophoresis.
19. The recommended time is approximate and is to illustrate the fact that manipulation of the samples should be kept to a minimum in order to prevent the physical shearing of DNA strands, which could artificially raise the SSB content in a sample.
20. The tops of microcentrifuge tubes should be secured using parafilm or other method in order to prevent the inadvertent opening of the tubes due to the built up pressure during the heat denaturation.

21. During the last 15 min of denaturation, submerge the solidified agarose gels into the electrophoresis chamber, making sure that the wells are oriented towards the negative electrodes.
22. In order to minimize any mechanical stress on the DNA, we simply add the loading buffer to each sample, with no mixing at this step. We find that the components sufficiently mix as they are transferred from the microcentrifuge tube to the gel well in the subsequent steps.
23. Setting up the electrophoresis chamber can be done while DNA samples are denaturing, in order to increase the efficiency of the assay. This will also allow the gel(s) to stay hydrated and chill uniformly to the cold room temperature. Carefully remove any air bubbles that may have formed in the wells. We typically run two gels per electrophoresis chamber and line them up side by side.
24. We typically avoid using specially designed thin gel loading pipette tips when loading DNA samples in order to avoid any shearing of the DNA. Exercise caution when loading a DNA sample to a well, so that samples in different wells do not mix and that air bubbles are not introduced to the wells. These potential problems can be avoided by delivering a DNA sample to a well slowly, so that the sample settles at the bottom of the well.
25. Using a gel with ten wells, we typically analyze only eight samples in total: six DNA samples, the size marker DNA, and the positive control DNA. In order to minimize any effects that the end lanes may have on DNA migration, we load only the inner eight wells with DNA. A typical loading scheme would be: empty well, marker DNA, DNA samples under investigation 1–6, positive control DNA, and empty well.
26. We found that 16 h of electrophoresis allowed the dye front from the loading buffer and the 2.3-kb size marker to migrate to the edge of the gel. We have also found that this duration provides sufficient migration of the DNA samples to allow for visualization. Electrophoresis time may need to be adjusted according to the apparatus used.
27. We find that 4 h of destaining removes enough acridine orange from the gel so that DNA smears are visually more prominent (i.e., they are brighter) than the background staining. For the experiments where we use two gels for electrophoresis, we take images of each gel separately.
28. We base our qualitative assessment of SSB content by visually comparing the extent (i.e., the length) of DNA migration between samples. The presence of SSBs in double-stranded DNA produces low molecular weight DNA fragments upon denaturation. Therefore, the greater the SSB content, the longer the DNA smear produced after sample processing and electrophoresis.

29. This assay produces an image similar to that obtained from the Comet assay, where high molecular weight DNA retained above the 23.1-kb marker is analogous to the high molecular weight DNA retained in the head of the comet (*see* ref. 12). The DNA smear produced in this assay is analogous to a comet tail, where the extent of DNA migration is ultimately determined by SSB content. In order to produce a numerical comparison between DNA smears, we use the CometScore software from Tritek to derive a metric (i.e., tail moment) similar to that used for the Comet assay (*see* refs. 10, 12). The tail moment for each lane with a DNA sample is determined by multiplying the tail length by the percentage of DNA in the tail. We define the tail length as the length of the DNA smear below the 23.1-kb marker, whereas the percentage of DNA in the tail is the image intensity of the DNA smear divided by the sum of the image intensities for the high molecular weight DNA (i.e., above the 23.1-kb marker) and the DNA smear. Therefore, we interpret a higher tail moment as an indicator of greater SSB content.

Acknowledgements

This work was supported in part by National Institute of Environmental Health Science grants (P42-ES05948, P30-ES10126).

References

1. Pitot HC, Dragan YP (1996) Chemical carcinogenesis. In: Klaassen CD (ed) Casarett and Doull's toxicology, 5th edn. McGraw-Hill, New York, pp 201–267
2. Caldecott KW (2006) Mammalian single-strand break repair: mechanisms and links with chromatin. DNA Repair 6:443–453
3. Georgiou C, Papapostolou I, Grizntzalis K (2009) Protocol for the quantitative assessment of DNA concentration and damage (fragmentation and nicks). Nat Protoc 4:125–131
4. Anderson D, Plewa MJ (1998) The international comet assay workshop. Mutagenesis 13:67–73
5. Luke AM et al (2010) Accumulation of true single strand breaks and AP sites in base excision repair deficient cells. Mutat Res 694:65–71
6. McMaster GK, Carmichael GG (1977) Analysis of single- and double-stranded nucleic acids on polyacrylamide and agarose gels by using glyoxal and acridine orange. Proc Natl Acad Sci U S A 74:4835–4838
7. Drouin R, Gao S, Homquist GP (1996) Agarose gel electrophoresis for DNA damage analysis. In: Pfeifer GP (ed) Technologies for detection of DNA damage and mutations. Plenum Press, New York, pp 37–43
8. Lin P-H et al (2003) Aldehydic DNA lesions induced by catechol estrogens in calf thymus DNA. Carcinogenesis 24:1133–1141
9. Rodriguez H et al (2003) Mapping of peroxyl radical induced damage on genomic DNA. Biochemistry 38:16578–16588
10. Pachkowski BF et al (2009) Cells deficient in PARP-1 show an accelerated accumulation of DNA single strand breaks, but not AP sites, over the PARP-1-proficient cells exposed to MMS. Mutat Res 671:93–99
11. Nakamura J, La DK, Swenberg JA (2000) 5′-Nicked apurinic/apyrimidinic sites are resistant to beta-elimination by beta-polymerase and are persistent in human cultured cells after oxidative stress. J Biol Chem 275: 5323–5328
12. Olive PL, Banath JP (2006) The comet assay: a method to measure DNA damage in individual cells. Nat Protoc 1:23–29

Chapter 9

Denaturing Gradient Gel Electrophoresis (DGGE)

Fiona Strathdee and Andrew Free

Abstract

Denaturing Gradient Gel Electrophoresis (DGGE) is a technique used to separate short- to medium-length DNA fragments based on their melting characteristics. It has been used frequently for identifying single-nucleotide polymorphisms without the need for DNA sequencing and as a molecular fingerprinting method for complex ecosystem communities, in particular in conjunction with amplification of microbial 16S rRNA genes. Here, the principles of DGGE, based on partial DNA strand separation at a given position in a gradient of chemical denaturant, are described, and an example protocol, optimized for fingerprinting of 200–300 bp fragments of bacterial 16S rRNA genes, is given.

Key words Denaturing gradient gel electrophoresis, DGGE, Single-nucleotide polymorphisms, Fingerprinting, 16S rRNA gene analysis, Microbial communities

1 Introduction

Denaturing Gradient Gel Electrophoresis was originally designed to detect single point mutations (single-nucleotide polymorphisms; SNPs) in genes associated with particular diseases [1]. It relies on the fact that a single-stranded DNA molecule migrates more slowly than the equivalent double-stranded molecule during electrophoresis, due to increased interaction of the unbonded nucleotides in the single-stranded molecule with the gel matrix. By contrast, the stacked, hydrogen-bonded nucleotides in the double-stranded molecule pass much more easily through the gel. DGGE employs a polyacrylamide gel containing an increasing gradient of chemical denaturants (usually urea and formamide) through which the DNA molecules pass by electrophoresis. As double-stranded DNA passes through the gradient, each molecule will begin to denature at a particular concentration of denaturant dependent upon its %GC content and the exact arrangement of bases in the sequence. Denaturing molecules are retarded at the point of denaturation

Svetlana Makovets (ed.), *DNA Electrophoresis: Methods and Protocols*, Methods in Molecular Biology, vol. 1054,
DOI 10.1007/978-1-62703-565-1_9, © Springer Science+Business Media New York 2013

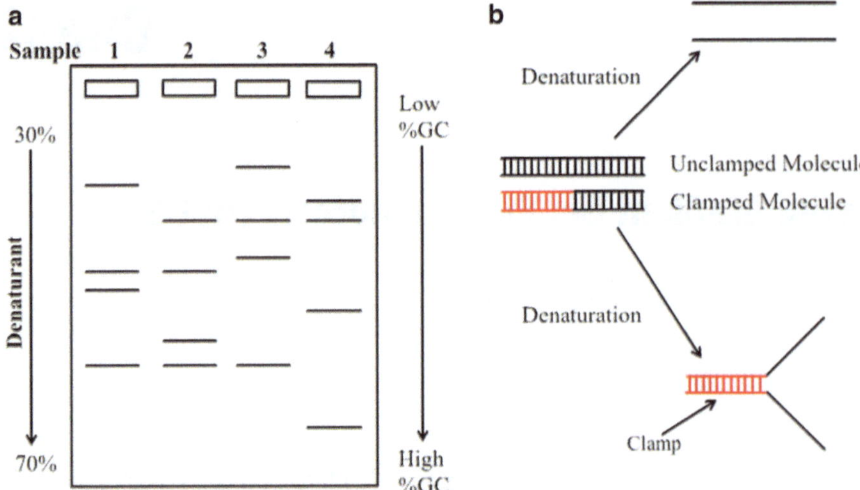

Fig. 1 Principle of denaturing gradient gel electrophoresis. (**a**) Typical parallel DGGE gel, in which the gradient of denaturant (typically 30–70 %; *see* Subheading 3) runs parallel to the direction of electrophoresis. Bands are observed at the positions in the gradient at which individual molecules partially denature (**b**) and hence cease to migrate. Bands of the same mobility correspond to identical sequences (in a fingerprinting assay) or polymorphisms, homo- or heteroduplex forms (in an SNP assay). Where sequences of varying %GC content are separated, the bands lower down the gel correspond to higher %GC sequences, although other sequence-specific factors determine exact melting behavior. (**b**) Partial denaturation of a double-stranded DNA molecule at a particular point in the gradient; the GC-rich "clamp" region (*red*), which is added to the PCR amplicons, remains base paired to prevent full denaturation to single DNA strands. A molecule with no clamp denatures fully to single strands and does not form a clear band in the gel

due to their newly acquired branched structure and can be visualized following electrophoresis (Fig. 1a), therefore achieving separation of molecules based on %GC content and DNA sequence rather than by size.

If the two strands of DNA become totally separated into single-stranded molecules, mobility will increase again. DGGE studies therefore normally employ PCR primers in which a 35–40 nt GC-rich tail, called a GC clamp, is engineered onto one of the primers [2]. This results in a partially melted structure in which the GC clamp remains double-stranded (Fig. 1b), which is retarded and effectively stops migrating at the point at which it is formed. For optimal formation of a single, clean band corresponding to each sequence being electrophoresed, the GC clamp is usually positioned adjacent to the highest melting domain of the native sequence, thus forcing melting to proceed in a single direction along the molecule towards the clamp and avoiding the possibility of multiple partially melted conformations which could cause a smear or multiple bands. DGGE gels can be run with the gradient of denaturant either parallel to the direction of electrophoresis, in which case single bands form at the position of denaturation (Fig. 1), or perpendicular to the electrophoresis direction, in which case denaturation curves corresponding to each molecule are obtained. A gradient of temperature instead of chemical denaturant can also be employed, in which case the technique is called Temperature Gradient Gel Electrophoresis (TGGE). Here, we focus on the methodology, applications, and analysis of parallel DGGE gels.

To detect SNPs associated with disease by DGGE, wild-type, and mutant fragments, the latter containing a polymorphism are run as separate samples on a gel; the mutant fragment will denature at a different gel position from the wild-type fragment dependent upon the effect of the polymorphism on melting behavior. It is also possible to mix the fragments, denature the strands, and allow them to re-anneal prior to loading onto the gel; this will allow the detection of additional bands due to the formation of heteroduplex molecules containing one wild-type and one mutant strand [3]. More recently, DGGE has become an important tool in microbial ecology, where it is used as a molecular fingerprinting method to assess microbial diversity in complex mixed communities [4]. It is most commonly applied to analyses using the 16S rRNA gene, which is universally conserved and has a mosaic structure of highly conserved regions separated by regions of sequence variation. PCR primers binding the conserved regions are used to amplify the intervening 200–400 bp variable region(s) from total community DNA; the mixed PCR products are then separated by parallel DGGE, with the melting behavior determined by the sequence of the variable region [5]. In theory each band obtained corresponds to a single microbial species, although the presence of multiple sequence-variant 16S rRNA genes in certain species, together with occasional co-migration of amplicons with different sequences, means that this correlation is not absolute. Although most commonly applied to rRNA genes, DGGE can be used to study any locus which exhibits interspecies sequence variation in regions of suitable size flanked by conserved sequences to act as primer binding sites.

2 Materials

Prepare all solutions using ultrapure distilled water and electrophoresis-grade chemicals for optimal gel quality. Store solutions at either room temperature or 4 °C as indicated. Follow local waste disposal regulations for all reagents, in particular solutions containing acrylamide and silver compounds. Gloves must be worn.

2.1 Denaturing Gradient Gel Reagents

Two concentrations of acrylamide stock are needed for mixing to achieve the gradient in the gel, one at 0 % denaturant and the second at 80 % denaturant. Chemical denaturant consists of a mixture of urea and formamide. These solutions should be made up in the fume hood in glass bottles, then wrapped in aluminum foil and stored at 4 °C.

1. Low (0 %) denaturant stock, 100 mL: Mix 2 mL 50× TAE stock (see below), 27 mL 30 % acrylamide/*N,N′*-methylenebisacrylamide solution (37.5:1 acrylamide: *N,N′*-methylenebisacrylamide), 71 mL dH_2O. Store at 4 °C.
2. High (80 %) denaturant stock 100 mL: Dissolve 33.6 g urea in 32 mL formamide, 2 mL 50× TAE stock (see below), and 27 mL 30 % acrylamide/*N,N′*-methylenebisacrylamide solution. Extensive stirring and some heating may be required. Add dH_2O to 100 mL final volume (a very small volume of dH_2O is required). Store at 4 °C.
3. 10 % APS (ammonium persulfate): Dissolve 0.1 g of ammonium persulfate powder in 1 mL of dH_2O. Once mixed, keep on ice. APS solution can be kept at 4 °C for up to 1 week; for longer time intervals, make up a fresh 1 mL stock.

2.2 Gel Loading and Running Solutions

1. 50× TAE stock, 1 L: Dissolve 242 g Tris base in 57.1 mL glacial acetic acid, 100 mL 0.5 M EDTA (pH 8), and dH_2O up to 950 mL. (Dispense glacial acetic acid in a fume hood.) Adjust pH to 8.0 as necessary and make up to 1 L with dH_2O. Store at room temperature.
2. 1× TAE 10 L: Mix 200 mL 50× TAE with 9.8 L dH_2O. Store at room temperature.
3. Gel loading buffer: Dissolve bromophenol blue and xylene cyanol to final concentrations of 0.05 % (w/v) in 10 mL of 70 % glycerol.

2.3 Silver Staining Solutions

These solutions should be made up fresh on the day. Quantities given are per DGGE gel.

1. Fix solution: Mix 2.5 mL glacial acetic acid, 50 mL ethanol, and 447.5 mL dH_2O in a fume hood.
2. Stain solution: Dissolve 0.3 g silver nitrate powder in 300 mL of the fix solution made in **item 1**.
3. Developing solution: Mix 2.7 mL formaldehyde and 197.3 mL dH_2O in a fume hood, then dissolve 6 g sodium hydroxide in this solution.

2.4 Other Solutions and Chemicals

For setting up gel: *N,N,N′,N′*-tetramethylethylenediamine (TEMED), water-saturated butanol, ethanol, distilled water.

2.5 Specialist Equipment

1. DGGE gel apparatus, including glass plates, clamps, spacers, casting stand and tank with heater, stirrer, and buffer recirculation (e.g., Bio-Rad DCode™ system).
2. Twin-well gradient mixer, magnetic stirrer, stir bars, peristaltic pump and tubing.

3. GelBond® PAG Film hydrophobic/hydrophilic membrane (Lonza Corp.; *see* **Note 1**).
4. Plastic trays with lids for silver staining (*see* **Note 2**).
5. Scanner for gel image acquisition, plus appropriate software for analysis.

3 Methods

The detailed methodology described below is for the Bio-Rad DCode™ gel system. For other systems, the details of setting up the gel apparatus, the quantities of solution required per gel and the electrophoresis conditions may differ; please refer to the manufacturer's instructions for guidance. All procedures should be carried out at room temperature.

3.1 Setting Up the Gel Apparatus

Lay out the glass plates, clamps, stand, sponges, and alignment card. The procedure involves making a gel "sandwich" with the plates, plastic spacers, and clamps prior to pouring (Fig. 2). The procedure for a single DGGE gel (*see* **Note 3**) is as follows:

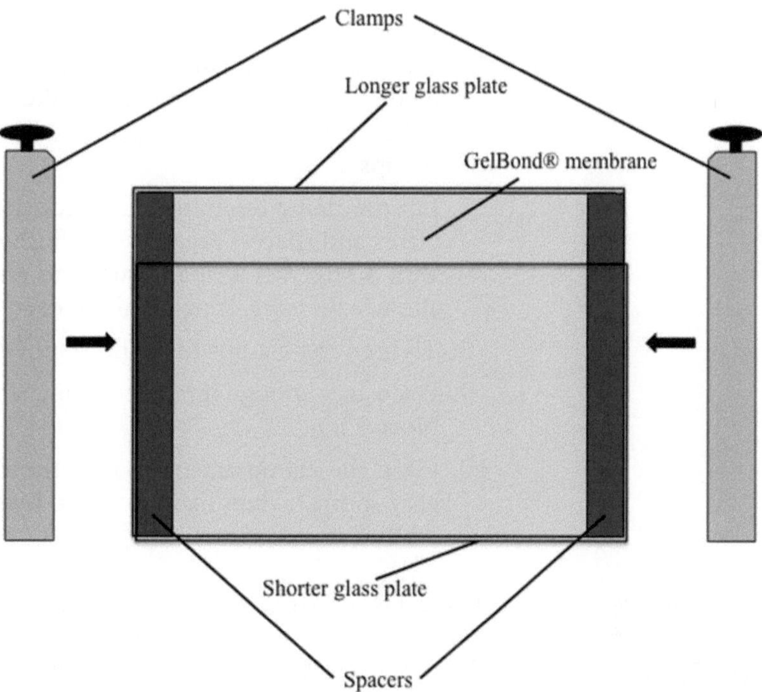

Fig. 2 Assembly of the gel sandwich. Arrange the components in the order longer glass plate → GelBond® membrane (hydrophilic side upwards) → spacers → shorter glass plate, with the cleaned faces of the glass plates facing inwards. Then slot the gel sandwich into the clamps and tighten slightly before transferring to the casting stand

1. Lay one long glass gel plate and one shorter glass gel plate on the bench with the faces to be positioned on the inside of the gel sandwich uppermost. Clean the upper sides thoroughly with water first and then ethanol, polishing until dry (*see* **Note 4**).
2. Using the shorter glass plate as a template, cut the GelBond® membrane to a width 2–3 mm narrower than the plate and of the same length. Drop a little dH_2O onto the cleaned side of the long glass plate and orientate the membrane with the hydrophobic side adjacent to the glass (the gel will bind to the hydrophilic side; *see* **Note 5**). Press the hydrophobic side of the membrane down, wiping from center to outside to remove excess water. The water will cause the hydrophobic side of the membrane to adhere to the glass.
3. Lay two spacers on top of the membrane flush with the sides of the plate (*see* **Note 6**).
4. Lay the shorter glass plate on top of the spacers, with the cleaned side facing the spacers.
5. Lift the entire gel sandwich and fit into the clamps, loosely tightening, orientating as in Fig. 2, with the spacers and clamps running vertically. The gel sandwich will only fit into the clamps in one orientation.
6. Insert the alignment card between the spacers and glass panels, and stand the assembled, clamped gel sandwich in the alignment slot of the casting stand. Press gently from the top and either side to ensure optimal alignment, then tighten the clamps.
7. Lift the clamp assembly and check that the bottoms of the glass plates and spacers are perfectly flush with each other, checking with a finger as well as visually to ensure there are no gaps to allow leaks later. If necessary, loosen the clamps and readjust.
8. Tighten the clamps fully and remove the alignment card.
9. Place grey sponge into the casting slot of the casting stand (*see* **Note 7**).
10. Place the clamp assembly into the casting slot on top of the grey sponge, then fix in position by sliding in and turning the catches at either side.

3.2 Preparing the Gel Mixes from Stock Solutions

The gel is prepared with a plug along the bottom (to prevent leaks during casting of the main gel), a main gradient gel where the DNA separates, and a stacking gel into which the samples are loaded (Fig. 3). Quantities given below are given for a 30–70 % denaturant gradient across the gel from top to bottom (*see* **Note 8**), for one gel in the Bio-Rad DCode™ system with 1.0 mm spacers. Different gradient endpoints can be obtained by adjusting the quantities of the two stock solutions appropriately. The gradient is

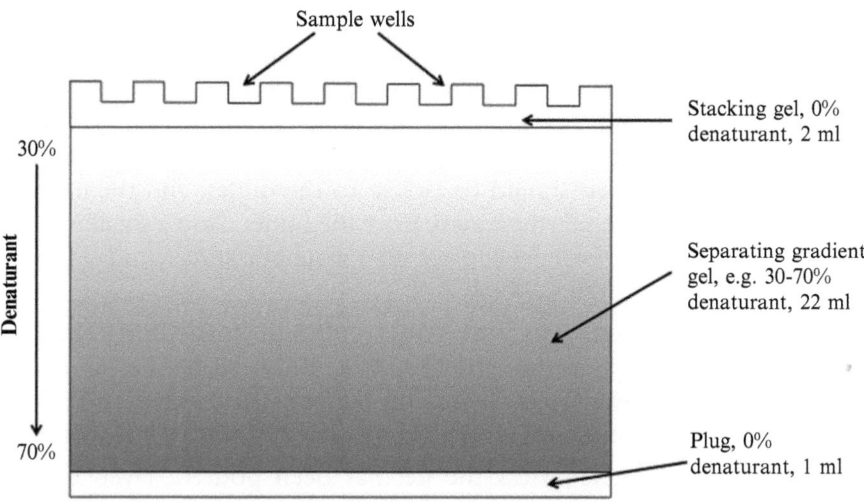

Fig. 3 Schematic diagram of a parallel DGGE gel. The plug at the base of the gel, which prevents leaks during the pouring of the gradient gel, is poured first. The gradient gel is then poured via a gradient mixer and peristaltic pump on top of the plug, and allowed to set. Finally, the stacking gel, which ensures that the sample is concentrated into a fine band before entering the gradient, is poured on top, and a comb inserted to form the sample wells

formed as the two gel solutions mix continuously during the pouring process. (The quantity of gel solutions required for other gel sizes or different spacer thicknesses will need to be ascertained beforehand.)

Into four labelled disposable plastic tubes, add the following:

Plug: 1 mL of 0 % DGGE solution.

Main gel, 30 %: 6.9 mL of 0 % DGGE solution + 4.1 mL of 80 % DGGE solution = 11 mL total.

Main gel, 70 %: 1.4 mL of 0 % DGGE solution + 9.6 mL of 80 % DGGE solution = 11 mL total.

Stacker: 4 mL of 0 % DGGE solution.

Do not add the catalysts (APS and TEMED) to each tube until that portion of the gel is ready to be poured, as polymerization will start immediately.

3.3 Pouring the Gel

1. The plug: Into the 1 mL of 0 % DGGE solution, add 1 μL/mL TEMED and 10 μL/mL APS.
2. With a pipette and tip, pour the plug solution gently down one side of the space between the glass plates then tip the whole stand and glass sandwich to ensure even coverage of solution along the bottom. Plug sets in 10–15 min.
3. The main (gradient) gel: Connect the twin-well gradient mixer (*see* **Note 9**) to a length of tubing and the tubing via a peristaltic

pump to the gel assembly. Insert the end of the tubing between the gel plates at one side adjacent to a spacer (*see* **Note 10**). Close the taps on the gradient mixer and pour the 11 mL of the appropriate low- and high-concentration gel solutions (*see* Subheading 3.2) into the two wells. The higher concentration used should be closest to the outlet, and the lower concentration farthest away from the outlet. Add a small magnetic stir bar to each well, place on a magnetic stirrer, and set the stirrer bars rotating slowly.

4. Add 5 μL/mL APS and 0.5 μL/mL TEMED into each well. Open the outlet tap first, followed by the tap connecting the two wells. Switch on the peristaltic pump and run at ~5 mL/min until all the gel mix has run through (*see* **Note 11**).
5. As soon as the gel has been poured, layer 1 mL of water-saturated butanol across the top of the gel with a pipette. This ensures a level gel edge. Sets in about 35–45 min.
6. As soon as the gel is poured, run dH_2O through the gradient mixer, taps, and tubes, and leave the taps on the gradient mixer open.
7. Once the gel is set, pour off the butanol and remove the last few drops by capillary action with a tissue.
8. Ease the comb between the glass plates, positioning 1 cm above the top of the gradient gel.
9. The stacking gel: Add 10 μL/mL APS and 1 μL/mL TEMED to the stacking gel solution and mix. With a pipette tip, pour down both sides gently and tip the whole casting stand to ensure even coverage of solution around comb.
10. When set (approximately 20 min), undo the catches and remove the gel sandwich unit with clamps from the casting stand, keeping the comb in place.
11. Mount the gel sandwich assembly unit with the longer glass plate outwards onto the (yellow) core, sliding the pegs on the core into the slots on the clamps and clicking into position with the catches (*see* **Notes 12** and **13**).

3.4 Preheating and Running the Gel

1. The standard electrophoresis time for the DCode™ system is 16 h. To allow time to pour and load the gel, we therefore recommend preheating 7 L of 1× TAE running buffer in the gel tank (*see* **Note 14**) to 65 °C while the stacking gel is setting, as this can take 1–2 h (*see* **Note 15**).
2. Before taking lid off tank, switch the heater and pump off and wait 30 s (*see* **Note 16**).
3. Remove ~500 mL 1× TAE buffer from the DGGE tank and transfer the assembled core to the tank (*see* **Note 17**). The core needs to be orientated so that the red dot is on the right-hand side, to ensure correct connection of the electrodes.

4. Add the 1× TAE buffer removed in **step 3** to the upper buffer reservoir. Ensure the combs are submerged under the buffer before easing them out very carefully, from both ends. The gel is only 1 mm thick and can tear easily (*see* **Note 18**).
5. Flush wells with buffer to remove unpolymerized acrylamide using a syringe and medium-gauge needle.
6. Ensure that the buffer in the tank is at the "run" level.
7. Load samples consisting of 2 μL of loading dye plus 10–15 μL of DNA sample with a fine-tipped gel loading tip (*see* **Note 19**).
8. Refit the lid, adjust the thermostat to 60 °C, and allow the temperature to equilibrate.
9. Run the gel for 960 min (16 h) at 75 V (3.75 v/cm for a 20 cm gel; *see* **Note 20**).
10. Stop the gel run, open the gel tank, remove the core, and dismantle the gel sandwich for staining. Keep the 1× TAE buffer as this can be used 3–4 times.

3.5 Staining the Gel

Instructions are given for silver staining. Other methods are possible but are generally less sensitive. Fixative, staining, and destaining solutions, described in Subheading 2.3, should be prepared fresh for each use.

1. Transfer the gel to a suitable staining tray, cover with 200 mL of fixative solution, and shake gently for 30 min.
2. Remove the fixative solution, retaining for later use.
3. To the remaining 300 mL of fixative solution, add 0.3 g silver nitrate (*see* Subheading 2.3).
4. Incubate the gel in this silver staining solution for 20 min with shaking.
5. Rinse with plenty of dH_2O (*see* **Note 21**); remove the water.
6. Add 200 mL of developing solution to the gel and incubate with shaking for 30–40 min. Dark bands should appear in the gel lanes.
7. When developed, rinse as in **step 5** and return to the fixative solution kept back from the first step. Keep the gel in fixative for 30 min or longer (*see* **Note 22**) prior to scanning.
8. For long-term storage, the gel can be transferred to a preservative solution of 25 % ethanol and 10 % glycerol.

3.6 Gel Analysis

The complex banding patterns which can be obtained from DGGE gels, in particular when used for fingerprinting of complex mixtures of DNA sequences, are best analyzed using automated band matching software to generate matrices of band intensity versus position in the gradient. The DGGE gel shown in Fig. 4a was

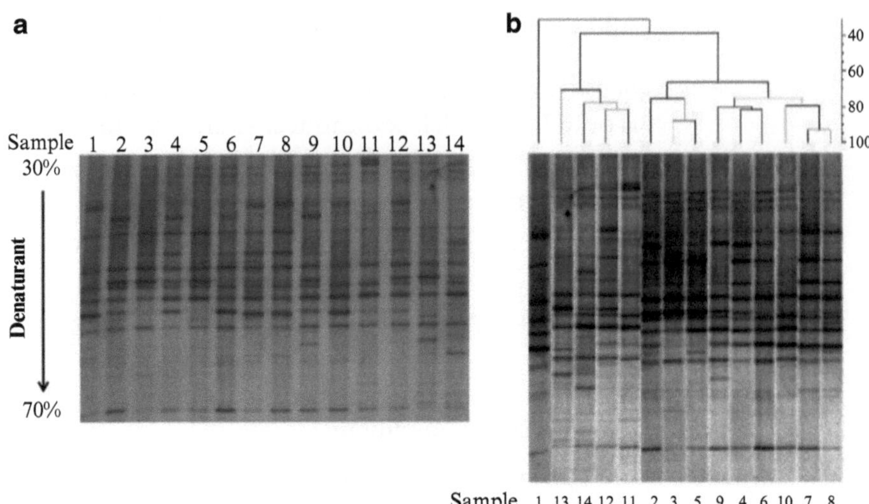

Fig. 4 Example of a 30–70 % parallel DGGE gel used to study the diversity of microbial communities using 16S rRNA gene amplicons. (**a**) The ~200 bp V3 region of the bacterial 16S rRNA gene was amplified by PCR with GC clamp-containing primers, run on the DGGE gel for 16 h, and silver stained. Moderate bacterial diversity and a significant percentage of shared species are seen within the samples. (**b**) Reordering of the gel lanes from the *left panel* by similarity following band matching in BioNumerics 6.0 and determination of the cosine similarity indices from the matrix of matched band intensities. The dendrogram was constructed using the UPGMA joining method; similarity between the communities is indicated by the scale on the *right*

generated by amplifying a short, variable region (the V3 region) of the bacterial 16S rRNA genes from total DNA extracted from 14 intermediate-complexity microbial communities. GC clamp-containing "universal" primers, which are designed to bind to the conserved DNA sequences flanking the V3 region of the 16S rRNA gene in all bacterial species, were used for the amplification.

The gel image was scanned and imported into the gel analysis package BioNumerics 6.0 (Applied Maths), which was used to detect automatically bands in the lanes and match equivalent bands in different lands, generating a matrix of band position versus abundance in each sample. A multivariate comparison between the samples was then carried out to generate cosine similarity scores between the different samples, and a dendrogram of community relatedness was constructed based on these similarity scores (Fig. 4b). The use of automated band detection and matching software avoids the potential biases associated with manual comparison. Alternative methods of analysis of the band abundance matrix include calculation of Bray-Curtis similarity coefficients followed by nonmetric multidimensional scaling in two dimensions, for which we use the Primer 6 package (Primer-E Ltd.).

4 Notes

1. Although not essential, we recommend using GelBond® membrane to support the gel, as it makes handling and storage much simpler.
2. A set of trays should be dedicated specifically to silver staining, as contaminants can cause excessive background staining.
3. Two gels can be poured and run simultaneously in the DCode® system. However, problems can occur with uneven running of the rear gel, due to its proximity to the heating element. If such problems occur, placing the gel tank on a magnetic stirrer and stirring gently during electrophoresis can ensure a more even temperature distribution.
4. Use paper hand towels with few loose fibers in order to minimize fibers sticking to the glass plates, which can affect gel quality.
5. If unsure which side of the GelBond® membrane is the hydrophobic side, test by dropping water onto the membrane: The water will bead on the hydrophobic side.
6. Vaseline smeared down the sides of the spacers helps prevent leaks. This therefore reduces frowning caused by DNA running faster down the sides which can affect automated band matching. Vaseline should be smeared onto the sides of the spacers in contact with the gel.
7. Covering the grey sponge with a piece of Parafilm can help prevent leaks.
8. We recommend starting with a broad gradient (30–70 %). If the bands are clustered too tightly, the gradient can be altered to separate the bands further on a subsequent gel run. For instance, if the bands are clustered at the bottom, a second run should be carried out with an increased denaturant concentration at the lower end of the gradient range.
9. We do not recommend use of the Model 475 gradient former supplied with the Bio-Rad DCode® system, as it is more complex to use and less reliable than a simple twin-well gradient mixer coupled with a magnetic stirrer and a peristaltic pump.
10. A fine tip on the end of the tube ensures it remains in place while the gel is being poured.
11. When the gel mix runs low, tip the gradient mixer to ensure the remaining quantity of lower concentration gel mix flows through for a complete gradient.
12. Check that the white rubber gasket is sitting flush with the top of the short glass plate.
13. If only one gel is to be run, use a blank made up by clamping a longer and a shorter glass plate together (without the spacers, gel, etc.) to clip into the second side of the core.

14. The tank must be filled to the "run" level to ensure that the float is in the buffer for preheating. The safety mechanism will detect a lack of buffer and switch off the heating if the float drops.
15. The buffer needs to be at 60 °C for the electrophoresis, so heating to 65 °C allows for cooling while combs are taken out and samples loaded.
16. If it is removed immediately, the thermal fuse in the heating element can blow.
17. Remove 500 mL of buffer from tank before inserting the core as this will displace liquid; save this buffer to add to the upper buffer well.
18. The gel walls at the edges of the wells can be straightened out with a small spatula inserted into the wells if they become bent.
19. Exact volume of DNA sample depends on the concentration and the staining method to be used.
20. As the buffer recirculates via the pump, it can generate transient short circuits between the upper and lower buffer reservoirs. If the power pack has a current fluctuation detection system, this needs to be deactivated or it may trigger and halt the electrophoresis. On Bio-Rad PowerPac Basic, this is achieved by holding down the Stop key in "current" mode until "dE9" appears in the display.
21. Several rinses with ~500 mL dH_2O are recommended.
22. After the final incubation in fixative, we wash and leave the gel in water until scanning. We then wrap the gel in cling film for storage as it is backed with GelBond®.

Acknowledgements

This work was supported by Leverhulme Trust a grant no. F/00158/BX, by EPSRC grant no. EP/E030173, and by Darwin Trust Research Fellowship to A.F.

References

1. Fischer SG, Lerman LS (1979) Length-independent separation of DNA restriction fragments in 2-dimensional gel electrophoresis. Cell 16:191–200
2. Abrams ES, Murdaugh SE, Lerman LS (1990) Comprehensive detection of single base changes in human genomic DNA using denaturing gradient gel-electrophoresis and a GC clamp. Genomics 7:463–475
3. Bio-Rad Laboratories (1996) The DCode™ universal mutation detection system (manual). Bio-Rad Laboratories, Hemel Hempstead

4. Muyzer G, Dewaal EC, Uitterlinden AG (1993) Profiling of complex microbial populations by denaturing gradient gel electrophoresis analysis of polymerase chain reaction-amplified genes coding for 16S ribosomal RNA. Appl Environ Microbiol 59:695–700

5. Felske A, Osborn AM (2005) DNA fingerprinting of microbial communities. In: Osborn AM, Smith CJ (eds) Molecular microbial ecology. Taylor and Francis, Abingdon, pp 65–96

Chapter 10

Polyacrylamide Temperature Gradient Gel Electrophoresis

Viktor Viglasky

Abstract

Temperature Gradient Gel Electrophoresis (TGGE) is a form of electrophoresis in which temperature gradient is used to denature molecules as they move through either acrylamide or agarose gel. TGGE can be applied to analyze DNA, RNA, protein–DNA complexes, and, less commonly, proteins. Separation of double-stranded DNA molecules during TGGE relies on temperature-dependent melting of the DNA duplex into two single-stranded DNA molecules. Therefore, the mobility of DNA reflects not only the size of the molecule but also its nucleotide composition, thereby allowing separation of DNA molecules of similar size with different sequences. Depending on the relative orientation of electric field and temperature gradient, TGGE can be performed in either a parallel or a perpendicular mode. The former is used to analyze multiple samples in the same gel, whereas the later allows detailed analysis of a single sample. This chapter is focused on analysis of DNA by polyacrylamide TGGE using the perpendicular mode.

Key words Temperature gradient gel electrophoresis, DNA, RNA, Protein, Polyacrylamide, Gel staining, Dye

1 Introduction

Temperature Gradient Gel Electrophoresis (TGGE) was first described by Roger Wartell from Georgia Institute of Technology, and extensive initial work was carried out by the Riesner laboratory in Germany [1]. TGGE is a special type of electrophoresis where nucleic acids undergo denaturation during gel electrophoresis as a result of being exposed to a temperature gradient. A similar effect could be achieved by subjecting nucleic acids to a pH (chemical) gradient during Denaturing Gradient Gel Electrophoresis (DGGE). Both TGGE and DGGE utilize melting behavior of molecules as the primary principle for separating them in a gel. Both forms of electrophoresis are useful for analyzing nucleic acids, both DNA and RNA, as well as for analyzing proteins. However, the chemical gradients are harder to reproduce accurately and consistently [1], while TGGE offers a quick gradient setup and reproducible separation conditions [2].

Svetlana Makovets (ed.), *DNA Electrophoresis: Methods and Protocols*, Methods in Molecular Biology, vol. 1054, DOI 10.1007/978-1-62703-565-1_10,

TGGE is a particularly powerful technique when separation of RNA or DNA fragments of identical length is required. In contrast to the conventional electrophoresis method, molecules are separated according to their melting behavior, and therefore it becomes possible to separate identical-length DNA fragments different in their primary sequence by as little as a single base pair. Not surprisingly, the most common application of TGGE is analysis of single nucleotide polymorphism (SNP).

Two fundamental points should be taken into consideration in order to understand the basic principles of TGGE. The first is how the structure of a biomacromolecule changes with a temperature increase or decrease; the second is how these structural changes affect the movement of the macromolecule through a gel matrix during electrophoresis. A temperature increase may lead to melting of nucleic acid molecules which would either slow down or speed up their migration through gel. When a molecule increases or decreases its volume without any change of surface charge, then its mobility decreases or increases, respectively. For example, G-quadruplex DNA dimers move more slowly than monomers after denaturation induced by a temperature increase [3], and vice versa. In contrast, the mobility of monomolecular G-quadruplex DNA is higher at lower temperature and decreases during the unfolding process [4]. However, unfolding is often accompanied by a change of the surface charge of a molecule, which in turn significantly influences migration speed.

While the same TGGE equipment can also be used for analyzing proteins and the gel runs can generate similar-looking patterns, the fundamental principles for protein analysis are quite different [5]. Proteins are either polyanionic or polycationic molecules depending on the pH and therefore they move either towards a cathode or towards an anode. An increase in temperature induces protein unfolding. As a result, changes in the friction volume are observed as well as changes in the isoelectric point and protein aggregation—all leading to a decrease in protein mobility in gel. Thus, TGGE not only separates molecules but also provides additional information about their melting behavior and stability [2, 6].

The temperature gradient is created on a metal plate adjacent to the electrophoretic apparatus, normally, by using circulating water to cool the plate on one side and to heat it on the other side: The stable linear gradient is maintained by two circulating thermostats. The lower temperature edge of the apparatus is typically set at 10 °C and the higher temperature one is at 80 °C. As a result, the linear temperature gradient is uniform within the range of 15–75 °C throughout the thermostated plate (Fig. 1).

The electric field can be oriented either parallel or perpendicular to the temperature gradient, thereby defining the two corresponding modes of TGGE. The parallel mode is very similar to standard electrophoresis, with the addition of a temperature

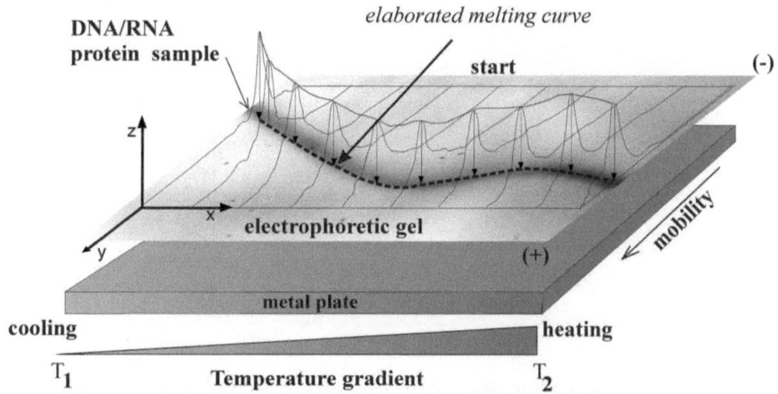

Fig. 1 Principles of TGGE. The linear perpendicular gradient of temperature is set up on a metal plate where one edge is cooled and the other one is heated. The electrophoresis apparatus is connected to this metal plate. The range of the temperature gradient is defined by the temperatures T_1 and T_2

gradient. In this mode, dsDNA molecules migrate along a gradient of temperature without any structural change until they reach the point of their melting temperature where the DNA duplexes begin to denature. After denaturation, the complementary DNA strands migrate separately as they normally differ from each other in their molecular mass. The dsDNA melting point, which is dependent on the DNA sequence, determines the position in the gel where the dsDNA band splits into two ssDNA bands. The parallel mode TGGE allows analyzing many different DNA samples in a single experiment and is limited only by the number of wells in a gel.

In the perpendicular mode, DNA samples move in the electric field perpendicularly to the temperature gradient. The DNA electrophoretic mobility increases linearly with the increase of temperature as long as no conformational change is induced. Therefore, any nonlinear changes in DNA mobility, either reduction or increase, reflect structural changes such as DNA duplex melting. A homogeneous population of DNA molecules produces a clear melting profile as a single curve (Fig. 2), whereas a sample of nonhomogeneous DNA yields multiple curves, each representing a separate subpopulation of DNA molecules. In the perpendicular mode, only one sample can be loaded into a gel. The perpendicular mode of TGGE is the main focus of this chapter.

Commercial equipment for TGGE is available from Biometra. This company offers two different versions of the apparatus for TGGE. The smaller system provides advantage of shorter running time and therefore is convenient for quick optimization of running conditions as well as for routine analysis of simpler samples. The larger system provides longer separation distance and is thus ideally suited for analysis of complex samples. Instead of the circulating

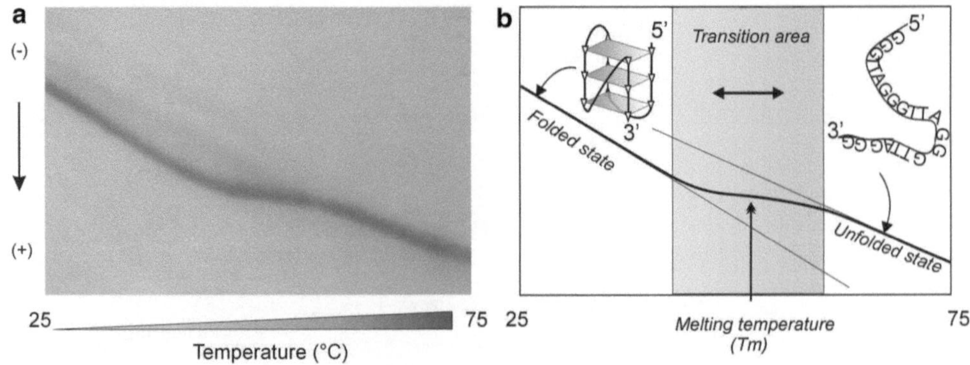

Fig. 2 A representative TGGE of the DNA oligomer $(GGGTTA)_3GGG$ analyzed in 10 % polyacrylamide in 0.5× TBE in the presence of 2.5 mM KCl. The gel was electrophoresed for 3 h at 5 V/cm and stained with Stains-All. (**a**) An image of the original TGGE gel; (**b**) a graphic interpretation of the gel in **a**. The oligomer folds into a G-quadruplex structure at lower temperatures (*left*), and an increase in temperature to above 50 °C (*right*) induces its unfolding [4]. The *vertical double arrow* represents the position of melting points. The mobility shows a nonlinear shape for the temperature region where the unfolding occurs (transition area)

thermostats described above, the Biometra TGGE instruments are powered by Peltier thermoblocks which allow precise temperature control by a microprocessor.

2 Materials

2.1 Polyacrylamide Gel Components

1. 40 % acrylamide/bisacrylamide stock solution. For optimal TGGE results, the ratio between these two components should be no higher than 19:1 (*see* **Note 1**). Commercially available acrylamide stock solutions are used in many research laboratories. If this is not an option, weigh 38 g of acrylamide monomer and 2 g of bisacrylamide (cross-linker) and transfer to a 150 mL graduated cylinder containing ~ 40 mL of dH_2O and mix for about 30 min. Solubility of acrylamide is facilitated by heating up the solution to 70 °C for the first 10 min. Bring the volume to 100 mL with dH_2O and filter through Whatman filter paper. Store at 4–10 °C, in a glass bottle away from light. Prior to use, it is recommended to adjust the temperature of stock solution to room temperature and to mix the solution vigorously for about 3 min.
2. Ammonium persulfate (10 % w/v). Dissolve 1 g of ammonium persulfate in dH_2O and bring the volume to 10 mL. Store at 4°C. Make fresh every couple weeks.
3. TEMED (Bio-Rad or Sigma). Store at 4 °C.

2.2 Buffers and Solutions

1. 0.5× TBE: 45 mM Tris–borate, 2.0 mM EDTA. TBE is usually made and stored as a 5× stock solution (*see* **Note 2**). Mix 54 g Tris base, 27.5 g boric acid, 20 mL of 0.5 M EDTA in 800 mL dH_2O. Bring up the volume to 1 L with dH_2O. The pH of the concentrated stock buffer should be ~8.3.
2. 100 mM Tris–HCl, pH 8.3. Dissolve 12.1 g Tris base in 800 mL dH_2O and adjust pH with HCl. Bring up the volume to 1 L with dH_2O.
3. 6× Gel-loading buffer: 0.25 % (w/v) bromophenol blue, 0.25 % (w/v) xylene cyanol FF, 0.10 % (w/v) Orange G, 40 % (w/v) sucrose in water. Gel-loading buffers are mixed with the samples before loading into the slots of the gel (*see* **Note 3**).
4. Ethidium bromide 10 mg/mL in dH_2O. Store at room temperature in dark bottles or bottles wrapped in aluminum foil. The dye is usually added to molten agarose and electrophoresis buffers at a concentration of 0.5 μg/mL.
5. Thiazole orange solution: 10 mg/mL in DMSO. Store at 4–14 °C in dark bottles or 1.7 mL microcentrifuge tubes wrapped in aluminum foil.
6. SYBR Gold staining solution (0.001–0.0025 %). Dilute the stock SYBR Gold stain 10,000-fold to make a 1× staining solution (0.001–0.0025 %) as follows: dilute 20 μL of stock solution in 200 mL 1× TBE buffer.
7. Methanol–acetic acid solution. Mix methanol, dH_2O, and glacial acetic acid at the v/v/v ratio of 5:4:1. For 1 L, add 500 mL of methanol and 100 mL of acetic acid to 400 mL of dH_2O. Mix and store at room temperature.
8. Coomassie Brilliant Blue (CBB) R-250: 0.25 % CBB in methanol–acetic acid. Add 0.25 g of CBB R-250 to 100 mL of methanol–acetic acid solution (*see* **Note 4**).
9. Stains-All stock solution: 0.1 % Stains-All (Sigma) w/v in formamide. Store at 4 °C and protect from light (use a dark bottle). To make Stains-All working solution, dilute the stock 1:9–1:4 in 15 mM Tris pH 8.0, 25 % isopropanol. The staining develops faster at the higher stain concentrations.

2.3 Other Materials

1. Gel temperature-monitoring strips, BioWhittaker (optional, *see* **Note 5**).
2. Electrophoresis apparatus, glass plates, comb, and spacers. Spacers (usually silicone rubber) vary in thickness from 0.5 to 2.0 mm.
3. Peristaltic pump.

3 Methods

3.1 Polyacrylamide TGGE

In order to obtain sufficiently accurate thermodynamic parameters characterizing the thermal unfolding of biopolymeric molecules (protein, RNA, DNA) from perpendicular TGGE, reproducible and controllable gel preparation is essential. The main requirement for the gels used in TGGE is the standardization of gel polymerization at specific pH values, in addition to chemical inertness, thermal stability, and homogeneity. The homogeneity of a polyacrylamide gel, i.e., its characteristic porosity throughout the whole gel, depends on the polymerization conditions, which can be manipulated effectively at slightly alkaline pH values. The polymerization of acrylamide to gel at acidic conditions is not well defined; higher concentrations of TEMED and ammonium persulfate are required in this case. The time of gel polymerization at low pH (pH <5.0) is lengthened, and for pH values below 2.8, polyacrylamide does not polymerize [2].

3.1.1 Gel Preparation

1. Assemble a casting mould as shown in Fig. 3a. Casting mould does not have to be provided directly with the apparatus. Two standard 20 × 20 cm glass plates and four 0.5–2 mm spacers can be used for this purpose.
2. Prepare gel solution of the desired polyacrylamide concentration (8–15 %) by mixing the appropriate amount of 40 % acrylamide/bisacrylamide stock solution, 100 mM Tris–HCl, and dH_2O. For example, to make 100 mL of 10 % acrylamide solution, mix 10 mL of 100 mM Tris–HCl, 25 mL of 40 % acrylamide/bisacrylamide stock solution, and 65 mL dH_2O (*see* **Note 6**).
3. Add ammonium persulfate and TEMED to the final concentrations of 0.05 % each, mix well, and pour immediately into the gel casting mould (Fig. 3a).
4. After the polymerization, disassemble the casting mould and place the gel in a clean tray and soak it in 2–3 L of dH_2O for 2 h at 85 °C. Repeat this step at least twice. The soaking can be done at room temperature, if the time of the soaking procedure is extended (*see* **Note 7**).
5. Soak the gel in the gel running buffer overnight (preferred) or for at least 2 h; *see* Fig. 3b.

3.1.2 Gel Electrophoresis

1. Place the gel into the electrophoresis tank, between the two thermostable glass plates (Fig. 2c). Press the gel down between the plates and try to avoid air bubbles between the gel and the glass plates. A paperweight can be placed on the upper glass plate to help in pressing the gel down.

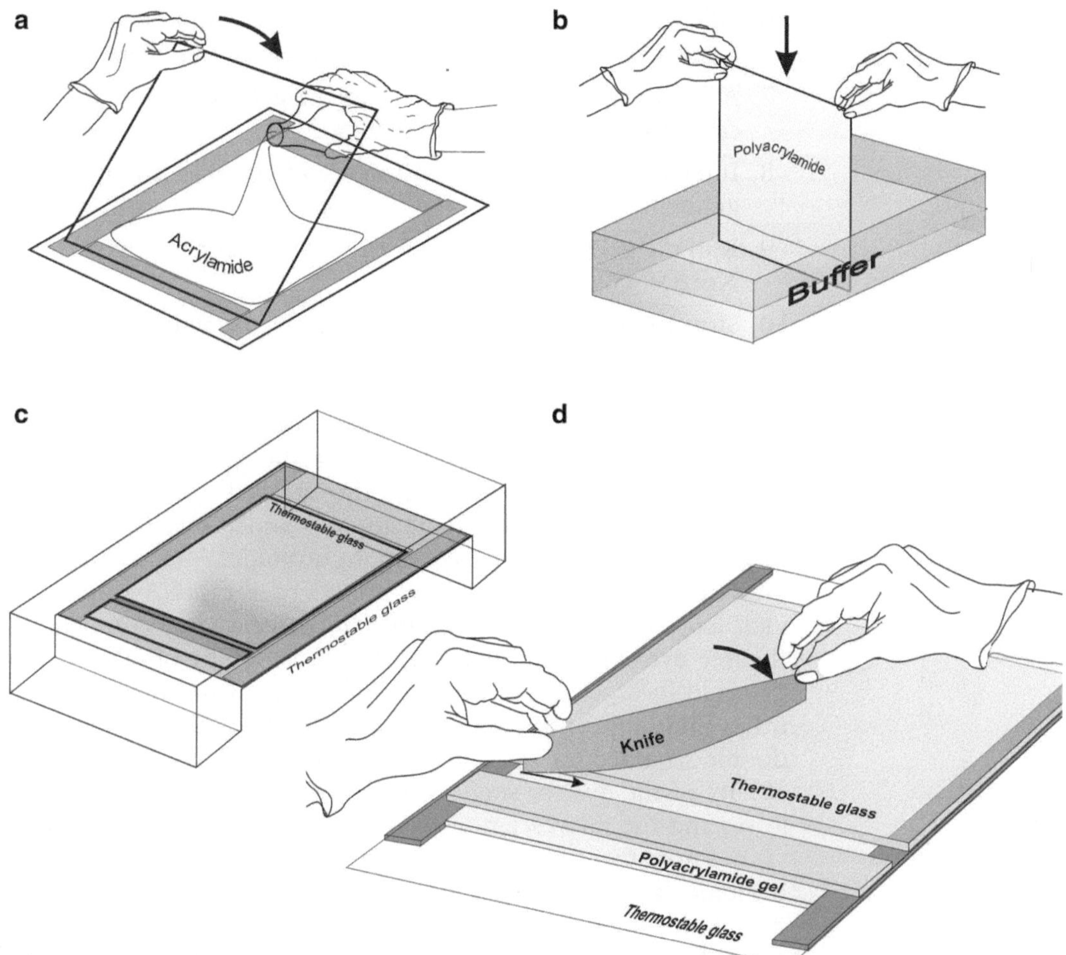

Fig. 3 Preparing a gel for TGGE, perpendicular mode. (**a**) Pouring the acrylamide gel into the gel casting mould, (**b**) transferring gel into a tank containing electrophoresis buffer, (**c**) assembling the gel in the apparatus, and (**d**) the well cutting technique

2. Heat the TGGE apparatus to 60–70 °C for 5–10 min. This improves the adherence of the gel to the glass plates.
3. Cool the apparatus down to room temperature.
4. Cut a narrow, 1.5–2.0 mm, electrophoretic well across the top of the gel in one of the exposed edges of the gel between the two glass plates (Fig. 3d). Press the rest of the exposed gel edge down with another glass plate. Use a thin sharp knife to adjust the size of the polyacrylamide gel to fit into the apparatus.
5. Remove the cut strip of the gel from the well with a sharp scalpel. Use a syringe filled with electrophoresis buffer to rinse the well(s).

6. Fill the gel tank with the 0.5× TBE gel running buffer and apply a DNA sample (μg scales) to the well (*see* **Note 8**).
7. Allow the sample to migrate into the gel for 15 min at ~5 V/cm, without a temperature gradient.
8. Interrupt the run for 30–40 min and establish a temperature gradient.
9. Run the gel at 5–10 V/cm for 2–7 h with constant voltage; *see* **Note 9**.
10. Stop the run and proceed with one of the staining protocols described below.

3.2 DNA/RNA Staining

The most convenient and commonly used method to visualize DNA/RNA in agarose gels is staining the gels with fluorescent dyes which intercalate between the stacked bases in double-stranded and, to a lesser extent, single-stranded nucleic acids. The most commonly used dye is ethidium bromide. After intercalating into the double helix, the dye bound to the DNA displays an increased, ~20- to 30-fold greater, fluorescent yield compared to the free dye in gels and buffers. As little as ~10 ng of DNA can be detected in a gel in the presence of ethidium bromide (0.5 μg/mL). Ethidium bromide can be used to detect both single- and double-stranded nucleic acids, both DNA and RNA. However, the affinity of the dye for single-stranded nucleic acids is significantly lower and the fluorescent yield is comparatively poor. It is important to note that polyacrylamide gels cannot be cast with ethidium bromide, as well as some alternative stains, because it inhibits polymerization of acrylamide. Similarly, SYBR Gold should not be added before the gel run is complete because its presence in the gel will cause severe distortions in the electrophoretic properties of DNA and RNA. Polyacrylamide gels are therefore stained after the gel has been run. Staining is accomplished by immersing the gel in electrophoresis buffer or dH_2O containing ethidium bromide at 0.5 μg/mL for 30–45 min at room temperature.

There are alternatives to ethidium bromide which are advertised as being less dangerous and providing a better performance. For example, several SYBR-based cyanine dyes, OliGreen, GoldView, thiazole orange, GelRed, and GelGreen are from the new generation of fluorescent in-gel stains for nucleic acids. Influorescent Stains-All, methylene blue, and silver nitrate have also been used. The advantages of these dyes can be summarized as follows: (1) they all are less mutagenic than ethidium, (2) they are as sensitive as ethidium is, and (3) the same simple staining procedure, as the one used for ethidium, may provide additional advantages. For example, the GoldView dye allows distinguishing between ds and ssDNA/RNA as it emits green fluorescence when bound to dsDNA and red fluorescence when bound to ssDNA or RNA. Similarly, thiazole orange can be helpful to assay for triplex and quadruplex DNA as complexes of thiazole orange with DNA triplexes and G-quadruplexes are very stable and do not dissociate during chromatography and gel electrophoresis. Thiazole orange binding to either triple- or quadruple-stranded DNA structures results in a >1,000-fold increase in dye fluorescence. A slightly smaller increase of fluorescence is observed for single-stranded and duplex DNA [7]. Therefore, it may be advantageous to consider alternatives to ethidium for detecting nucleic acid in agarose and polyacrylamide gels.

The general, the staining procedure is very simple and similar for all the mentioned dyes. For example, SYBR Gold is used to stain DNA by soaking the gel, after separation of the DNA fragments, in a 1:10,000 dilution of the stock in dH_2O or gel running buffer. The greatest sensitivity is obtained when the gel is illuminated with UV light at ~300 nm and visualization is carried out with green or yellow filters; *see* **Note 10**.

1. Switch off the electric current and heating/cooling of the apparatus.
2. Leave the apparatus to equilibrate to room temperature.
3. Gently remove the upper glass plate. Carefully hold one edge of the gel and transfer it to the appropriate staining solution (*see* **Note 11**).
4. Stain the gel for 30–45 min at room temperature. Use just enough staining solution to cover the gel completely. Avoid any bubbles under the gel as it will interfere with the homogeneity of gel staining. Occasionally, gently shake the container with the gel.
5. Remove the staining solution using a water vacuum exhauster. Add dH_2O and rinse the gel with at least 10 gel volumes of dH_2O for 5 min. Repeat the wash step twice.
6. Take the gel out of the staining container and place it on a transilluminator plate and take a picture.

Usually, no additional destaining is required. However, detection of very small amounts (<10 ng) of DNA is made easier if the background fluorescence caused by unbound ethidium bromide is reduced by soaking the stained gel in dH_2O or 1 mM $MgSO_4$ for 20 min at room temperature.

3.3 Staining DNA–Protein Complexes in Polyacrylamide Gels

Staining of protein–DNA complexes is based on staining the protein molecules.

1. Immerse the gel after electrophoretic separation in at least 5 gel volumes of CCB staining solution, and place the container with the gel on a slowly rocking or rotating platform for a minimum of 2–4 h at room temperature (*see* **Note 12**).

2. Remove the stain and save it for future applications (can be reused for at least 30 times).
3. Destain the gel by soaking it in the methanol–acetic acid solution on a slowly rocking or rotating platform for 4–8 h. Change the destaining solution at least three times during this incubation. The more thoroughly the gel is destained, the smaller the amount of protein can be detected by staining with CBB. Destaining for 24 h usually allows as little as 0.1 μg of protein to be detected in a single band (*see* **Note 13**).
4. After destaining, store the gel in dH_2O, in a sealed plastic bag. Gels may be stored indefinitely without any diminution in the intensity of staining; however, fixed polyacrylamide gels stored in dH_2O will swell and may become distorted during the storage. To avoid this problem, store fixed gels in dH_2O containing 20 % glycerol. Stained gels should not be stored in destaining buffer, because the stained protein bands will fade.
5. Photograph the stained gel to make a permanent record.

3.4 Staining Both Proteins and Nucleic Acids

Both proteins and nucleic acids can be visualized by Stains-All, a highly sensitive stain which can also provide additional information about analyzed samples; *see* **Note 14**.

1. Carefully remove the gel from glass plates and place it into a container a bit larger than the gel.
2. Add at least 20 gel volumes of Stains-All working solution and make sure the gel is fully submerged into the stain. Place the container on a slowly rotating platform and incubate for 20–40 min (*see* **Note 15**).
3. Remove the Stains-All solution and rinse the gel twice with dH_2O.
4. Fill the container with dH_2O and place it under direct light. Destain the gel for 30–60 min. Check the gel often to monitor the progress of destaining as over-destained gels will produce suboptimal results. Destaining is complete when the gel is clear, with no purple shade in the background, and the bands appear sharp (*see* **Note 16**).
5. Photograph the gel to make a permanent record.

3.5 Digital Recording of Electrophoretic Gels

A gel stained with a fluorescent dye or another contrast molecule (Stains-All) can be recorded and digitalized. The present level of technology allows image capturing using integrated systems containing light sources, fixed-focus digital cameras, and thermal printers. The CCD cameras of these systems use a wide-angle zoom lens that allows the detection of very small amounts of fluorescently stained DNA (0.01–0.5 ng). Photographs of stained gels can be made using transmitted UV or visible light.

Most commercially available transilluminators emit UV light at 254/302 nm. Digital image processing of an electrophoretic gel can be exploited to find the skeleton of deformed electrophoretic bands representing the dependence of protein/nucleic acid mobility on temperature. The skeleton curve coordinates are obtained by the algorithm which tracks the darkest pixels of the analyzed part of the image. The darkest pixels are obtained as a transversal line profile of electrophoretic band at the corresponding temperatures (*see* Fig. 1). The curve represents the most abundant population of conformation state of proteins/nucleic acids.

4 Notes

1. Gels can be cast with acrylamide solutions containing different acrylamide/bisacrylamide (cross-link) ratios, such as 29:1 and 37.5:1, in place of the 19:1 ratio recommended here. The mobility of DNA and dyes in such gels will be different from those given in this protocol, and therefore the final concentration of acrylamide in the gel should be adjusted accordingly. In order to maintain the elastic properties of gel, the concentration of acrylamide in gel should be greater than 9 %. WARNING: Wear gloves while working with acrylamide.
2. More concentrated stock solutions of TBE tend to precipitate during longer storage, therefore dilute the concentrated buffer stock just prior to use. Make the gel solution and the appropriate electrophoresis buffer from the same stock solution to ensure the same concentration of buffer components in both electrophoretic gel and buffer.
3. Gel-loading buffers contain glycerol which increases the density of the sample solution, ensuring that the DNA settles into the well. The colored dyes allow visualizing the loading process directly and estimating the sample migration in a gel during electrophoretic separation.
4. If you prepare a fresh stock solution, filter the solution through a Whatman No.1 filter to remove any particulate matter. After the staining is complete, the band intensity may be further enhanced by destaining the gel in 30 % methanol and 10 % acetic acid. CBB has a sensitivity of staining 8–10 ng of protein per band (e.g., 8–10 ng of BSA is visible in 4–20 % SDS acrylamide gels). CBB may be harmful when inhaled, ingested, or absorbed through direct contact with the skin. Appropriate gloves and safety glasses should be worn when handling these substances.
5. These strips are thermochromic liquid crystal (TLC) indicators that change color as the temperature of the gel increases during electrophoresis. Temperature-monitoring strips are not needed

if your electrophoresis apparatus that has a built-in thermal sensor is used.

6. To prepare 100 mL of 8, 12, or 15 % acrylamide gel solutions, mix 10 mL of 100 mM Tris–HCl and 20, 30, or 37.5 mL of 40 % acrylamide/bisacrylamide stock solution, respectively. Add dH_2O to bring the total volume to 100 mL.
7. After this treatment, the gel and its thermal stability are suitable for TGGE experiments. Gels may be stored in dH_2O for 3–5 days at 4 °C. In addition, it is advisable to pre-electrophorese each gel for about 1.5–2 h prior to sample applications. These steps help to remove residual, reactive, charged compounds and low molecular weight impurities from the gel as they may cause artifacts.
8. Concentrated nucleic acid samples used for electrophoresis should be exhaustively dialyzed against dH_2O prior to being diluted with the gel running buffer containing the gel-loading buffer.
9. Continuous recirculation of the electrophoresis buffer from anode/cathode to cathode/anode chamber with a peristaltic pump is recommended for electrophoretic separation longer than 4 h.
10. Ethidium bromide staining requires an orange filter, but for SYBR Gold and thiazole orange, a yellow or a green filter works better. The maximum fluorescence yield and contrast with SYBR Gold and thiazole orange staining are obtained by epi-illuminating the gel with ~254 nm UV light rather than ~310 nm. However, routinely used standard transillumination can work as well. Gels stained with methylene blue and Stains-All can be photographed with white light; no filter required.
11. Using gloves is strongly recommended!
12. Instead of CBB, another dye, Amido Black, can be used.
13. If destaining is prolonged, there may be some loss in the intensity of staining of protein/nucleic acid bands. A more rapid rate of destaining can be achieved by the following adjustments: (a) destaining at higher temperatures (45 °C), (b) including a few grams of an anion exchange resin or a piece of sponge in the destaining solution as these absorb the stain as it leaches from the gel.
14. The staining solution is typically made by dissolving the dye in a buffer containing formamide. It has been reported that Stains-All is 50–100 % more intense than CCB and 30–70 % more sensitive than ethidium bromide. As little as 3 ng of 125 bp DNA fragment and ~90 ng tRNA can be detected in a polyacrylamide gel. The intensity of Stains-All can be increased by adding silver nitrate. The Stains-All also allows to differenti-

ate between phosphorylated and non-phosphorylated proteins. After staining, RNA obtains a bluish purple shade, DNA becomes blue, and proteins are of a reddish shade. It is recommended to stain polyacrylamide gels in the dark and then destain them by removing the gel from the staining solution and exposing it to light from a light box until sufficient destaining has occurred.

15. The amount of time necessary to stain effectively is somewhat empirical, but it will get longer as the staining solution ages. Make fresh staining solution every few months.
16. Destaining can be facilitated by the following procedure. Place the gel in a Pyrex glass pan for destaining. Put the pan on a sheet of white blotting paper and place a desk lamp above the pan for about 10–12 min. Watch closely, because it takes only ~15 min to destain the gel and then the bands will start disappearing. This method is not recommended if the visualization of faint bands is the main goal of the procedure.

Acknowledgements

This work was supported by grants from the Slovak Research and Development Agency under the contract No. APVV-0280-11 and by the Slovak Grand Agency (No. 1/0504/12).

References

1. Rosenbaum V, Riesner D (1987) Temperature-gradient gel electrophoresis. Thermodynamic analysis of nucleic acids and proteins in purified form and in cellular extracts. Biophys Chem 26:235–246
2. Viglasky V, Antalík M, Bagel'ová J, Tomori Z, Podhradský D (2000) Heat-induced conformational transition of cytochrome c observed by temperature gradient gel electrophoresis at acidic pH. Electrophoresis 21:850–858
3. Poniková S, Tlučková K, Antalík M, Víglaský V, Hianik T (2011) The circular dichroism and differential scanning calorimetry study of the properties of DNA aptamer dimers. Biophys Chem 155:29–35
4. Víglasky V, Bauer L, Tlucková K (2010) Structural features of intra- and intermolecular G-quadruplexes derived from telomeric repeats. Biochemistry 49:2110–2120
5. Creighton TE, Shortle D (1994) Electrophoretic characterization of the denatured states of staphylococcal nuclease. J Mol Biol 242:670–682
6. Riesner D, Steger G, Zimmat R, Owens RA, Wagenhöfer M, Hillen W, Vollbach S, Henco K (1989) Temperature-gradient gel electrophoresis of nucleic acids: analysis of conformational transitions, sequence variations, and protein-nucleic acid interactions. Electrophoresis 10:377–389
7. Lubitz I, Zikich D, Kotlyar A (2010) Specific high-affinity binding of thiazole orange to triplex and G-quadruplex DNA. Biochemistry 49:3567–3574

Chapter 11

Separation of DNA Oligonucleotides Using Denaturing Urea PAGE

Fiona Flett and Heidrun Interthal

Abstract

Denaturing urea polyacrylamide gel electrophoresis (PAGE) allows the separation of linear single-stranded DNA molecules based on molecular weight. This method can be used to analyze or purify short synthesized DNA oligonucleotides or products from enzymatic reactions.

In this chapter we describe how to prepare and how to run these high concentration polyacrylamide gels. We detail how to transfer a gel onto Whatman paper and how to dry it. Radiolabelled oligonucleotides are visualized by PhosphorImager technology.

Key words Urea, Formamide, Sequencing gel, DNA oligonucleotide, Denaturing PAGE, Tdp1

1 Introduction

Urea polyacrylamide gel electrophoresis (PAGE) employs 6–8 M urea in combination with heat (45–60 °C) to denature secondary DNA structures. This allows the separation of single-stranded DNA molecules based on their molecular weight [1]. Before loading onto the gel, DNA oligonucleotide samples are first denatured by heating in the presence of formamide. Both formamide and high concentrations of urea disrupt the hydrogen bonding necessary for DNA to adopt secondary or double-stranded structures. Therefore, DNA molecules can be resolved according to molecular weight up to single nucleotide resolution. The size of DNA molecules which can be resolved is dependent on the acrylamide percentage chosen [1]. The smaller the DNA fragment, the higher the concentration of acrylamide that should be used.

Denaturing PAGE was developed originally for DNA sequencing applications; however, due to the resolution it provides, it is now also commonly used for enzymatic activity assays such as nuclease digestion assays [2].

Svetlana Makovets (ed.), *DNA Electrophoresis: Methods and Protocols*, Methods in Molecular Biology, vol. 1054,
DOI 10.1007/978-1-62703-565-1_11,

Table 1
Concentrations of acrylamide giving optimum resolution of DNA fragments using denaturing PAGE [3]

DNA fragment size (in nucleotides)	Acrylamide concentration (%)
>200	4
80–200	5
60–150	6
40–100	8
10–50	12
<20	20

Before beginning the experiment you should consider the percentage of acrylamide gel required for your application. The chosen acrylamide concentration should allow migration of a single-stranded oligonucleotide approximately one-half to three-fourths of the way through the gel which will resolve oligonucleotides from $n+1$ and $n-1$ species, as well as oligonucleotides with various modifications. Table 1 provides recommendations for the percentage of acrylamide that should be used, which are dependent upon the length of oligonucleotide that is to be resolved [1, 3] (*see* Table 1).

Here we describe the use of denaturing PAGE for an enzymatic activity assay using tyrosyl DNA phosphodiesterase 1 (Tdp1) [4]. Tdp1 can remove various 3′ adducts, including a biotin group, from the 3′ end of DNA oligonucleotides to leave a 3′ phosphate at the end [2]. In this example, substrate 20-mer oligonucleotides with a 3′ biotin moiety were radiolabelled with ^{32}P on the 5′ end. The labelled DNA substrate was mixed with the enzyme Tdp1 to allow the reaction to occur, and then both substrate and product were resolved on a 15 % denaturing PAGE (see Fig. 4).

2 Materials

2.1 Equipment

1. Sequencing gel apparatus (Model S2, Gibco BRL, distributed by Biometra) with glass plates, 0.4 mm spacers, and matching combs. Self-adhesive temperature strip for monitoring gel temperature.
2. Power supply (2,000 V, 60 W).
3. Large gel dryer, e.g., Hybaid Gel Vac or Bio-Rad Gel Dryer Model 583 with vacuum pump.
4. Phosphor storage screen (430 mm × 350 mm) with a cassette (GE Healthcare).

5. PhosphorImager (Storm or similar).
6. Heat block at 95 °C for 1.5 mL reaction tubes.
7. Microtube cap locks for 1.5 mL tubes (Fisherbrand TUL-105-020Y).
8. Kim wipes.
9. 250 mL glass beaker.
10. 50 mL syringe.
11. 19 gauge needle.
12. Black clamps.
13. Plastic pipettes.
14. Razor blade.
15. Metal spatula.
16. 3MM CHR blotting paper (Whatman—731-2505).
17. Polyvinylchloride film, e.g., Saran wrap or Clingfilm.
18. Disposable absorbent bench liner.
19. Suitable protection for radioactive work (shields).

2.2 Reagents

All solutions should be prepared using deionized water:

1. UreaGel System (National Diagnostics—EC-833).

   Contains: UreaGel Concentrate 1 L contains 237.5 g of acrylamide, 12.5 g of methylene bisacrylamide, and 7.5 M urea in water.

   UreaGel Buffer7.5 M urea in deionized water.

   UreaGel Diluent 0.89 M Tris-Borate–20 mM EDTA buffer, pH 8.3 (10× TBE), and 7.5 M urea.
2. A scouring powder (Ajax, Vim, Comet).
3. 100 % Ethanol.
4. Gel Slick (Lonza Bioscience Rockland—catalogue number 50640).
5. 10 % ammonium persulfate. Dissolve 1 g of ammonium persulfate (Sigma-A3678-25G) in deionized water so that the total volume becomes 10 mL. The solution can be stored at 4 °C for a few months.
6. *N*, *N*, *N*, *N'*-tetramethyl-ethylenediamine (TEMED—Fisher—EC-503). Store at 4 °C.
7. Formamide (Sigma-F9037).
8. Formamide loading dye: 96 % formamide, 20 mM EDTA, 0.03 % xylene cyanol, 0.03 % bromophenol blue. Make the loading dye by mixing 9.6 mL of formamide with 400 μL 0.5 M EDTA (pH 8.0), 60 μL of 5 % xylene cyanol, and 60 μL of 5 % bromophenol blue. Aliquot and store at −20 °C.

9. 10× TBE: 890 mM Tris base, 890 mM boric acid, 20 mM EDTA. Dissolve 107.8 g Tris base (FW = 121.14) and 55.0 g boric acid (FW = 61.83) in approximately 900 mL deionized water. Add 20 mL of 0.5 M EDTA (pH 8.0), and adjust the solution to a final volume of 1 L. Autoclave for long-term storage and dilute tenfold with water before use.

3 Methods

3.1 Preparing Glass Plates for Denaturing PAGE

1. *First Use*: Begin by preparing a set of glass plates, one large and one small, for denaturing PAGE. The first time a set of brand new glass plates are used, they should be thoroughly scrubbed to roughen the surface, using a scrubbing brush and Ajax to scrub both plates well on both sides using circular motions. Then thoroughly wash off the Ajax with water and rinse the plates with 100 % EtOH. Dry off the plates using a lint-free tissue such as Kim wipes (*see* **Note 1**).
2. Treat one side of the smaller plate with 1 mL gel slick to silanize the surface and prevent the gel from adhering to the surface and facilitate post-electrophoretic separation (*see* **Note 2**). Use Kim wipes to quickly distribute the Gel Slick solution evenly. Then rinse the plate with water and polish the small plate by using a circular motion (*see* **Note 3**). Ensure that absolutely no gel slick comes into contact with the larger plate (*see* **Note 4**).
3. Following a gel run, wash the plates with water followed by 100 % EtOH to ensure that no dust, grease, salt, or acrylamide is stuck to the plates. Dry off the plates using a lint-free tissue such as Kim wipes. Repeat **steps 2** and **3** for each use.

3.2 Preparing the Gel Solution

1. Prepare the sequencing gel mix. We recommend to use the UreaGel System from National Diagnostics (19:1) [3] which is supplied as three solutions, UreaGel Concentrate, UreaGel Diluent, and UreaGel Buffer, which are to be combined in different amounts, depending on the final volume of the gel mix (Vt) and the desired percentage of acrylamide in the gel (*X*) (*see* **Note 5**). Use the following formulas to calculate the required volumes of UreaGel Concentrate (Vc), UreaGel Diluent (Vd), and UreaGel Buffer (Vb).

Vc = [(Vt) (*X*)]/25	Vc = UreaGel Concentrate volume Vb = UreaGel Buffer volume
Vb = 0.1 (Vt)	Vd = UreaGel Diluent volume Vt = Total casting solution volume
Vd = Vt − (Vc + Vb)	*X* = % Acrylamide desired

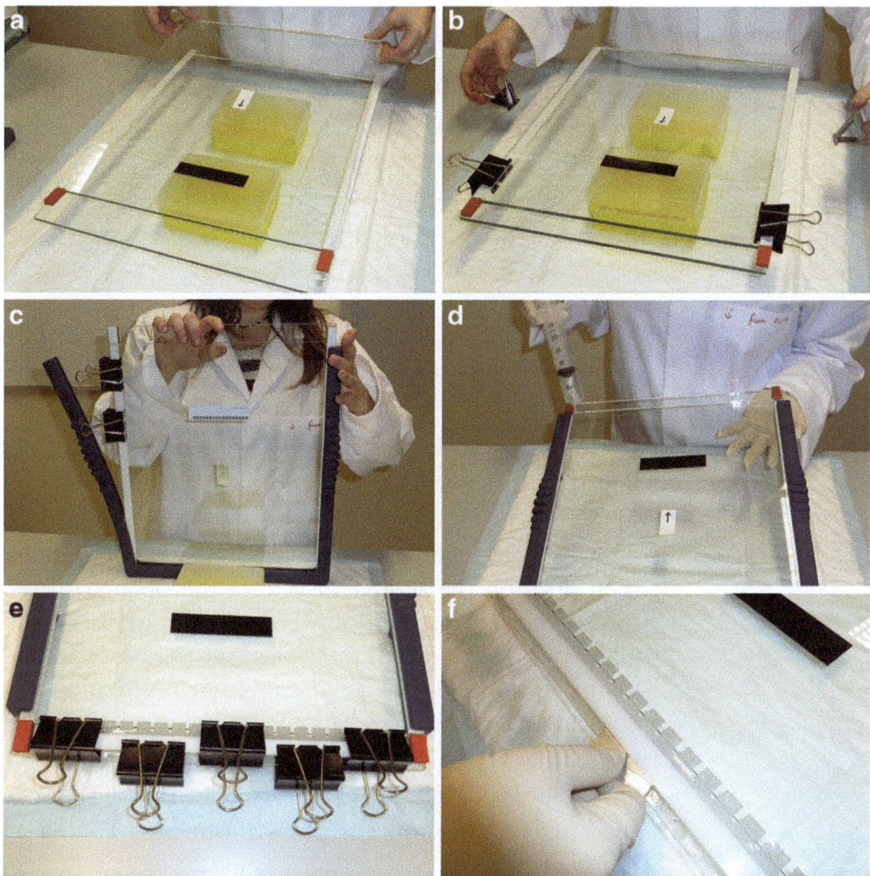

Fig. 1 Assembling of the gel cassette and casting of the gel

2. For example, for a 15 % denaturing PAGE, cast within a gel cassette of dimensions 0.4 mm × 440 mm × 350 mm gel, 70 mL of gel solution is required. Thus, mix 42 mL UreaGel Concentrate, 21 mL UreaGel Diluent, and 7 mL UreaGel Buffer.
3. Mix the UreaGel components in a glass beaker by swirling gently in both directions to minimize air bubbles, and chill on ice before use (*see* **Note 6**). While the gel mix is chilling, assemble the gel cassette.

3.3 Assembling the Gel Cassette

1. Lay the larger glass plate (outer face down) on a raised surface which will support the plate but allow access to the edges (*see* **Note 7** and Fig. 1a). Place the spacers along both side edges of the glass plate, with the foam pad at the top of the spacer lying two-thirds of the way up the plate, leaving the ends of the spacers sticking out at the bottom.
2. Place the small plate (outer face up) on top of the larger glass plate and spacers then push the top edge of the smaller plate up

against the foam pads of the spacers. The spacers should easily slide up the larger glass plate until the bottom edges of both glass plates and the spacers all align (Fig. 1a).

3. Fix two black clamps on each side edge of the gel cassette to hold the glass plates in place (Fig. 1b).
4. Seal the side and bottom edges of the gel cassette with the silicon sleeve provided with the sequencing gel apparatus (*see* **Note 8** and Fig. 1c).

3.4 Casting the Gel

1. To the UreaGel components on ice, add 800 μL 10 % ammonium persulfate for every 100 mL gel solution and swirl gently to mix.
2. Add 40 μL TEMED for every 100 mL gel solution and swirl gently to mix thoroughly and initiate polymerization of the gel (*see* **Note 9**).
3. Draw the gel solution into a 50 mL syringe avoiding air bubbles, then position the syringe at one of the open corners along the top of the gel cassette. Slowly pour the gel between the glass plates by pressing lightly on the syringe (*see* **Note 10** and Fig. 1d). Continue to fill up the gel cassette while watching the gel front ensuring no air bubbles become trapped (*see* **Note 11**). Overfill the gel slightly so that a small pool of gel lies on the top lip of the larger, bottom plate.
4. Insert the comb into the gel without introducing air bubbles (*see* **Note 12**).
5. Clamp along the top edge of the cassette using three black clamps to clamp the glass plates together and alternating with two black clamps to clamp the comb and larger glass plate together (Fig. 1e).
6. Check if the gel is still full to the top edge of the cassette. If not pour more gel into the cassette.
7. Leave gel to polymerize at room temperature for 30–60 min (*see* **Note 13**).
8. If the gel is to be left overnight, place wet paper towels on the glass plate but do not allow the paper towel or water to come into contact with the gel. Then seal the open end of the gel cassette and paper towels with Saran wrap to prevent the gel from drying out. The gel may be stored in this way for up to 48 h at room temperature.
9. Once the gel has polymerized, remove the black clamps from the gel cassette (*see* **Note 14**).
10. In order to remove the comb, add a little 1× TBE to the top of the gel cassette. Run a razor blade under the comb to raise it from the large glass plate and allow liquid to run beneath the

comb. With the razor blade sitting under the comb, slowly and gently pull the comb out, applying even pressure along the length of the comb (*see* **Note 15** and Fig. 1f).

11. Remove the gel from the silicon sleeve surrounding the cassette. Use a razor blade to remove any small pieces of acrylamide from above the wells, to avoid them falling into and getting stuck in the wells. Clean outer faces of the gel cassette with water using Kim wipes to ensure the surfaces are clean and free from pieces of acrylamide, so that the gel cassette can lie flush against the electrophoresis apparatus.

3.5 Pre-running the Gel

1. Position the gel cassette in the tank, with the smaller plate lying against the cooling plate and forming a seal with the upper buffer chamber. Hold in place using the upper and lower clamps (*see* **Note 16**).
2. Ensure the upper buffer chamber draining knob is closed. Make 1 L of 1× TBE and then use to fill the upper and lower buffer chambers, ensuring the upper buffer chamber seal is intact and no buffer is leaking from the upper to lower buffer chamber behind the glass plates.
3. Using a 19 gauge needle attached to a syringe, flush out the wells with 1× TBE to remove urea which accumulates in the wells and get rid of air bubbles (Fig. 2a).
4. Pre-run the gel setting the power as a limiting factor at 60 W constant for at least 1 h to bring gel up to temperature ~50 °C (*see* **Note 17**). While the gel is pre-running, you can prepare the samples so that they are ready to be loaded by the time the pre-run is finished.

   Once pre-run, rinse the wells of the gels again using 1× TBE. Flush the wells one last time immediately before loading samples onto the gel (*see* **Note 18**).

3.6 Loading and Running Denaturing PAGE

To run oligonucleotides on denaturing PAGE, the DNA should be radioactively labelled in order to allow it to be visualized. In our example we use a 20-mer oligonucleotide which is 5′ end labelled using ^{32}Pγ-ATP. The amount of DNA used depends on the amount required for your enzymatic reactions and how radioactive your DNA is. The volume of the sample loaded also depends on the comb used during the casting of the gel. Standard combs have 32 or 20 wells, which can accommodate up to 6 and 12 μL, respectively.

1. Mix DNA samples with formamide loading dye. Ensure the final concentration of formamide is at least 50 % in each sample.
2. Secure the lids of the tubes containing your samples, then denature the DNA at 95 °C for 2 min, spin down briefly, and

Fig. 2 Gel running and disassembly of the gel cassette

load immediately. Alternatively, samples can be stored at −20 °C until future use, and reheated at 95 °C for 2 min immediately before loading.

3. Using a 20 μL pipette, load samples into the gel carefully (*see* **Note 19** and Fig. 2b).
4. Run the gel at 60 W constant power until the bromophenol blue dye has migrated 3/4 to 4/5 the length of the gel, which should take approximately 2 h. Check the dye migration graph (Fig. 3) for an estimate of where the blue dyes should run. Gel running times will vary depending on the percentage of acrylamide used and the length and thickness of the gel.
5. To resolve a 20-mer oligonucleotide from similar species on a 15 % denaturing gel (dimensions 0.4 mm × 440 mm × 350 mm) (Fig. 4), the gel should be run until the bromophenol blue marker (dark blue) is 8–10 cm from the bottom of the gel (Fig. 2c). The gel running time should be approximately 2 h.

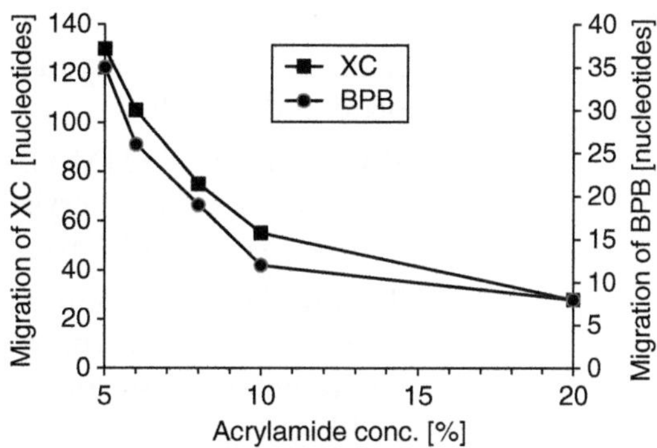

Fig. 3 This graph indicates the positions at which the bromophenol blue (BPB) and xylene cyanol (XC) dyes run at depending on the acrylamide concentration of the urea PAGE gel. Data are from ref. 3

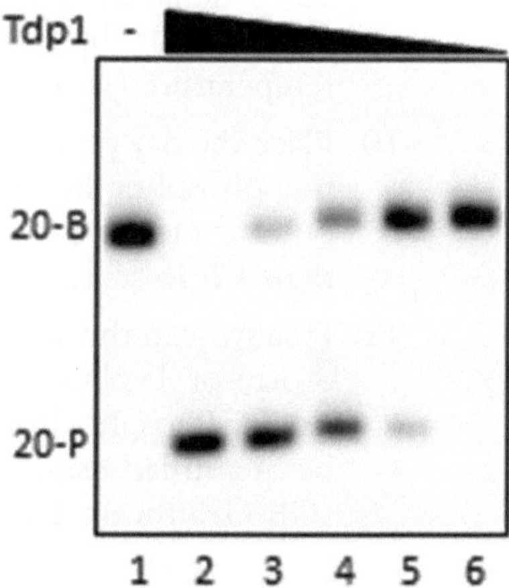

Fig. 4 15 % Denaturing PAGE of a 20-mer oligonucleotide labelled on the 5′ end with P^{32} following electrophoresis at 60 W for 2 h. Gel was dried for 5 h at 80 °C then image retrieval and quantitation were carried out using a Storm PhosphorImager and ImageQuant software (GE Healthcare). Gel shows a Tdp1 enzymatic activity assay: Substrate 20-mer oligonucleotide with a 3′ Biotin (20-B Lane 1) was incubated for 10 min at 37 °C with fivefold dilutions of Tdp1 (Lanes 2–6), capable of cleaving the 3′ Biotin and leaving a product of a 20-mer oligonucleotide with a 3′ Phosphate (20-P) which is clearly resolved from the substrate on this gel system

3.7 Gel Cassette Disassembly and Gel Drying

1. Stop the run by turning off the power and remove the power leads from the gel box.
2. Open the upper buffer chamber draining knob to empty the running buffer into the lower chamber.
3. Release the gel cassette from the tank by loosening the upper and lower clamps, then place the cassette on a flat surface with the larger plate on the bottom.
4. Remove the side spacers by pulling up and out holding on to both the spacer and foam pads on the ends (*see* **Note 20** and Fig. 2d).

5. Slowly pry open the gel plates using a spatula. The smaller silanized plate should lift off easily, leaving the gel on the larger plate underneath.
6. Place a sheet of 3MM CHR blotting paper on top of the gel. Depending on the acrylamide percentage of the gel, the gel may or may not stick to the paper.
7. To transfer the gel onto the paper, starting at one corner, very carefully peel back the gel and the paper together, until the whole gel is transferred onto the blotting paper (*see* **Note 21** and Fig. 2e, f).
8. Cover the gel with a layer of Saran wrap and use a razor blade to carefully cut away excess Saran wrap around the edges of the gel (*see* **Note 22**).
9. Dry the gel in a gel dryer using vacuum and high heat. For a 15 % denaturing PAGE, we recommend drying for 5 h at 80 °C, with the temperature rising quickly to the maximum temperature (*see* **Notes 23** and **24**).
10. Place the dry gel into a cassette and expose it for a suitable time to a phosphor storage screen. The exposure time is dependent on how radioactive your samples were and can range from less than 1 h to several days.
11. Finally, scan the screen with an appropriate scanner such as the Storm or Typhoon PhosphorImagers from GE Healthcare or similar PhosphorImagers available from Fuji. The gel can then be quantified using an appropriate software such ImageQuant (GE Healthcare Life Sciences).

4 Notes

1. Due to the thinness of the gels, air bubbles can easily become trapped during the casting of the gel. Therefore, it is important for the glass plates to be cleaned well to minimize the risk of bubbles.
2. Each plate can be used in four different orientations. Before running a gel with samples, it is advised to run loading dye only in order to find the best orientation for the set of plates to give nice, straight bands across the width of the gel. Once determined, for each following gel run, the plates should be used in the same orientation, thus mark the plates on the outer side with reference to top and bottom of each plate. Use a small piece of strong tape with an arrow drawn on it.
3. If over-polished, the plate will become static. If this happens add more water. Plates should be treated before each run. This treatment will also help to minimize the risk of air bubbles accumulating during the casting of the gel.

4. Even the smallest droplets of Gel Slick on the larger can result in big problems in pouring a gel, causing air bubbles to form in the gel. In this light the same Kim wipe should never be used on both the large and small plates. Store the plates in a rack, keeping the smaller silanized plate separate from the larger glass plate to avoid Gel Slick coming into contact with the larger plate.
5. Commercially prepared polyacrylamide solutions are highly recommended as this avoids inhaling neurotoxic acrylamide powder. However, unpolymerized acrylamide is highly toxic and should be handled accordingly.
6. Cooling the gel mix while assembling the gel apparatus will slow down polymerization when ammonium persulfate and TEMED are added, allowing more time to pour the gel, which otherwise would set very quickly.
7. Setting the plates on top of two pipette tip boxes or a thick book works well.
8. If using a silicon sleeve ensure the feet of the sleeve are underneath which will give the gel the slight slope required for casting and setting the gel. If using gel tape to seal the cassette, ensure the edges are well sealed, then put a small block or box underneath the gel cassette to raise the open end of the gel cassette to create the small slope which will help during casting and setting the gel.
9. Once ammonium persulfate and TEMED have been added, polymerization begins, thus the casting of the gel must be done immediately and efficiently. In warmer temperatures, for example, in the summer, polymerization can happen very quickly therefore the volume of ammonium persulfate and TEMED can be reduced to increase the polymerization time.
10. While casting the gel it is recommended to raise the plate to a 45° angle to encourage flow to the bottom. It is useful to rest the left corner of the gel cassette on your hip slightly higher than the right corner, then inject the gel from the right corner using the syringe.
11. Air bubbles may become trapped during the casting of the gel. If this happens they may be shifted by knocking on the glass plates above the air bubble.
12. The wells should be 3–4 mm deep. The comb should not be fully inserted, as the pieces of gel separating the wells are very fragile and may easily be pulled out by the comb.
13. The gel should polymerize in 30–45 min. To ensure the gel is polymerized, after casting, a small volume may be drawn into a glass Pasteur pipette and also left to polymerize. Once polymerized it will be stuck in the pipette.

14. While the comb is still in position, it is recommended to mark the position of the wells on the larger glass plate with a Sharpie pen, otherwise the wells may be harder to see.
15. Pulling the comb out too quickly or with uneven pressure can distort the shape of the wells.
16. The upper and lower clamps should be tightened just enough to hold the gel cassette in place and form a seal with the upper buffer chamber. Over tightening the clamps can cause the glass plates to crack and the springs to wear out.
17. Attach a temperature strip in the center of the outer face of the larger glass plate to monitor the temperature during the gel running. Temperature should be between 45 and 60 °C. If the temperature is too low, the DNA may not be fully denatured. If the temperature is too high, the gel may run with diffuse or distorted bands.
18. If urea is not flushed out of the wells before loading, it will cause diffuse bands.
19. Check manufacturer's recommendations for volume of sample to load, as the volume is dependent on well size. For example, when using a 32 well comb, 6 μL can be loaded, or for a 20 well comb, 12 μL can be loaded. A smaller sample volume will produce sharper bands. Samples should be pipetted into the wells slowly to allow the samples to run down to the bottom of the wells and to avoid introducing air bubbles. Pipetting too quickly can force the sample down to the bottom of the well then cause it to rise up the side of the well and out of the well. When pipetting the sample in, the well will be too narrow to accommodate the 20 μL pipette tip. Therefore, the pipette tip must be carefully pushed into the well which will squash it slightly but still allow adequate control for loading the sample. In our experience, this is more appropriate than any specialist gel loading tips.
20. Pulling on the foam pads can cause them to be detached from the spacers, and they are hard to reattach.
21. If a higher percentage of acrylamide is used (>12 %), the gel will not stick to the blotting paper. Thus, the transfer can be tricky and requires patience and practice to avoid the gel ripping. If the gel does rip, it may be best to use another piece of blotting paper to transfer the remaining gel.
22. The Saran wrap will not lie perfectly flat, but a little bit of folding won't affect gel drying. Do not try to remove and reposition Saran wrap as the gel will stick to it and come off the blotting paper. There is a risk here of ripping the gel or the gel folding back and sticking to itself. Do not have Saran wrap overlapping the edges of the blotting paper as this may cause problems during the gel drying.

23. Urea gels can be hard to dry. If the gel is not completely dry at the end of the drying cycle, it will fragment when the vacuum is released or even after it has been taken off the gel dryer.

24. Care should be taken when transferring the gel to the gel dryer. Keep the gel flat to avoid the gel falling off the filter paper or folding over on itself. Also while positioning the gel in the gel dryer, hold on to the saran wrap so that it remains stuck to the gel. Saran wrap will stick to the cover of the gel dryer, and careless handling may cause the gel to be lifted off the filter paper potential causing the gel to fold or rip. Therefore, it is also important to hold onto the saran wrap after the gel has been dried, so the gel remains covered.

References

1. Maniatis T, Jeffrey A, van deSande H (1975) Chain length determination of small double- and single-stranded DNA molecules by polyacrylamide gel electrophoresis. Biochemistry 14:3787–3794
2. Interthal H, Chen HJ, Champoux JJ (2005) Human Tdp1 cleaves a broad spectrum of substrates, including phosphoamide linkages. J Biol Chem 280:36518–36528
3. Sequagel ureagel system protocol. https://www.nationaldiagnostics.com/electrophoresis/protocol/sequagel-ureagel-system-protocol. Accessed 11 Dec 2012
4. Interthal H, Pouliot JJ, Champoux JJ (2001) The tyrosyl-DNA phosphodiesterase Tdp1 is a member of the phospholipase D superfamily. Proc Natl Acad Sci U S A 98: 12009–12014

Chapter 12

Pulsed-Field Gel Electrophoresis of Bacterial Chromosomes

Julia S.P. Mawer and David R.F. Leach

Abstract

The separation of fragments of DNA by agarose gel electrophoresis is integral to laboratory life. Nevertheless, standard agarose gel electrophoresis cannot resolve fragments bigger than 50 kb. Pulsed-field gel electrophoresis is a technique that has been developed to overcome the limitations of standard agarose gel electrophoresis. Entire linear eukaryotic chromosomes, or large fragments of a chromosome that have been generated by the action of rare-cutting restriction endonucleases, can be separated using this technique. As a result, pulsed-field gel electrophoresis has many applications, from karyotype analysis of microbial genomes, to the analysis of chromosomal strand breaks and their repair intermediates, to the study of DNA replication and the identification of origins of replication. This chapter presents a detailed protocol for the preparation of *Escherichia coli* chromosomal DNA that has been embedded in agarose plugs, digested with the rare-cutting endonuclease NotI, and separated by contour-clamped homogeneous field electrophoresis. The principles in this protocol can be applied to the separation of all fragments of DNA whose size range is between 40 kb and 1 Mb.

Key words *Escherichia coli*, Agarose gel electrophoresis, Pulsed-field gel electrophoresis (PFGE), Contour-clamped homogeneous electric field (CHEF), Pulse time, Agarose plugs

1 Introduction

Standard agarose gel electrophoresis is an extremely helpful technique used for the separation of DNA fragments. However, the upper limit of DNA size for this technique is between 20 and 40 kb, meaning that DNA fragments above this molecular weight cannot be resolved. In order to overcome this limitation, pulsed-field gel electrophoresis (PFGE) was developed. By using PFGE, DNA fragments as big as 12 Mb have been successfully resolved [1]. There are many different apparatus and separation parameters that can affect the size range of DNA molecules resolved by PFGE, and for a detailed review of these, *see* ref. 2. This chapter focuses on the use of contour-clamped homogeneous electric field (CHEF) for the separation of *Escherichia coli* chromosomal DNA that has

Svetlana Makovets (ed.), *DNA Electrophoresis: Methods and Protocols*, Methods in Molecular Biology, vol. 1054,
DOI 10.1007/978-1-62703-565-1_12,

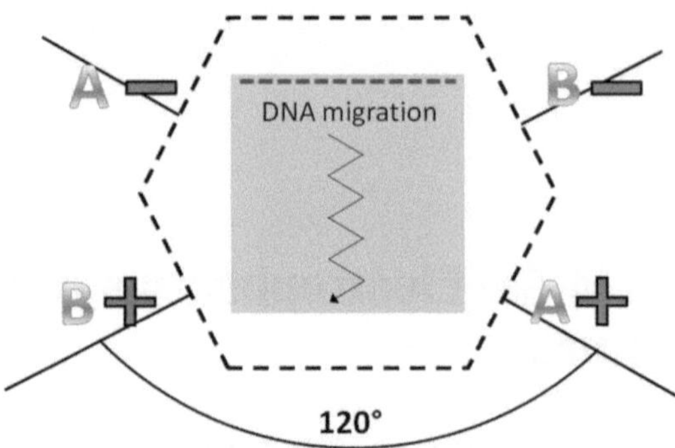

Fig. 1 Contour-clamped homogeneous electric field (CHEF) apparatus. A diagrammatic representation of a hexagonal tank with the plugs (*dark grey* rectangles at the *top* of the gel), which contain the chromosomal DNA, that are loaded into the wells of the gel. Electrodes are represented by the *dashed black lines*. Here two electric fields, the A channel and the B channel, which generate the inclusion angel of 120° required for the separation of the DNA, are also represented in the diagram. During the run, the electric field is switched from the A channel to the B channel for a given period of time (pulse time), resulting in a zigzag migration of the DNA. The period of time in which the DNA moves in any particular direction is the same, resulting in a net forward migration of the DNA

been digested with the rare-cutting endonuclease, NotI, to generate chromosomal fragments ranging from 15 kb to 1 Mb in size.

CHEF differs from standard agarose gel electrophoresis in that the electric field is regularly switched between two orientations that run diagonally across the gel [3]. As the electric field is switched, the DNA fragments need to be reoriented through an angle of 120° (known as the inclusion angle) before they can restart migrating. Larger fragments require more time to be reoriented than smaller fragments, and it is this reorientation time that allows for their separation. Figure 1 depicts a diagrammatic representation of a CHEF apparatus and the snakelike migration undertaken by the DNA molecules during electrophoresis.

The time during which the electric field is set to any particular channel is the single most important factor that determines the size range of the fragments that can be resolved in a given run. This time is referred to as the pulse time. Short pulse times result in the successful separation of small fragments, while longer pulse times allows for the separation of big fragments. To further improve the separation, a ramp can be added to the pulse time. This means that the pulse time will be slowly, and regularly, increased over the course of the run. For example, during a 10 h run the initial pulse time may be set to 5 s and the final pulse time may be set to 14 s. After every hour, the pulse time will increase by 1 s until 14 s are reached in the final hour of the run. The benefit of adding a ramped pulse time to the electrophoresis is that it increases the linear relationship between the migration rate and the molecular weight of the DNA.

As with standard agarose gel electrophoresis, altering other parameters such as agarose concentration, buffer temperature, or field strength can also affect the size range of the DNA fragments resolved. Nevertheless, in order to retain band sharpness, it is advised to use a high agarose concentration (1 % when the fragments to resolve are smaller than 1 Mb) and to maintain a buffer temperature below 14 °C. Additionally, it is advised not to use high field strengths (no more than 6 V/cm when separating *E. coli* DNA fragments digested with NotI) as the larger fragments of DNA can become trapped in the wells under these conditions [4]. A lower field strength not only overcomes this problem but also promotes band sharpness resulting in a clearer separation of the DNA fragments.

Finally, another key factor for the successful separation of large chromosomal fragments by CHEF is the integrity of the chromosomal DNA prepared for analysis. Conventional techniques, which isolate the DNA in solution, are not appropriate for analysis by PFGE as the DNA is heavily damaged during the extraction. To overcome this problem, it is advised to embed the cells in ultrapure low melting-temperature agarose prior to their lysis. Upon cell lysis, the chromosomal DNA is released into the agarose plug, which serves to trap and protect it from shearing forces.

2 Materials

2.1 Bacterial Culture and Preparation of Chromosomal DNA Embedded in Agarose Plugs

1. L-broth: 1 % Bacto-tryptone (Difco) (w/v), 0.5 % yeast extract (Difco) (w/v), and 1 % NaCl (w/v). Adjust the pH to 7.5 using NaOH, and then autoclave. Stored at room temperature.
2. 0.5 M EDTA, pH 8.0. Stored at room temperature.
3. 55 °C incubator or water bath.
4. DNA size marker (NEB; low range PFG marker, yeast chromosome PFG marker).
5. TEN buffer: 50 mM Tris–HCl, pH 8.0, 50 mM EDTA, and 100 mM NaCl. Stored at room temperature.
6. Low melting point agarose (Invitrogen UltraPure™ LMP Agarose Cat. no. 16520–050).
7. NDS solution: 0.5 M Na_2 EDTA, 10 mM Tris base, 0.6 mM NaOH, and 34 mM *N*-Lauroylsarcosine. Dissolve the Na_2 EDTA, Tris base, and NaOH in 350 mL of deionized water. In a separate container, completely dissolve the *N*-Lauroylsarcosine in 50 mL of distilled water. Once dissolved, add it to the main solution. Adjust the pH to 8.0 using NaOH and bring the total volume up to 500 mL. Store at room temperature.
8. Proteinase K solution: 50 mg/mL in water. Stored at −20 °C.
9. Agarose plug molds (Bio-Rad).

10. 2 mL microcentrifuge tubes.
11. Heat block set at 37 °C.
12. Lab tape. 250 ml flask.
13. 50 mL conical tubes.

2.2 Digestion of DNA Set in Agarose Plugs

1. 1× restriction buffer, with or without DTT (1 mM) and BSA (typically 1 mg/mL).
2. Restriction enzyme.
3. 2 mL microcentrifuge tubes.
4. Rocker at room temperature and 37 °C.

2.3 Pulsed-Field Gel Electrophoresis

1. Ultrahigh gel strength agarose (AquaPor™ ES; Fisher catalogue number ELR-300-040F).
2. 0.5 M EDTA, pH 8.0. Stored at room temperature.
3. 5× TBE, pH 8.0: 0.89 M Tris base, 0.89 M Boric acid, 10 mM EDTA.
4. TE buffer: 10 mM Tris–HCl pH 7.4, 1 mM EDTA.
5. Pulsed-field gel apparatus (Bio-Rad CHEF-DR™ II PFGE).
6. Power supply (Bio-Rad Pulsewave® 760 switcher).
7. Cooling system (VWR Scientific 1166 Refrigerated Circulating Waterbath).
8. Ethidium bromide 0.5 μg/mL in water.
9. Gel documentation system.

3 Methods

3.1 Bacterial Culture and Preparation of Chromosomal DNA Embedded in Agarose Plugs

1. Inoculate a single bacterial colony into 5 mL of L-broth. Grow the cells overnight at 37 °C with agitation.
2. Measure the OD_{600} of the overnight culture and dilute the cells in 25 mL of fresh L-broth (in a 250 mL conical flask) to an OD_{600} of 0.05–0.08. Grow the cells at 37 °C with agitation until they reach an OD_{600} of 0.2–0.4. During this time, chill some TEN buffer on ice (21 mL for each set of plugs to be prepared).
3. While the cells are growing, prepare 2 % LMP agarose in TEN buffer (500 μL for each set of plugs to be prepared) and store at 55 °C to prevent it from setting.
4. Once the cells have reached the desired OD_{600}, harvest a volume of cells that corresponds to an OD_{600} of 4.0 (if cells have reached an OD_{600} of 0.4, then 10 mL of culture needs to be harvested; *see* **Note 1**).
5. Wash your cell pellet with 10 mL of ice-cold TEN buffer by pipetting the cells up and down. Centrifuge the cells at 4 °C to

re-pellet them again, and repeat the washing step. Resuspend the cells in 1 mL of chilled buffer TEN to obtain your cells at an OD_{600} of 4.0. Place the cells on ice until all samples to be processed have been collected.

6. For the preparation of plugs, pre-warm 2 mL microcentrifuge tubes in a heat block set at 37 °C (one microcentrifuge tube for each set of plugs to be prepared). Once warmed, add 500 μL of the 2 % LMP agarose, prepared previously, to each microcentrifuge tube and place the tubes back into the heat block.
7. Briefly warm the cells to 37 °C in the heat block, and then add 500 μL of the cell suspension to the 500 μL of 2 % LMP agarose to give you a cell mix in 1 % LMP agarose. Briefly vortex each tube and pipette the cell-agarose solution into agarose plug molds that have had their bottom sealed with tape.
8. Allow the plugs to set at 4 °C for 30 min and then extrude each set of plugs into a 50 mL conical tube.
9. Add 10 mL of NDS solution, with Proteinase K added, to a final concentration of 1 mg/mL.
10. Incubate at 55 °C overnight to allow for cell lysis (*see* **Note 2**).
11. Once the cells have been fully lysed and the DNA within released into the plugs, the plug samples can be stored indefinitely at 4 °C in NDS solution (500 μL per plug).

3.2 Digestion of DNA Set in Agarose Plugs

1. Place a plug into a 2 mL microcentrifuge tube and add 1.5 mL of 1× restriction buffer corresponding to the enzyme of your choice (*see* **Note 3**). Place the tube on a rocker and rock for 1 h at room temperature.
2. Remove the restriction buffer and replace with another 1.5 mL of fresh 1× restriction buffer and place back on the rocker for 1 h. Repeat the washes for a total of 6 h.
3. To digest the DNA, add 500 μL of 1× restriction buffer and the appropriate amount of restriction enzyme (*see* **Note 4**). Incubate at 37 °C (or at the optimum temperature for the enzyme of choice) overnight with gentle rocking.

3.3 Pulsed-Field Agarose Gel Electrophoresis

1. Briefly cool the plugs to 4 °C. Meanwhile, prepare 1 % ultra-high gel strength agarose in 0.5× TBE and cool to 55 °C.
2. Briefly rinse the plugs in 1 mL TE buffer to remove any residual enzyme. Using 10 μL of the warmed agarose, stick the plugs, and the DNA size marker of your choice, to the gel comb (*see* Fig. 2). Incubate the comb at 4 °C for 30 min.
3. Place the comb with the attached plugs in the pulsed-field casting tray and gently pour the 1 % melted agarose (*see* **step 1**) around it until all the plugs are covered by the agarose. Allow the gel to set for 30 min.

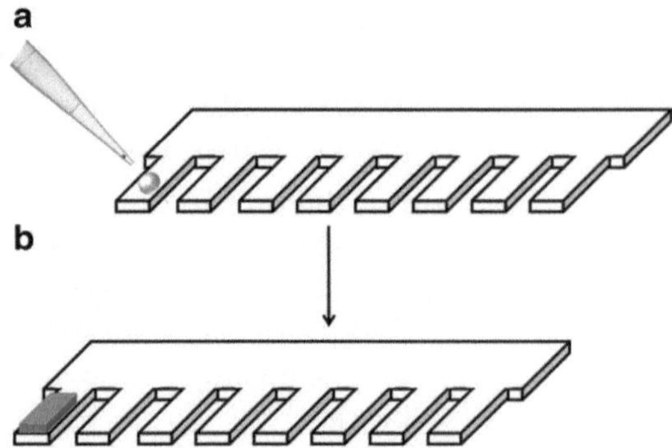

Fig. 2 Loading plugs onto a PFGE comb. (**a**) Pipette 10 μL of the melted 1 % agarose onto one of the teeth of the comb. (**b**) Position a plug onto the agarose. Repeat until all the plugs have been attached to the comb and put the comb at 4 °C to allow the plugs to set

4. Remove the comb and run the gel at 6 V/cm at 4 °C, with an appropriate pulse time (5–30 s for resolving DNA molecules ranging from 100 kb to 1 Mb and 20–120 s for resolving DNA molecules ranging from 1 to 2 Mb; *see* Fig. 3 and *see* **Note 5**).
5. Once the run has stopped, move the gel into a tray.
6. Stain the gel with 0.5 μg/mL of ethidium bromide in water, for 20 min.
7. Photograph the gel using a gel-documentation system.
8. If needed, process the gel for Southern blotting (*see* **Note 6**).

4 Notes

1. Do not exceed an OD_{600} of 6.0 as high concentrations of DNA do not run as a sharp band in a pulsed-field gel.
2. If a temperature of 55 °C is not appropriate for the experiments being carried out, for example, if the DNA species of interest are sensitive to high heat, it is possible to carry out two overnight incubations with Proteinase K at 37 °C. Make sure to replace the NDS/Proteinase K solution with fresh NDS/Proteinase K solution for the second overnight incubation. Incubate with gently rocking of the plugs if possible.
3. If you plan to make the restriction buffer from scratch there is no need to add BSA or DTT for the duration of the washes as this step is only designed to wash out excessive NDS and Proteinase K and equilibrate the plugs in restriction buffer. BSA and DTT should be added to the restriction buffer just prior to the addition of the restriction enzyme.
4. For the ethidium bromide stained gels depicted in Fig. 3, 10 U of NotI were used for the digestion of each plug.

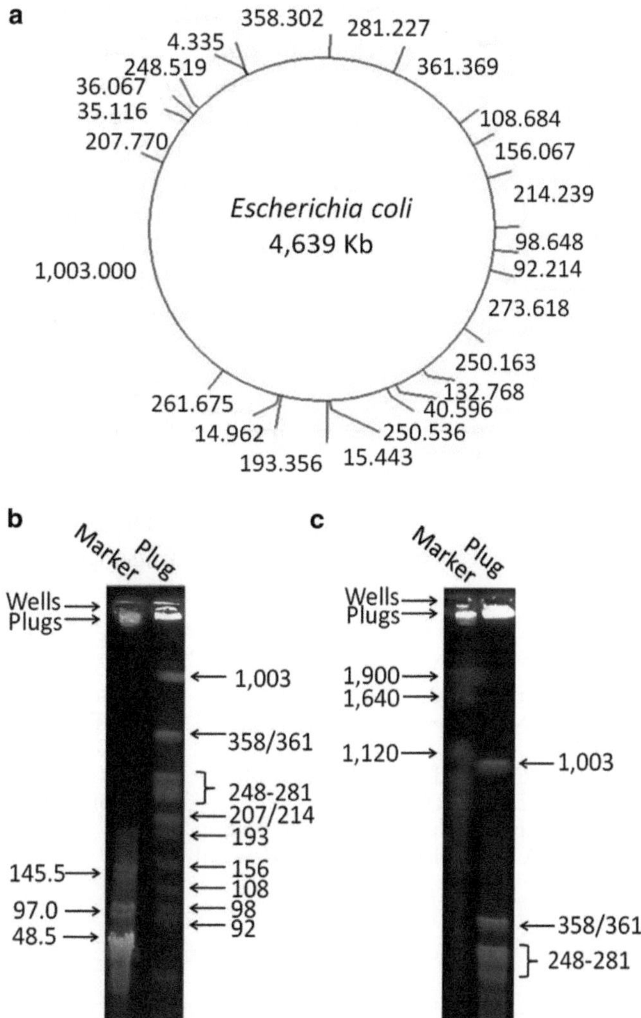

Fig. 3 CHEF of *E. coli* chromosomal DNA following digestion with NotI. (**a**) A NotI restriction map of the *E. coli* chromosome showing the sizes (kb) of the 23 fragments generated. (**b**) Ethidium bromide staining of NotI digested chromosomal DNA separated on a 1 % agarose gel. The gel was run at 6 V/cm, 4 °C, for 17 h with an initial pulse time of 5 s and a final pulse time of 30 s. (**c**) Ethidium bromide staining of NotI digested chromosomal DNA separated on a 1 % agarose gel. The gel was run at 6 V/cm, 4 °C, for 17 h with an initial pulse time of 20 s and a final pulse time of 120 s. NEB yeast chromosome PFG ladder was used as a marker

5. It is important to note that plasmids can migrate at a high molecular weight in a pulsed-field gel and may migrate at the same rate as a chromosomal band of interest.
6. When processing a PFGE for Southern blotting, it is detrimental to depurinate the DNA run in the gel. Depurination is typically used to fragment stretches of DNA that are bigger than 1 kb in order to increase the efficiency of their transfer to a membrane. This process occurs through the depurination of

the DNA by 0.25 M HCl. This leaves behind apurinic sites, which are converted to DNA strand breaks when exposed to NaOH during the denaturation step. Nevertheless, experiments in the lab have shown that depurination following CHEF reduces the amount of DNA that is transferred to the membrane (data not shown). This may be due to natural breakage of the DNA strands that occurs during electrophoresis as the DNA is forced to "snake" through the agarose matrix. The addition of depurination probably results in over-fragmentation of the DNA, which can result in poor DNA retention on the membrane.

Acknowledgements

We would like to thank Dr. Martin White and Ewa Okely for helpful suggestions to the manuscript.

References

1. Obrach MJ et al (1988) An electrophoretic karyotype of *Neurospora crassa*. Mol Cell Biol 8:1469–1473
2. Maule J (1998) Pulsed-field gel electrophoresis. Mol Biotechnol 9:107–126
3. Chu G, Vollrath D, Davis RW (1986) Separation of large DNA molecules by contour-clamped homogeneous electric fields. Science 234: 1582–1585
4. Gunderson K, Chu G (1991) Pulsed-field electrophoresis of megabase-sized DNA. Mol Cell Biol 11:3348–3354

Chapter 13

Resolution of Budding Yeast Chromosomes Using Pulsed-Field Gel Electrophoresis

Aziz El Hage and Jonathan Houseley

Abstract

Pulsed-field gel electrophoresis (PFGE) is a technique that resolves chromosome-sized DNA molecules in an agarose gel. As well as DNA mapping and karyotyping applications, PFGE techniques are well adapted to follow DNA rearrangements over time in a quantitative manner. Because of the very large sizes of the DNA molecules analyzed, DNA preparation, electrophoresis, and Southern blotting processes present unique challenges in PFGE experiments. In this chapter, we describe a robust PFGE protocol covering the preparation of intact *Saccharomyces cerevisiae* chromosomal DNA, specific running conditions for the resolution of small, medium- and large-sized chromosomes and their by-products, and basic Southern blotting and hybridization instructions for the analysis of these molecules.

Key words *Saccharomyces cerevisiae*, Pulsed-field gel electrophoresis (PFGE), Contour-clamped homogeneous electric field (CHEF), Agarose embedded yeast DNA, Southern blot, Hybridization

1 Introduction

In conventional gel electrophoresis, DNA moves through the gel at a rate proportional to the logarithm of its size, leading to an exponential decline in resolution with increasing DNA length. This imposes a natural limit to the size of molecules that can be resolved up to the point where the DNA is so large that changes in length have no detectable effect on mobility. The best one-dimensional agarose gel electrophoresis methods have an effective resolution limit of ~50 kb in size, for which long running times and fragile low percentage gels are required. To circumvent the problems inherent to the resolution of large DNA molecules ingenious pulsed-field gel electrophoresis (PFGE) techniques were developed in the 1980s [1, 2], culminating in contour-clamped homogeneous electric field (CHEF) devices that are capable of reproducibly resolving molecules up to 12 Mb [3, 4].

Svetlana Makovets (ed.), *DNA Electrophoresis: Methods and Protocols*, Methods in Molecular Biology, vol. 1054,
DOI 10.1007/978-1-62703-565-1_13, © Springer Science+Business Media New York 2013

During electrophoresis in agarose gels DNA molecules are "sieved" through the gel matrix in a size-dependent manner. Large linear molecules like DNA exist in solution as floppy coils but are thoroughly unravelled as they squeeze through the much smaller pores in the agarose gel. Like a thread passing through the eye of the needle, the DNA is stretched out and therefore "reptates" (moves in a "snake-like" manner) through the gel. In PFGE techniques, instead of applying a constant electrical field as for conventional DNA electrophoresis, the direction of the electric field is periodically changed forcing the DNA molecules to rotate to align with the new field before progressing through the gel. Realignment occurs at a rate that decreases with increasing length of the molecule, so if the direction of the field is changed periodically ("pulsed"), a larger molecule must spend more time realigning to each change of field than a smaller one and will have less time to progress through the gel before the next field change. The realignment rate varies linearly with DNA length, ensuring that large and small molecules are resolved to a similar extent for a given change in size, avoiding the resolution limit in conventional DNA electrophoresis. In a standard CHEF protocol the direction of the electric field is switched between –60° and +60°, causing the DNA to zigzag through the gel while undergoing 120° angle changes [3]. Gels were originally run with a constant switching time, providing linear resolution within a given range; however, because longer switch times resolve longer DNA molecules, switch times are now routinely increased through the run ("ramped"), greatly improving the linear range of resolution [5].

PFGE methods have been traditionally applied to relatively simple tasks such as transgene mapping and karyotyping as well as analyzing the products of recombination reactions. PFGE also provides a powerful quantitative method for following DNA rearrangements over time. Examples include formation and resolution of meiotic recombination intermediates [6], damage and repair of chromosomes [7, 8], copy number changes [9, 10], and progression of DNA replication [11, 12].

DNA isolation by standard techniques is unsuitable for isolating intact chromosomes as pipetting and vortexing lead to shearing of long DNA molecules; even with very careful handling, it is almost impossible to obtain DNA molecules larger than 100 kb. In order to avoid DNA shearing, cell samples are lysed and the DNA is digested within a solid agarose plug. A large number of factors including running conditions and Southern blotting can also affect the quality of PFGE experiments. Here we describe in detail a robust PFGE protocol well suited for the separation of relatively small (as required for meiotic recombination analysis) and large DNA molecules (as required for ribosomal DNA recombination analysis). Careful following of this protocol should yield results of high quality and reproducibility.

2 Materials

Prepare all solutions using deionized water.

2.1 DNA Preparation and Gel Running Components

1. Agarose for DNA plug preparation (SeaKem LE agarose, Lonza, catalogue number 50001) (*see* **Note 1**).
2. Plug molds (Sample CHEF Disposable Plug Mold, Bio-Rad, catalogue number 1703706).
3. Lyticase (Sigma, catalogue number L2524) (*see* **Note 2**). Prepare 1 mL of lyticase 17,000 U/mL stock by dissolving 17,000 U lyticase powder (the activity concentration in U/mg is supplied by the manufacturer) in 490 μL water, 500 μL glycerol 100 %, 5 μL 1 M K_2HPO_4, and 5 μL 1 M KH_2PO_4. Store at −20 °C, stable for a few months.
4. Large orifice 200 μL tips (Starlab, catalogue number E 10118400).
5. Water bath (50 °C) and heating block (37, 42, 50 and 100 °C).
6. Proteinase K (Roche, catalogue number 03115801001). Prepare 20 mg/mL solution in water. Store at −20 °C.
7. Proteinase K buffer (PK buffer): 100 mM EDTA, 0.2 % sodium deoxycholate, and 1 % *N*-lauroylsarcosine sodium. Prepare fresh on day of use.
8. Wash buffer: 10 mM Tris–HCl pH 7.6 and 50 mM EDTA.
9. Gel rocker.
10. Certified Megabase agarose (Bio-Rad, catalogue number 161-3109).
11. Chromosome size marker: for low to medium range, we use Midrange II PFG marker (NEB, catalogue number N3552S) or lambda ladder PFG marker (NEB, catalogue number N0340S), and for high range we use *H. wingei* chromosomes (Bio-Rad, catalogue number 170-3667).
12. TBE buffer: 0.5× or 1× depending on run conditions (*see* Table 1), prepared fresh (*see* **Note 3**). To make 2.5 L of 1× TBE: 27 g Tris base, 13.25 g boric acid, 10 mL 0.5 M EDTA pH 8. To make 2.5 L of 0.5× TBE: 13.5 g Tris base, 6.6 g boric acid, 5 mL 0.5 M EDTA pH 8.
13. PFGE machine: CHEF-DRII or DRIII System with chiller unit and casting stand (Bio-Rad, *see* **Note 4**).
14. SYBR Safe DNA gel stain (Molecular Probes, Invitrogen, catalogue number S33102) or ethidium bromide (10 mg/mL in water).

2.2 Southern Blotting and Hybridization Components

1. Blotting membrane: positively charged nylon transfer Hybond-N^+ membrane (GE Healthcare, catalogue number RPN 303B).
2. Blotting paper: Whatman 3MM.

Table 1
PFGE running conditions for three ranges of DNA, *see* example results in Fig. 2

Molecular weight range resolved	Small (40–400 kb)	Medium (200–1,500 kb)	Large (1,000 to >3,500 kb)
Gel percentage (%)	1.3	1	0.8
TBE	0.5×	0.5×	1×
Run time (h)	24	24	68
Switch times (ramped) (s)	15–25	60–120	300–900
Voltage (V/cm)	6	6	3

3. Blotting solutions:
   (a) Depurinating solution—0.25 M HCl (freshly diluted from concentrated HCl, usually 37 %/12.1 M).
   (b) Denaturing solution—1.5 M NaCl, 0.5 M NaOH.
   (c) Neutralizing solution—0.5 M Tris–HCl pH 7.5, 1.5 M NaCl.
4. Paper towels for blotting (C-Fold hand towel 2 ply white, Scientific Laboratory Supplies, catalogue number FC 5804).
5. Transfer buffer (20× SSC): 3 M NaCl, 0.3 M Sodium Citrate, pH 7.0.
6. UV-cross-linking box (UV Stratalinker 1800, Stratagene).
7. Phenol:chloroform:isoamylalcohol (25:24:1) (PCI) (*see* **Note 5**).
8. 1.5 mL tight-cap (Safe-Lock, Eppendorf, catalogue number 0030 120.086) and 2 mL screw-cap (Anachem, catalogue numbers 2330-00 and 2001-52) microcentrifuge tubes.
9. 425–600 μm glass beads (Sigma, catalogue number G8772).
10. Break buffer: 10 mM Tris–HCl pH 8, 2 % Triton X-100, 1 % SDS, 100 mM NaCl, and 1 mM EDTA.
11. 1× TE solution plus RNAse A: 10 mM Tris–HCl pH 8, 1 mM EDTA, and 0.25 μg/μL RNAse A (RNase A DNase-free, AppliChem, catalogue number A3832).
12. QIAquick Gel Extraction Kit (Qiagen, catalogue number 28704).
13. DECAprime II Random Primed DNA Labelling Kit (Ambion, Invitrogen, catalogue number AM1455).
14. Liquid nitrogen (*see* **Note 6**).
15. [α-^{32}P]dATP or [α-^{32}P]dCTP (3,000 Ci/mmol, 10 mCi/mL) (Perkin Elmer).
16. Mini Quick Spin DNA columns (Roche, catalogue number 11814419001).

17. Church hybridization buffer: 0.5 M sodium phosphate pH 7.2, 1 mM EDTA, 1 % BSA and 7 % SDS. For preparation of 500 mL Church buffer, dissolve 5 g BSA (bovine serum albumin fraction V) by small amounts in 100 mL water with stirring. In a separate 1 L beaker, mix 171 mL of 1 M Na_2HPO_4 with 79 mL of 1 M NaH_2PO_4. Add 1 mL 0.5 M EDTA pH 8 and water to ~350 mL then 35 g SDS. Stir and warm gently to dissolve, and then add water to 400 mL total volume. Add the BSA solution slowly with stirring then filter sterilize. Store at room temperature for short term, or aliquot and store at −80 °C for long term.
18. Hybridization bottles, nylon mesh (VWR, catalogue number BARN308-6A), and rotary hybridization oven.
19. 10 or 20 % SDS stock solution.
20. Phosphor Storage Screen with Cassette and PhosphorImager.

3 Methods

3.1 Preparation of Agarose Embedded Yeast DNA

1. Weigh 0.16–0.2 g SeaKem LE agarose in a 50 mL conical tube and add 10 mL dH_2O, seal and place in a beaker containing ~200 mL water (*see* **Note 7**). Bring agarose to the boil using a microwave, and then equilibrate to 50 °C in a water bath.
2. Harvest 0.25–1 × 10^8 cells (1.25–5 OD_{600}) from an appropriate growth phase by centrifuging at 2,000 × *g* 5 min at 4 °C (*see* **Note 8**). Resuspend cells in 1 mL PFGE wash buffer, transfer to a microcentrifuge tube, and spin again. Decant the supernatant and resuspend cells in 50 μL wash buffer (*see* **Note 9**).
3. Warm cell suspension to 50 °C in a water bath. Just prior to mixing the cells with agarose, add 1 μL lyticase and immediately proceed with **step 4**.
4. Combine cell suspension with an equal volume of agarose (from **step 2**) using a large orifice 200 μL tip. Mix by pipetting or vortex for 5–10 s and transfer the mixture to a plug mold (*see* **Note 10**). Let agarose polymerize.
5. Use the snap off tool provided on the plug mold to push the solidified agarose plug into a microcentrifuge tube containing 500 μL of wash buffer with 10 μL lyticase. Incubate plugs for 1 h at 37 °C in a heating block.
6. Remove the wash buffer. Add 0.5 mL of PK buffer containing 1 mg/mL proteinase K. Incubate the plugs overnight at 50 °C in a heating block.
7. Wash the plugs four times for 30 min each wash in 1 mL wash buffer, at room temperature with gentle agitation on a rocker. Store the plugs at 4 °C in PFGE wash buffer or 0.5 M EDTA (*see* **Note 11**).

3.2 Casting the Gel and Loading the DNA Plugs

Casting the gel requires the following components provided with the CHEF system: a casting stand with removable end gates, a platform on which the gel is cast and a comb (positioned on a comb holder).

1. Assemble the casting stand (*see* **Note 12**) and place the comb holder into one of the two positioning slots on each side of the casting stand. Make sure the bottom of the comb is approximately 1 mm above the surface of the platform. Prepare 120 mL of pulsed-field Megabase agarose solution in TBE (*see* Table 1 for recommended gel percentages and TBE concentrations) by microwaving, and ensure the agarose is fully dissolved (*see* **Note 13**). Cool to 60 °C. Pour molten agarose into the casting stand and let solidify at room temperature.
2. Remove the comb holder gently. Fill the wells with buffer from the tank (*see* **Note 14**). Using clean tweezers place the DNA plug on a smooth surface, e.g., a Petri dish and cut the plug in half with a clean razor. Load one half of the plug into the sample well (*see* **Note 15**). Include a chromosome size marker preferably in the first well of the gel.

3.3 Electrophoresis

1. Remove the gates from the casting stand. Place the gel and platform assembly in the frame in the center of the electrophoresis cell (*see* **Note 16**).
2. Program the CHEF system with appropriate run conditions (*see* Table 1 for example). Start the run and make sure that the buffer is circulating through the cooler set to 12 °C (*see* **Note 17**).

3.4 Gel Staining and Imaging

1. When the run is finished (*see* **Note 18**) place the gel in 200 mL 1× TBE with 1× SYBR Safe or 10 μg/mL ethidium bromide in a plastic box.
2. Allow the gel to stain for 1 h with gentle agitation.
3. Wash two times 15 min each wash with 1× TBE to reduce the staining background.
4. Image the gel (*see* **Note 19**, example gel images are shown in Fig. 2).

If needed process the gel for Southern blotting (Subheadings 3.5 and 3.6).

3.5 Non-electrophoretic DNA Transfer

1. Place the gel in a plastic box on a shaker with gentle agitation. Cover the gel with a solution of 0.25 M HCl and leave for 20–30 min to depurinate the DNA (*see* **Note 20**).
2. Discard the depurinating solution and soak the gel in the denaturing solution for 30 min to unwind the DNA strands and introduce breaks at the abasic sites generated by depurination.

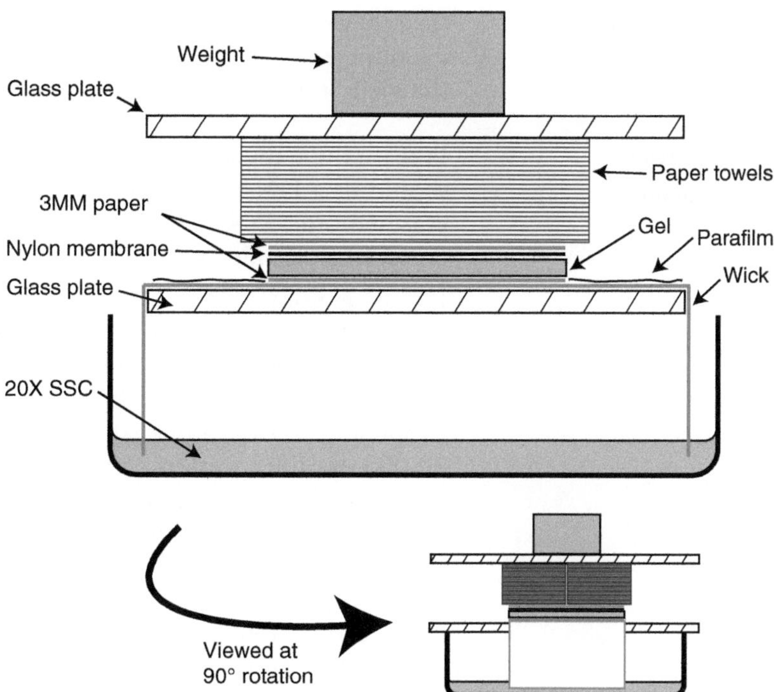

Fig. 1 Schematics of a basic Southern blotting assembly

This will assist DNA transfer by breaking long DNA molecules into small pieces.

3. Discard the denaturing solution and soak the gel in the neutralizing solution two times 15 min.
4. Cut a piece of Hybond membrane the same size as the gel. Wet with water then soak with 20× SSC. Cut two pieces of blotting paper the same size as the gel, and one piece the same width as the gel but ~50 cm long. Soak all blotting papers in 20× SSC.
5. Set up the capillary transfer as shown in Fig. 1. Fill the plastic box (24 cm × 24 cm × 7 cm) with ~1 L 20× SSC and balance a long glass plate on top. Drape the long sheet of blotting paper over the glass plate, touching the liquid on each side to form a wick, and place a sheet of blotting paper on top. Place the gel on the blotting papers then carefully lay the membrane on the gel (*see* **Note 21**). Place a sheet of blotting paper on the membrane then role a pipette over the assembly to remove bubbles. Cover exposed areas of the wick with Saran Wrap or Parafilm to avoid short circuits, then add a stack of paper towels, a glass plate and a weight.
6. Transfer the DNA for at least 24 h.
7. Disassemble the transfer setup: carefully pull the membrane off the gel and place it on a sheet of dry blotting paper so that the

surface with the bound DNA is facing upwards. Place the membrane inside the Stratalinker and immobilize the DNA to the membrane by UV cross-linking (120 mJ/cm^2).

3.6 Southern Hybridization

1. Prepare the template for random primed probe synthesis. This should be a DNA fragment of 0.2–1 kb derived by PCR from yeast genomic DNA (Subheading 3.6, **step 2**) or by restriction digest of a plasmid (skip **step 2**, Subheading 3.6).
2. If yeast genomic DNA is required as a PCR template: break a yeast cell pellet (2–5 mL overnight saturated culture) in the presence of 200 μL break buffer, 200 μL PCI and 200 μL glass beads in a screw-cap tube, by vortexing for 5–10 min at room temperature. Centrifuge at room temperature for 10 min 20,000 × *g* (full speed in a microcentrifuge). Remove carefully ~140 μL from the upper phase and precipitate DNA with 0.5 mL of 100 % ethanol. Spin in a microcentrifuge at full speed for 10 min at 4 °C, discard the supernatant, and wash the DNA pellet with 0.5 mL of 70 % ethanol. Centrifuge at full speed for 10 min at 4 °C, discard the supernatant, dry the DNA pellet, and resuspend it in 25–40 μL of 1× TE solution plus RNAse A. Incubate 10–15 min at 42 °C to degrade the RNA. Set up a standard PCR reaction using 1 μL of 1/10 dilution of the genomic DNA.
3. Load the PCR or restriction digest on a standard 1 % agarose gel stained with 1× SYBR Safe or 10 μg/mL ethidium bromide, run until your band of interest is in the middle of the gel, cut the DNA band, and purify using QIAquick Gel Extraction kit. Elute the DNA from the QIAquick column with 30 μL water.
4. Synthesize a random primed probe using DECAprime II Random Primed DNA Labelling Kit following the manufacturer's procedure (*see* **Note 6**). You can use either [α-32P] dATP or [α-32P]dCTP (3,000 Ci/mmol) (*see* **Note 22**).
5. Meanwhile, place the membrane from Subheading 3.5 in a hybridization bottle with 10 mL Church buffer (*see* **Note 23**). Attach the bottle to the rotisserie of a hybridization oven set to 65°C and balance with an empty bottle if needed. Pre-hybridize at 65 °C for at least 1 h while rotating.
6. Purify the probe using a Sephadex G50-based column. Ensure that more activity is present in the flow-through than remains on the column (this can be assessed by holding the column or tube a fixed distance from a Geiger counter using tweezers). Discard the column, the purified probe is in the flow-through.
7. Denature the radiolabelled probe by heating at 100 °C for 10 min, snap chill on ice, and add to the hybridization bottle. Hybridize for 12–24 h at 65 °C with rotation.

8. Pour off the probe (*see* **Note 24**). Wash the membrane two times 10 min in 2× SSC, 0.1 % SDS at 65 °C and two times 20 min in 0.5× SSC, 0.1 % SDS at 65 °C (*see* **Note 25**).
9. Wrap the membrane in Saran Wrap and place in a cassette with a phosphor storage screen.
10. Scan with a PhosphorImager and if necessary perform longer exposures (*see* Fig. 2).
11. You can strip the ^{32}P-labelled probe by placing the nylon membrane in a plastic box containing a boiling solution of 0.1× SSC, 0.1 % SDS. Continuous shaking at room temperature on a rocker helps the stripping process. Repeat the procedure several times until the radioactive signal disappears completely. You can re-probe the membrane as many times as you wish providing that the stripping is successful (*see* **Note 26**).

4 Notes

1. Although most PFGE protocols require low melting point agarose, in our hands plugs made using SeaKem LE perform equivalently and are much easier to handle.
2. Contaminants in lower grades of lyticase affect PFGE resolution.
3. We have had variable results with TBE diluted from 10× stocks. We therefore make fresh 1×/0.5× each time.
4. Bio-Rad CHEF systems are by far the most widely used PFGE systems and are compatible with the running conditions described here. PFGE units are available without controlled refrigeration units; however, PFGE runs capable of separating chromosome-sized molecules take 24 h or more at relatively high electrical field and must therefore be carried out with precise temperature control to avoid gel distortions.
5. Use phenol pre-equilibrated with 10 mM Tris–HCl pH 8, 1 mM EDTA (Sigma, catalogue number P4557).
6. Liquid nitrogen will be used in Subheading 3.6, **step 4** for random primed probe synthesis. Note that tight-cap tubes should be used to avoid bursting in liquid nitrogen.
7. Microwaving the conical tube in a small beaker containing water prevents the agarose from boiling too vigorously.
8. For the simplest analysis, use stationary cells grown 1–2 days in YPD. For time-dependent processes such as replication, DNA damage, or recombination, cells should be in mid-log or other appropriate phase (e.g., meiosis). Note that cells can be fixed by resuspending in 70 % ethanol and stored at −20 °C for later processing if desired.

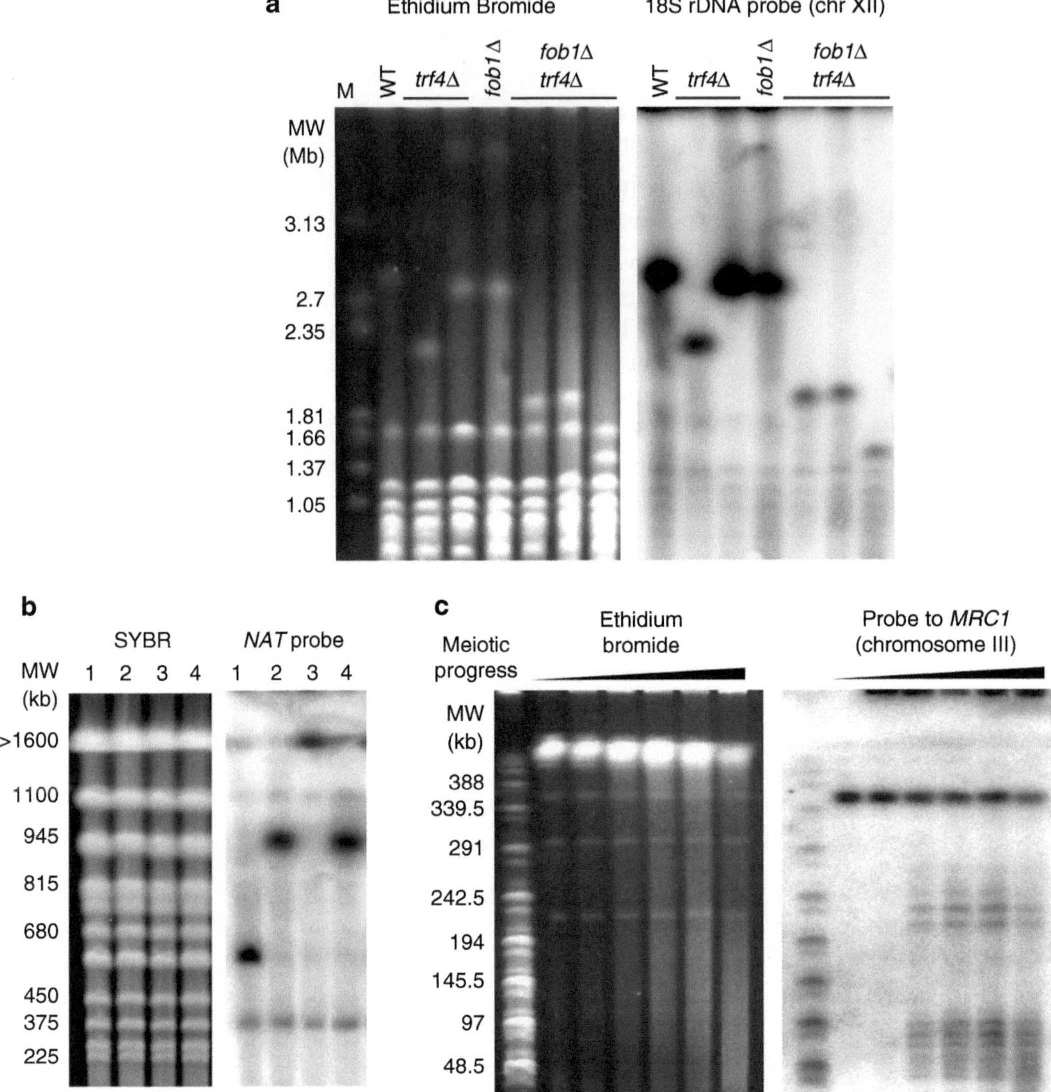

Fig. 2 Examples of separation of large-, medium-, and small-sized chromosomes with PFGE. In all panels, gel images are shown on the *left* and Southern blots on the *right*. (**a**) Separation of chromosome XII with large-sized running conditions. The length of chr. XII varies depending on the copy number of ribosomal DNA (for an rDNA array of 150–200 repeats chr. XII size can vary between 2.36 and 2.86 Mb). In some yeast mutants, here double *trf4*Δ *fob1*Δ strain [13], the rDNA array can be contracted to as little as 35 repeats (chr. XII size ~1.37 Mb). The probe on the Southern blot (*right panel*) is specific to the 18S rDNA. (**b**) Crude mapping of a randomly inserted subtelomeric transgene using medium-sized running conditions. A probe to the inserted NAT cassette (*right panel*) displays transgenes on different chromosomes. (**c**) Separation of meiotic recombination intermediates in a *sae2*Δ strain [14] using small-sized running conditions. Most chromosomes comigrate as a single strong band near the top of the gel, while the three smallest, chr. I, VI, and III, resolve in the gel at 225, 295, and ~350 Kb (*left panel*). A probe to the *MRC1* gene, which is located on chr. III, displays the full-length chromosome in all lanes and truncated DNA fragments (40–300 Kb) accumulating during meiosis (*right panel*). MW = Molecular weight, M = chromosome marker

9. Cells that flocculate (e.g., SK1 strain background) should be pipetted repeatedly until resuspended at this point.
10. It is advantageous to place the molds at 4 °C to rapidly solidify the agarose plugs.
11. The plugs are stable at 4 °C for at least a year.
12. Place the platform into the casting stand. Position one end gate over the screws protruding from the casting stand, insuring that the horizontal slot is facing towards the platform. Tighten the screws. Similarly, position the second end gate and tighten the screws. See the CHEF system manual (Bio-Rad) for diagrams. Use a spirit level to make sure the platform is sitting on a level surface. The casting stand provided with CHEF-DR systems is 14 cm wide × 13 cm long.
13. We usually prepare 2.5 L TBE (0.5× or 1×, Table 1), use 120 mL for the agarose gel, and pour the remaining in the electrophoresis cell. Both the pump and cooler should be switched on at this stage to allow the buffer to equilibrate to 12 °C.
14. Adding buffer in the wells helps avoiding air bubbles being trapped between the plug and the wall of the well sample.
15. Handling agarose plugs with tweezers requires practice; too tight a grip will squash the plug, but plugs are easily dropped and lost with tentative handling. The plug is often a little larger than the well but can easily be squeezed in the well. If the plug is much smaller than the well, ensure that it sits against the front edge of the well.
16. Make sure not to dislodge the gel from the platform otherwise the gel will float away during the run, obviously ruining the migration of the DNA molecules. Also ensure that the wells of the gel are facing the correct end (this information may change on different systems—check the CHEF system manual).
17. Because of the use of ramped switch times, optimization of novel PFGE run conditions is complex and beyond the scope of this review. The easiest approach is to use the conditions recommended for a PFGE marker that gives resolution in an appropriate range for your application. Alternatively, run conditions can be sourced [10, 13] from publications, and three widely applicable run conditions are given here in Table 1.
18. After each run, it is important to drain the running buffer from the electrophoresis cell and to allow circulation of 2.5 L distilled water for at least 30 min to ensure proper washing of the cell and the tubing.
19. Standard gel imaging systems can struggle to obtain good pictures of yeast PFGE gels as the staining intensity is often low. Laser scanning systems (we use FLA-5100 and FLA-7000 systems from FUJI/GE life sciences) provide much better results.

20. It is important to depurinate the DNA for at least 20 min and not more than 30 min otherwise the yield or quality of the transferred DNA can be greatly reduced.
21. Do not move the membrane after it has touched the gel. Small amounts of DNA transfer immediately, causing faint extra bands to appear on hybridization if the gel is moved.
22. Manipulate the radioisotope in a radiation protected area.
23. Hybridization mesh can be used if desired—this can help ensure even signal on large membranes.
24. The hybridization mix can be stored in a 50 mL conical tube at −20 °C for subsequent Southern hybridizations. Make sure to heat-denature the probe in a boiling water bath for at least 10 min before each use.
25. If washes have been performed in hybridization bottles, some background activity may be present where the membrane overlapped. To remove this background, wash the membrane for ~1 h with 0.1× SSC, 0.1 % SDS in a plastic box at room temperature.
26. Note that for repetitive regions like the rDNA (150–200 copies in yeast), it is not trivial to strip the probe from the membrane. Therefore, if a blot is to be probed for sequences of different copy number, we recommend that you probe for lower copy targets first.

Acknowledgements

The setup of the PFGE protocol was initially started in the laboratory of Professor David Tollervey.

A.E.H. is a research fellow in the laboratory of Professor David Tollervey who is funded by the Welcome Trust. J.H. is funded by Wellcome Trust grant 088335.

References

1. Carle GF, Olson MV (1984) Separation of chromosomal DNA molecules from yeast by orthogonal-field-alternation gel electrophoresis. Nucleic Acids Res 12:5647–5664
2. Schwartz DC, Cantor CR (1984) Separation of yeast chromosome-sized DNAs by pulsed field gradient gel electrophoresis. Cell 37:67–75
3. Chu G, Vollrath D, Davis RW (1986) Separation of large DNA molecules by contour-clamped homogeneous electric fields. Science 234:1582–1585
4. Orbach MJ, Vollrath D, Davis RW, Yanofsky C (1988) An electrophoretic karyotype of *Neurospora crassa*. Mol Cell Biol 8: 1469–1473
5. Clark SM, Lai E, Birren BW, Hood L (1988) A novel instrument for separating large DNA molecules with pulsed homogeneous electric fields. Science 241:1203–1205
6. Game JC (1992) Pulsed-field gel analysis of the pattern of DNA double-strand breaks in the Saccharomyces genome during meiosis. Dev Genet 13:485–497
7. Choy JS, Kron SJ (2002) NuA4 subunit Yng2 function in intra-S-phase DNA damage response. Mol Cell Biol 22:8215–8225

8. Lundin C, North M, Erixon K, Walters K, Jenssen D, Goldman AS, Helleday T (2005) Methyl methanesulfonate (MMS) produces heat-labile DNA damage but no detectable in vivo DNA double-strand breaks. Nucleic Acids Res 33:3799–3811
9. Kobayashi T, Ganley AR (2005) Recombination regulation by transcription-induced cohesin dissociation in rDNA repeats. Science 309:1581–1584
10. Kobayashi T, Heck DJ, Nomura M, Horiuchi T (1998) Expansion and contraction of ribosomal DNA repeats in *Saccharomyces cerevisiae*: requirement of replication fork blocking (Fob1) protein and the role of RNA polymerase I. Genes Dev 12:3821–3830
11. Kegel A, Betts-Lindroos H, Kanno T, Jeppsson K, Strom L, Katou Y, Itoh T, Shirahige K, Sjogren C (2011) Chromosome length influences replication-induced topological stress. Nature 471:392–396
12. Schollaert KL, Poisson JM, Searle JS, Schwanekamp JA, Tomlinson CR, Sanchez Y (2004) A role for *Saccharomyces cerevisiae* Chk1p in the response to replication blocks. Mol Biol Cell 15:4051–4063
13. Houseley J, Kotovic K, El Hage A, Tollervey D (2007) Trf4 targets ncRNAs from telomeric and rDNA spacer regions and functions in rDNA copy number control. EMBO J 26:4996–5006
14. McKee AH, Kleckner N (1997) A general method for identifying recessive diploid-specific mutations in *Saccharomyces cerevisiae*, its application to the isolation of mutants blocked at intermediate stages of meiotic prophase and characterization of a new gene SAE2. Genetics 146:797–816

Chapter 14

Analysis of DNA Damage via Single-Cell Electrophoresis

Diana Anderson and Julian Laubenthal

Abstract

The comet assay or single-cell gel electrophoresis assay is a relatively simple and sensitive technique for quantitatively measuring DNA damage and repair at the single-cell level in all types of tissue where a single-cell suspension can be obtained. Isolated cells are mixed with agarose, positioned on a glass slide, and then lysed in a high-salt solution which removes all cell contents except the nuclear matrix and DNA, which is finally subjected to electrophoresis. Damaged DNA is electrophoresed from the nuclear matrix into the agarose gel, resembling the appearance of a comet, while undamaged DNA remains largely within the proximity of the nuclear matrix. By choosing different pH conditions for electrophoresis, different damage types and levels of sensitivity are produced: a neutral (pH 8–9) electrophoresis mainly detects DNA double-strand breaks, while alkaline (pH ≥ 13) conditions detect double- and single-strand breaks as well as alkali-labile sites. This protocol describes a standard comet assay study for the analysis of DNA damage and outlines important variations of this protocol.

Key words Comet assay, Single-cell electrophoresis, DNA damage, Double-strand breaks, Single-strand breaks, Alkali-labile sites

1 Introduction

The comet assay or single-cell electrophoresis assay was introduced by Ostling and Johansson in 1984 [1] and significantly improved by Singh et al. in 1988, who established the alkaline version [2]; Olive et al. in 1990 [3], who established the neutral version; Klaude et al. in 1996 [4], who further developed the current alkaline version; and Tice et al. in 2000, who set up the first international guidelines for a standardized protocol which included important parameters such as slide preparation, electrophoresis and cell lysis conditions, DNA staining, and DNA damage quantification [5]. The basic principle of this assay is mixing isolated cells with melted agarose, followed by placing this mixture on a microscope slide already covered with a layer of agarose, subsequent cell lysis in a high-salt solution which removes all the cell contents except the DNA, and finally electrophoresis of the DNA from multiple single cells. However, even after cell lysis, which includes removal of all

Svetlana Makovets (ed.), *DNA Electrophoresis: Methods and Protocols*, Methods in Molecular Biology, vol. 1054,
DOI 10.1007/978-1-62703-565-1_14,

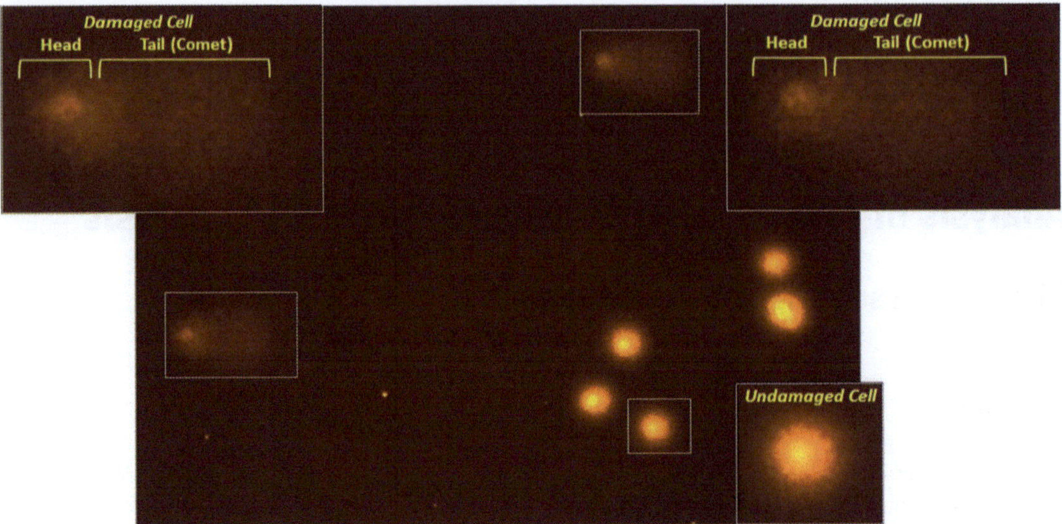

Fig. 1 Representative comet assay image of damaged and undamaged lymphocytes

the histone proteins, the DNA remains supercoiled and probably attached to the nuclear matrix. At the sites of DNA strand breaks, the supercoiling is lost and the relaxed DNA extrudes from the bulk of supercoiled DNA. This relaxed (i.e., damaged) DNA is electrophoresed away from the nuclear matrix towards the anode resulting in an extended, comet-shaped tail of the DNA after electrophoresis, while the undamaged DNA is confined as a head region (Fig. 1).

By choosing different pH conditions for electrophoresis and the preceding incubation, different DNA damage types and levels of sensitivity can be quantified. Electrophoresis can be performed under alkaline conditions, detecting double-strand breaks (DSBs) and single-strand breaks (SSBs; at pH 12.3), as well as alkali-labile sites (ALSs; at pH ≥ 13) [4, 6]. When using neutral (pH 8–9) electrophoresis, primarily DSBs are quantified, while only a few SSBs, due to the relaxation of supercoiled loops containing the breaks, contribute to the formation of the comet tail [3, 7]. Since toxicants induce primarily SSBs and ALSs, the alkaline comet assay is more commonly used. When combined with comprehensive image analysis, the comet assay can be used to identify apoptotic cells characterized by their faint and small head and a large dispersed comet tail, termed "ghosts" or "hedgehogs." Important variations of the standard alkaline comet assay protocol include further decondensation approaches prior to lysis, unwinding, and electrophoresis when assessing damage in condensed spermatozoal DNA [8, 9]; the addition of endonuclease III, formamidopyrimidine DNA glycosylases, or other DNA repair enzymes to convert oxidized pyrimidines or purines into SSBs, to subsequently detect oxidative

DNA damage [10, 11]; and the use of an electrophoresis buffer at neutral pH [8, 9] to detect primarily DSBs [7, 12].

The comet assay unites uniquely the advantages of an in situ assay with the simplicity of high-throughput techniques for the detection of DNA damage, on a low cost, simple, fast, and reproducible basis. That is why the comet assay constitutes the currently most widely applied method for detecting DNA damage at a single-cell level and became one of the reference methods in evaluating DNA damage and repair for biomonitoring in cohort as well as in genotoxicity studies [6, 13].

2 Materials

2.1 Slide Preparation

1. Microscope slides with frosted end.
2. Metal plate that can hold several microscope slides.
3. Polystyrene box that the metal plate can be fitted in.
4. Water bath or heat block set at 60 °C.
5. 1 % (w/v) normal melting point (NMP) agarose (Invitrogen, UK) in ddH_2O.

2.2 Agarose Embedding and Lysis of Cells

1. Phosphate buffer saline (PBS): 137 mM NaCl, 2.75 mM KCl, 1.45 mM KH_2PO_4, 15.25 mM Na_2HPO_4, pH 7.2.
2. Microcentrifuge (a speed of $200 \times g$ is required).
3. 0.5 and 1 % (w/v) low melting point (LMP) agarose (Invitrogen, UK) in PBS.
4. Water bath or heat block set at 60 °C.
5. Coverslips (22 mm × 50 mm).
6. Coplin jar.
7. Alkaline lysis solution: 2.5 M NaCl, 100 mM Na_2EDTA, 10 mM Tris base in ddH_2O. Store at room temperature for up to 2 weeks. Prior to use, add 10 % DMSO (v/v) and 1 % TritonX-100 (v/v) and chill this solution for 1 h at 4 °C (*see* **Notes 1** and **2**).

2.3 Electrophoresis

1. Horizontal electrophoresis tank. The size of this tank determines the number of slides used in a single experiment as well as the parameters of the power supply required.
2. Alkaline electrophoresis buffer: 300 mM NaOH, 1 mM Na_2EDTA in ddH_2O. Adjust pH to 13.2 with 10 M NaOH.
3. Power supply capable of reaching 20 V (0.8 V/cm on platform, 300 mA).
4. 50 mL plastic syringe.
5. Neutralization buffer: 0.4 M Tris base in ddH_2O. Adjust pH to 7.5 with 10 M HCl.

2.4 Staining

Staining solution: 20 μg/mL ethidium bromide in Tris-EDTA buffer (1 mM Na_2EDTA, 10 mM Tris–HCl, pH 8). Ethidium bromide can be replaced by either SYBRGold (Invitrogen, UK) or YOYO-1 iodide (Invitrogen, UK), both at the final dilution of 1:10,000 with similar results. Stock solutions are prepared at the 1:100 dilution using ddH_2O and can be stored at −20 °C for up to 6 weeks (*see* **Note 3**).

Caution: Ethidium bromide is mutagenic and possibly carcinogenic.

2.5 Quantification

1. Epifluorescence microscope equipped with a ×20 and a ×40 objectives; a filter capable for detecting an absorption of 510 nm and an emission of 595 nm (ethidium bromide fluorescence) as well as a 100 W mercury illumination (HBO 100) bulb (*see* **Note 3**).
2. Charged coupled device (CCD) camera. We are currently using an electron-multiplying CCD camera (Andor, UK) with 658 × 496 pixels.
3. Computer workstation with an image analysis software. We are currently using the comet 6.0 software from Andor.

2.6 Variations for the Neutral Comet Assay

1. Neutral lysis solution: 30 mM EDTA, 0.5 % SDS (w/v) in ddH_2O. Adjust pH to 8.0 with 1 M NaOH.
2. Neutral electrophoresis buffer: 90 mM Tris base, 90 mM Boric acid, 2 mM EDTA in ddH_2O. Adjust pH to 8.5 with 1 M NaOH.

2.7 Variations for the Sperm Cells Comet Assay

1. 2 % (w/v) low melting point (LMP) agarose (Invitrogen, UK) in PBS.
2. Alkaline lysis solution supplemented with 10 mM dithiothreitol (Sigma, UK).
3. Alkaline lysis solution supplemented with 0.05 g/mL proteinase K (Roche, UK).

3 Methods

See **Notes 4–6** for general requirements for the method.

3.1 Slide Preparation

1. Clean the microscope slides with 100 % ethanol and flame.
2. Melt 1 % (w/v) NMP agarose in dH_2O. Prevent liquid overflow or excessive evaporation.
3. Place melted agarose in a water bath at 60 °C.
4. Dip the slides briefly into the melted 1 % NMP agarose. The agarose should cover at least half of the slide. Remove excess agarose by wiping the back of the slide with a tissue.

5. Let the slides dry overnight at room temperature. Agarose pre-coated slides can be stored for up to 7 months in an airtight box at room temperature (*see* **Note 7**).

3.2 Agarose Embedding and Lysis of Cells

1. Place a metal plate on top of ice in an appropriate polystyrene box.
2. Melt LMP agarose in dH_2O to make 0.5 and 1 % agarose solutions. Keep agarose melted by placing the solutions on a hot plate or in a water bath at 60 °C.
3. Dilute cells obtained from cell culture, lymphocyte separation, or tissue extraction to a concentration of $1.5–3.0 \times 10^4$ cells/mL in PBS. For one comet slide, 1 mL of this solution is required, i.e., a total of $1.5–3.0 \times 10^4$ cells are required per experiment.
4. Spin down cell suspension in a microcentrifuge at $200 \times g$ for 5 min at room temperature.
5. Discard the supernatant and resuspend cells in 1 mL of fresh PBS. Repeat **steps 4** and **5**.
6. Discard the supernatant and resuspend the pellet in 100 μL of PBS.
7. Mix the cell suspension with an equal amount of 1 % LMP agarose solution by pipetting the mixture up and down. *Caution*: Avoid any air bubbles during mixing as this can introduce DNA damage.
8. Quickly apply 100 μL of the agarose-cell suspension in the center of a pre-coated agarose slide and cover the suspension with a coverslip.
9. Place the slide on the ice-cold metal plate and let the agarose set for about 10 min.
10. Once the agarose is set, remove the coverslip and apply an additional (third) layer of 100 μL of 0.5 % LMP agarose solution and cover with a coverslip.
11. Place the slide on the ice-cold metal place and let the agarose set for about 10 min (*see* **Note 8** and Fig. 2).
12. Once the agarose is set, remove coverslips and incubate slides at 4 °C for 1 h vertically in a Coplin jar filled with ice-cold alkaline lysis solution (*see* **Note 9**).
13. Remove the alkaline lysis solution carefully and wash slides without shaking three times for 15 min in Coplin jars filled with ice-cold ddH_2O at 4 °C. Ensure that the agarose layer is not damaged during these washing steps (*see* **Note 10**).
14. Transfer all slides carefully from the Coplin jar to the electrophoresis tank and position as close as possible to the anode to ensure lack of variability between the slides (*see* Fig. 2).

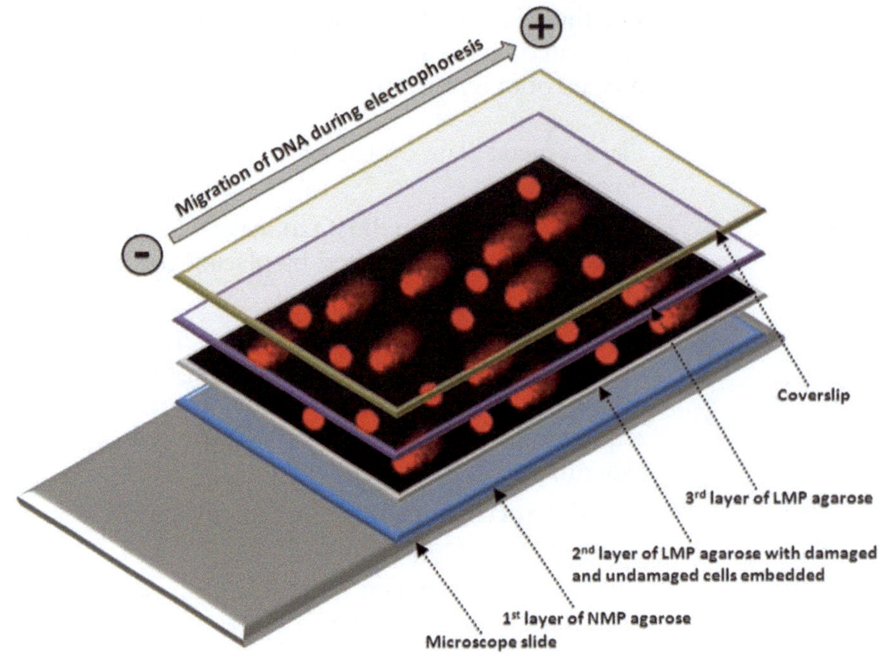

Fig. 2 A schematic of a comet assay slides structure, showing the three different agarose layers

When there are too few slides to fill the entire tank, fill gaps with empty slides to allow a homogenous current during electrophoresis.

3.3 Electrophoresis

1. Pour ice-cold alkaline electrophoresis buffer into the tank until all slides are covered with a thin layer (~4 mm) of buffer. Close the tank. Avoid bubbles on top of the slides. Place the tank into a cold room at 4 °C to prevent melting of LMP agarose during electrophoresis. Alternatively, the tank can be placed on ice, in a cold room or in a closed fridge at 4 °C during electrophoresis (*see* **Note 11**).
2. Preincubate slides for 30 min in alkaline electrophoresis buffer to unwind DNA and to convert alkali-labile sites into DNA breaks (*see* **Note 12**).
3. Electrophorese at constant voltage 0.75 V/cm with current reaching 300 mA, for 30 min. If the current is too high or low, remove or add, respectively, the alkaline electrophoresis buffer using a 50 mL syringe (*see* **Note 13**).
4. Remove all slides from the electrophoresis tank and drain the edges of the slides using a paper towel.
5. Wash slides three times, 5 min per wash, using neutralization buffer.
6. Rinse the slides carefully with ddH_2O, remove excess liquid from the slides, and dry at room temperature for 30 min.

3.4 Staining

Apply 60 μL of ethidium bromide staining solution on top of the slide and cover it with a coverslip. Generally, ethidium bromide can stain both DNA and RNA; however, under the comet assay alkaline conditions, all the RNA is degraded.

3.5 Quantification

Usually 50 cells in three repeats are scored in a comet experiment. There are three different ways to quantify the obtained results:

1. Evaluation of comet tail formation by eye. Usually cells are graded in five categories according to the following amounts of DNA in the tail, and each cell is assigned a score on the scale from 0 to 4: 0, <5 % of the DNA is in the tail (undamaged cell with a faint comet tail); 1, 5–20 % (low-level damage); 2, 20–40 % (medium level damage); 3, 40–95 % (high-level damage); and 4, >95 % (total damage). For 50 scored cells, the overall score (the sum of scores of the 50 cells) will be between 0 and 200 arbitrary units.
2. Semiautomated image analysis, in which cells are selected by the scorer manually. A CCD camera captures selected comets and commercial software quantifies the signal intensity and the diameter of the comet. The most common parameters detected with an (semi)automated system are comet tail length, % DNA in tail, % DNA in head, and the tail moment (the combination of tail length and % tail DNA). The most recent guidelines suggested % tail DNA as the international standard, covering the widest range of DNA damage, reviewed in ref. 14.
3. Automated image analysis, in which cells are automatically selected by the software. This is currently under development in different laboratories. However, major problems, amongst others, are overlapping of comets being scored as damage positive cells.

3.6 Variations for the Neutral Comet Assay

There are four variations compared to the alkaline comet assay protocol described in Subheadings 3.2 and 3.3:

1. In **step 12**, Subheading 3.2, use the neutral lysis solution instead of the alkaline lysis solution for the cell lysis.
2. In **steps 1–6**, Subheading 3.3, use the neutral electrophoresis buffer instead of the alkaline electrophoresis buffer for the incubations and electrophoresis.
3. In **step 3**, Subheading 3.3, use the voltage of 1 V/cm during the electrophoresis.
4. Skip the washes in the **step 5**, Subheading 3.3.

3.7 Variations for the Sperm Comet Assay

There are three alterations compared to the alkaline comet assay protocol described in Subheadings 3.1–3.5:

1. In **step 7**, Subheading 3.2, mix the sperm cell suspension with 2 % LMP agarose solution instead of a 1 % LMP agarose solution.
2. In **step 12**, Subheading 3.2, supplement the alkaline lysis solution with 10 mM dithiothreitol and incubate slides for 1 h at 4 °C. Then make alkaline lysis solution with 0.05 g/mL proteinase K and incubate slides for another hour at 4 °C.
3. In **step 3**, Subheading 3.3, electrophorese slides for 20 min.

4 Notes

1. Earlier experimental protocols added the anionic detergent sodium sarcosinate to the solution. This is not necessary for successful lysis and may induce precipitation of the lysis solution at 4 °C [2].
2. Addition of DMSO scavenges reactive radicals produced by iron which is released from hemoglobin during cell lysis. When using cells/tissue without hemoglobin, it is not necessary to add DMSO.
3. Other nucleic acids dyes which can be equally well used (with the appropriate filter on the microscope) include YOYO-1 iodide (absorption,491 nm; emission, 508 nm) and SYBR Gold (495 nm, 537 nm). Some laboratories use acridine orange (502 nm, 526 nm), but this tends to fade in our opinion.
4. This protocol works equally well on any eukaryotic cells except spermatozoa (*see* Subheading 3.6).
5. It is very important that the entire comet assay experiment (Subheadings 3.2–3.4) is performed on ice, only with cold (4 °C) solutions to avoid DNA repair taking place, in the dark or in a room with an orange filter to prevent UV light-induced DNA damage.
6. It is imperative that temperature, voltage settings, incubation times, lysis and electrophoresis time, electrophoresis buffer volume, and concentration/type of the nucleic acid dyes are not changed within matching experiments where more than one electrophoresis runs are used.
7. Longer storage of agarose pre-coated slides can increase the chance of detached agarose layers during lysis or electrophoresis.
8. The dimension of the agarose gel layer should match the dimension of the respective coverslip used. The gel should not fill the entire slide or contain any air bubbles.
9. Extending the lysis time for up to 8 h does not influence the success of the experiment. Longer lysis time variations are

reported in the literature, but this, in our opinion, significantly increases the chance of lost cells. Also, the pH should not be greater than pH 10.5 when increasing the lysis time [15].

10. Remaining salt traces from the lysis buffer can lead to an artificially expressed comet tail during electrophoresis. Hence, rinsing well with ddH_2O is essential.
11. An even, well-controlled layer of electrophoresis buffer covering the slides is essential for reproducible results, as even small differences in electrophoresis buffer volume result in variable electric fields strength, subsequently producing inconsistent comet tail lengths.
12. Incubation in electrophoresis buffer prior to electrophoresis depends on the type of cell suspension used, and incubation time can vary between 20 and 70 min. A 30 min incubation for cultured cells and blood samples is generally recommended and is the standard in our laboratory. Shorter incubation periods, however, can reduce the background DNA damage level found in control cells but at the same time will decrease the sensitivity of DNA damage detection in exposed cells.
13. Changing the current of the power supply during electrophoresis results in dramatic increases or decreases of comet tails. The settings described here are the international standards and yield the most reproducible results. As a guide, cells not being exposed to DNA damaging agents (control cells) should have around 5–12 % of their DNA in the tail and the tail lengths of no more than 15–25 μm [5].

References

1. Ostling O, Johanson KJ (1984) Microelectrophoretic study of radiation-induced DNA damages in individual mammalian cells. Biochem Biophys Res Commun 123:291–298
2. Singh NP, McCoy MT, Tice RR, Schneider EL (1988) A simple technique for quantitation of low levels of DNA damage in individual cells. Exp Cell Res 175:184–191
3. Olive PL, Banath JP, Durand RE (1990) Heterogeneity in radiation-induced DNA damage and repair in tumor and normal cells measured using the "comet" assay. Radiat Res 122:86–94
4. Klaude M, Eriksson S, Nygren J, Ahnstrom G (1996) The comet assay: mechanisms and technical considerations. Mutat Res 363:89–96
5. Tice RR, Agurell E, Anderson D, Burlinson B, Hartmann A, Kobayashi H, Miyamae Y, Rojas E, Ryu JC, Sasaki YF (2000) Single cell gel/comet assay: guidelines for in vitro and in vivo genetic toxicology testing. Environ Mol Mutagen 35:206–221
6. Fairbairn DW, Olive PL, O'Neill KL (1995) The comet assay: a comprehensive review. Mutat Res 339:37–59
7. Collins AR (2004) The comet assay for DNA damage and repair: principles, applications, and limitations. Mol Biotechnol 26:249–261
8. Anderson D, Schmid TE, Baumgartner A, Cemeli-Carratala E, Brinkworth MH, Wood JM (2003) Oestrogenic compounds and oxidative stress (in human sperm and lymphocytes in the Comet assay). Mutat Res 544:173–178
9. Sipinen V, Laubenthal J, Baumgartner A, Cemeli E, Linschooten JO, Godschalk RW, Van Schooten FJ, Anderson D, Brunborg G (2010) In vitro evaluation of baseline and induced DNA damage in human sperm

exposed to benzo[a]pyrene or its metabolite benzo[a]pyrene-7,8-diol-9,10-epoxide, using the comet assay. Mutagenesis 25: 417–425

10. Cemeli E, Baumgartner A, Anderson D (2009) Antioxidants and the Comet assay. Mutat Res 681:51–67
11. Collins AR, Azqueta A (2011) DNA repair as a biomarker in human biomonitoring studies; further applications of the comet assay. Mutat Res 736(1–2):122–129
12. Baumgartner A, Cemeli E, Anderson D (2009) The comet assay in male reproductive toxicology. Cell Biol Toxicol 25:81–98
13. Albertini RJ, Anderson D, Douglas GR, Hagmar L, Hemminki K, Merlo F, Natarajan AT, Norppa H, Shuker DE, Tice R et al (2000) IPCS guidelines for the monitoring of genotoxic effects of carcinogens in humans. International Programme on Chemical Safety. Mutat Res 463:111–172
14. Kumaravel TS, Vilhar B, Faux SP, Jha AN (2009) Comet assay measurements: a perspective. Cell Biol Toxicol 25:53–64
15. Collins AR, Oscoz AA, Brunborg G, Gaivao I, Giovannelli L, Kruszewski M, Smith CC, Stetina R (2008) The comet assay: topical issues. Mutagenesis 23:143–151

Chapter 15

Fluorescence In Situ Hybridization on Electrophoresed Cells to Detect Sequence Specific DNA Damage

Julian Laubenthal and Diana Anderson

Abstract

Fluorescence in situ hybridization (FISH) to label fragments of DNA with probes which can specifically locate a genomic region of interest, combined with the single cell electrophoresis (Comet) assay, also termed Comet-FISH, allows the quantification of DNA damage and repair at a specific genomic locus. While the Comet assay alone quantifies only the overall DNA damage of an individual cell, subsequent FISH on the electrophoresed single cell genome enables the coincidental localization of fluorescently labelled sequences (i.e., probes) to the respective damaged or undamaged genes or specific genomic regions of interest. In that way sequence specific DNA damage, global genomic and transcription coupled repair or the three dimensional ultrastructure of cells from any tissue can be comparatively investigated. This protocol provides a detailed description of the principles and basic methodology of a standard Comet-FISH experiment to study interphase cells of any tissue. Also important variations of the protocol (e.g., neutral conditions to detect double strand breaks) as well as the production of fluorochrome-labelled DNA probes via random priming are described.

Key words Fluorescence in situ hybridization (FISH), Single cell electrophoresis (Comet) assay, Sequence specific, DNA damage, DNA repair

1 Introduction

Fluorescence in situ hybridization (FISH) is a sensitive method for mapping and positioning DNA sequences with high resolution (up to 1 kb) in nearly any genome and is often used in research and clinical diagnostics to detect numerical and structural chromosomal aberrations. FISH utilizes the biological principle that denatured, i.e., single-stranded DNA stretches can identify a complementary single-stranded stretch of DNA, by using fluorescently labelled DNA as a probe to hybridize to its complementary sequences in a target cell [1, 2]. The experiment consists of labelling the probe DNA with fluorochrome-conjugated molecules via random hexanucleotide priming or nick-translation; heat-denaturing both the labelled probe DNA and target cell DNA;

Svetlana Makovets (ed.), *DNA Electrophoresis: Methods and Protocols*, Methods in Molecular Biology, vol. 1054, DOI 10.1007/978-1-62703-565-1_15, © Springer Science+Business Media New York 2013

hybridizing the denatured, single-stranded probe DNA to the denatured target cell DNA under specific conditions; washing under high stringency to eliminate the probe DNA that hybridized nonspecifically to regions of some similarity to the targeted locus; and counterstaining of the target cell chromatin. Using more than one FISH-probe allows the simultaneous detection of multiple targets in one cell in multiple colors. Repeat probes hybridize to repetitive DNA such as centromeric or telomeric sequences; single-copy probes detect specific chromosomal regions; and paint-probes detect whole chromosomes or chromosome-arms [3, 4].

The single cell electrophoresis (Comet) assay is a simple and sensitive technique for quantitatively measuring DNA damage and repair at the single cell level in virtually all types of tissue [5]. Isolated cells are mixed with melted agarose, placed on a glass slide, lysed in a high-salt solution which removes most cell contents except the DNA and finally electrophoresed. However, after proteolysis of most histones and disruption of the nucleosomes, the DNA remains supercoiled. That is likely caused by attachment of DNA stretches to leftovers of a nuclear matrix or framework [6, 7]. Hence, it is widely believed that these supercoiled DNA loops found after proteolysis represent domains, where each domain can be considered as a discrete topological unit. In this context, only DNA strand breaks are capable of relaxing and extend out of the supercoiled DNA bulk [8]. This relaxed (i.e., damaged) DNA is electrophoresed away from this supercoiled DNA bulk towards the anode resulting in an extended, Comet-shaped tail of DNA after electrophoresis, while the undamaged DNA is confined as a head region [8–10]. By choosing different pH conditions for electrophoresis and the preceding incubation (neutral or alkaline), different DNA damage types and levels of sensitivity can be quantified [11].

When hybridizing a fluorescently labelled DNA probe to this electrophoresed cell genome, the level of DNA damage or repair in a single cell can be measured on the level of single genes or specific genomic regions as small as 10 kb (Fig. 1). Santos et al. originally developed the first protocol for FISH on electrophoresed cells to investigate the nuclear ultrastructure of interphase cells as well as the three dimensional behavior of chromosomal domains and chromatin fibers in relation to the nuclear matrix [12]. Today this technique is frequently termed "Comet-FISH" [13] and usually used as a sophisticated biomarker for sequence specific DNA damage and repair in cancer cells to detect the preferential repair of actively transcribed genes, such as p53 in cancers of the bladder or breast [14, 15]; in surrogate cells, such as lymphocytes exposed to radiation [16, 17] and environmental toxicants [18, 19]; or cancer cells exposed to chemotherapeutical drugs [20, 21]. The technique can also be applied to distinguish between single- and double-strand breaks by adjusting the pH conditions for electrophoresis and incubations [22] in any tissue, including even plants [23].

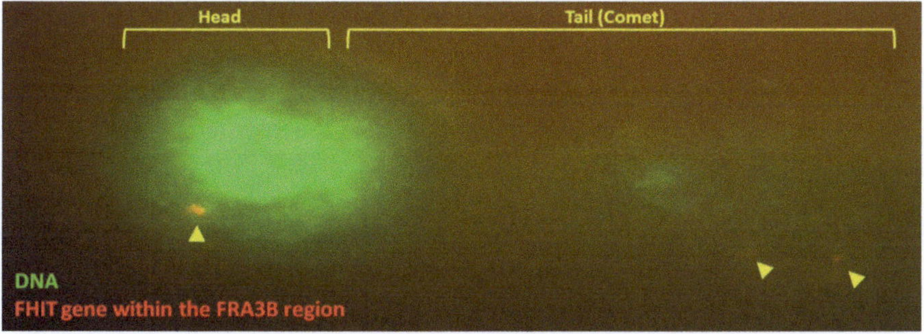

Fig. 1 Representative image of a damaged lymphocyte hybridized with a fluorescently labelled DNA probe, which spans the 1 Mb large FHIT gene within the most frequently expressed fragile site FRA3B. One FISH signal is seen in the (undamaged) DNA of the Comet head, whereas the second signal is split into two domains which appear in the (damaged) DNA of the Comet tail

However, compared to conventional FISH experiments where single cells are strongly attached with their cell membrane to a glass slide, during the Comet-FISH experiment, electrophoresed DNA is conserved in its three dimensional state within a heat sensitive agarose matrix. Therefore, this fragile electrophoresed DNA can only be chemically denatured using alkali solutions. Following hybridization, it should be washed with lower stringency buffers and very gentle handling. Longer incubation times to let the solutions penetrating the cells embedded three dimensionally in the gel matrix are required [5, 24–26].

2 Materials

2.1 DNA Probe Preparation

1. 20× saline-sodium citrate (SSC) buffer: 3 M NaCl, 300 mM Na citrate, pH 7.1.
2. For 10 mL of hybridization master mix (sufficient for more than 1,000 hybridization experiments), spread 1.43 g dextran sulfate (Sigma, UK) along the side of a 15-mL conical tube and add 7.86 mL of deionized Formamide (Sigma, UK), 1.55 mL of 20× SSC and 0.7 mL of ddH_2O. Heat up the mixture to 70 °C in order to dissolve the dextran sulfate and adjust the pH with 10 M NaOH to 7.0 with pH paper.
3. PN buffer: Prepare 1 L of 0.1 M Na_2HPO_4 (dibasic) and 500 mL of 0.1 M NaH_2PO_4 (monobasic). Titrate 0.1 M Na_2HPO_4 (pH > 9) by adding small volumes of 0.1 M NaH_2PO_4 (pH ~ 4.5) until a pH of 8.0 is reached. Add 0.05 % (v/v) Non-ident P-40.
4. Random primer mix: 125 mM Tris–HCl (pH 6.8), 12.5 mM $MgCl_2$, 25 mM 2-mercaptoethanol, 750 μg/mL oligodeoxyribonucleotide primers (Invitrogen, UK) (*see* **Note 1**).

5. 40 U/μL DNA polymerase Klenow fragment (Invitrogen, UK) (*see* **Note 1**).
6. Nucleotide buffer: 5 μL of each 100 mM dATP, 100 mM dGTP, and 100 mM dCTP (Promega, UK) are mixed with 2.5 μL of 1 M Tris–HCl, pH 7.7, 0.5 μL 0.5 M EDTA, pH 8.0, and 232 μL ddH_2O for a total of 250 μL nucleotide buffer. The final concentration of the nucleoside triphosphates is 2 mM each. Store at −20 °C.
7. 150 mM dTTP (Promega, UK).
8. Labelled dUTPs, such as Cy3-11-dUTP, Cy5-11-dUTP (Amersham, USA), fluorescein-11-dUTP, or Texas Red-11-dUTP (Roche, UK) at a concentration of 0.4 mM (*see* **Note 2**).
9. Human Cot-1 DNA (Invitrogen, UK).
10. Salmon sperm DNA (Invitrogen, UK).
11. Glycogen 20 mg/mL (Invitrogen, UK).
12. 100 % Ethanol.
13. 0.5 M EDTA, pH 8.0.
14. Heat block set at 100 °C.
15. A water bath or an incubator set at 37 °C.

2.2 Slide Preparation

1. Microscope slides with frosted ends.
2. Metal plate that can hold several microscope slides.
3. Polystyrene box in which the metal plate can be fitted.
4. Water bath or heat block set at 60 °C.
5. 1 % (w/v) normal melting point (NMP) agarose (Invitrogen, UK) in ddH_2O.

2.3 Agarose Embedding and Lysis of Cells

1. Phosphate-buffered saline (PBS): 137 mM NaCl, 2.75 mM KCl, 1.45 mM KH_2PO_4, 15.25 mM Na_2HPO_4, pH 7.2.
2. Microcentrifuge (a speed of 200 × *g* is required).
3. 0.5 and 1 % (w/v) low melting point (LMP) agarose (Invitrogen, UK) in PBS.
4. Water bath or heat block set at 60 °C.
5. Coverslips (22 mm × 50 mm).
6. Coplin jar.
7. Alkaline lysis solution: 2.5 M NaCl, 100 mM EDTA, 10 mM Tris base in ddH_2O. Store at room temperature for up to 2 weeks. Prior to use, add 10 % DMSO (v/v) and 1 % Triton X-100 (v/v) and chill this solution for 1 h at 4 °C (*see* **Notes 3** and **4**).

2.4 Electrophoresis

1. Horizontal electrophoresis tank. The size of this tank determines the number of slides used in a single experiment as well as the type of the power supply required.

2. Alkaline electrophoresis buffer: 300 mM NaOH, 1 mM EDTA in ddH_2O. Adjust pH to 13.2 with 10 M NaOH.
3. Power supply capable of reaching 20 V and 300 mA (0.8 V/cm on platform, 300 mA).
4. 50 mL plastic syringe.
5. Neutralization buffer: 0.4 M Tris–HCl in ddH_2O, pH to 7.5.

2.5 Denaturation and Hybridization

1. Glass Coplin jars.
2. 0.5 M NaOH.
3. 70, 85, 100 % Ethanol.
4. Incubator set at 37 °C.
5. Diamond pen (VWR, UK) to mark hybridization areas.
6. Rubber cement (Marabu, Germany) (*see* **Note 5**).
7. Parafilm© (Pechiney Plastic Packaging Company, USA).
8. Sealable plastic chamber lined with moist paper towels.
9. Iron block to be fitted in the plastic chamber.

2.6 Washing and Staining

1. Glass Coplin jars.
2. Water baths set at 37 and 45 °C.
3. 30 % deionized formamide in 2× SSC, pH 7.0.
4. 0.2× SSC, pH 7.0.
5. 2× SSC, pH 7.0.
6. Antifade solution: 0.223 g of 1,4-diazabicyclo[2.2.2]octane (Sigma, UK), 7.2 mL of Glycerol, 0.8 mL of ddH_2O, and 2 mL of 1 M Trizma base, pH 8.0. Store 1 mL aliquots at −80 °C and protect from light.
7. DAPI stock solution: 4′,6-Diamidine-2-phenylindole (DAPI; Vector Laboratories, USA) in Antifade solution to give a final concentration of 0.5 g/mL.
8. YoYo-1 iodide stock—please describe.
9. Coverslips, 22 mm × 22 mm.
10. Nail polish.

2.7 Microscopy and Image Analysis

1. Epifluorescence microscope equipped with ×40, ×63, and ×100 with oil objectives; a filter set capable for detecting 3–4 colors, such as Cy3 (excitation: 554; emission: 568), Fluorescein Isothiocyantate (490; 525), DAPI (350, 470) and infrared fluorochrome Cy5 (649; 666) visible only via a CCD camera; and a 100 W mercury illumination (HBO 100) bulb.
2. Charged coupled device (CCD) camera. We are currently using an electron multiplying CCD camera (Andor, UK) with 658 × 496 pixels.

2.8 Variations for Neutral Conditions to Detect Double-Strand Breaks Only

1. Neutral lysis solution: 30 mM EDTA, 0.5 % SDS (w/v) in ddH_2O. Adjust pH to 8.0 with 1 M NaOH.
2. Neutral electrophoresis buffer: 90 mM Tris base, 90 mM Boric acid, 2 mM EDTA in ddH_2O. Adjust pH to 8.5 with 1 M NaOH.

2.9 Variations When Using Spermatozoa

1. 2 % (w/v) LMP agarose (Invitrogen, UK) in PBS.
2. Alkaline lysis solution supplemented with 10 mM dithiothreitol (Sigma, UK).
3. Alkaline lysis solution supplemented with 0.05 g/mL proteinase K (Roche, UK).

3 Methods

3.1 DNA Probe Preparation via Random Priming

See **Notes 6** and **7** for general requirements for this section.

1. Set up a 20 μL random priming solution containing 100 ng/μL of probe DNA, 2.25 μL nucleotide buffer, 10 μL random primer mix, 1.75 μL labelled dUTP, and 3.5 μL dTTP. Mix the solution gently by pipetting up and down (*see* **Note 8**).
2. Boil the solution at 100 °C for 5 min.
3. Quickly cool the solution on ice to RT.
4. Spin briefly at 3,000 × *g*.
5. Add 0.5 μL of DNA polymerase I Klenow fragment (40 U/μL), gently mix by flicking the tube and spin briefly at 3,000 × *g*. *Caution*: Do not vortex the mixture as Klenow polymerase is sensitive to vortexing.
6. Incubate the reaction for 10–12 h at 37 °C in a water bath or an incubator.
7. Add 2 μL of 0.5 M EDTA, pH 8.0 to stop the reaction. Store at −20 °C until further use.

3.2 Slide Preparation

1. Clean the microscope slides with 100 % ethanol and flame.
2. Melt 1 % (w/v) NMP agarose in H_2O. Prevent liquid overflow or excessive evaporation.
3. Place melted agarose in a water bath at 60 °C.
4. Dip the slides briefly into the melted 1 % NMP agarose. The agarose should cover at least half of the slide. Remove excess agarose by wiping the back of the slide with a tissue.
5. Let the slides dry overnight at room temperature. Slides can be stored for up to 7 months in an airtight box at room temperature (*see* **Note 9**).

3.3 Agarose Embedding and Lysis of Cells

See **Notes 10–12** for general requirements for this section.

1. Place a metal plate on top of ice in an appropriate polystyrene box.
2. Melt the 0.5 and 1 % agarose solutions and keep melted on top of a hot plate or in a water bath at 60 °C.
3. Dilute cells obtained from cell culture, lymphocyte separation or tissue extraction to a concentration of $1.5–3.0 \times 10^4$ cells/mL in PBS. For one Comet slide 1 mL of this cell suspension is required, i.e., a total of $1.5–3.0 \times 10^4$ cells are required per experiment.
4. Spin down the cell suspension at $200 \times g$ for 5 min at room temperature.
5. Discard the supernatant and resuspend the cells in fresh PBS. Repeat **steps 4** and **5**.
6. Discard the supernatant and resuspend the pellet in 100 μL PBS.
7. Mix the cell suspension with an equal amount of 1 % LMP agarose solution by pipetting the mixture up and down. *Caution*: Avoid any air bubbles during mixing as this can introduce DNA damage.
8. Quickly apply 100 μL of the agarose-cell suspension in the center of a pre-coated agarose slide (from Subheading 3.2, **step 5**) and cover the suspension with a coverslip.
9. Place the slide on the ice-cold metal plate and let the agarose set for about 10 min (*see* **Note 13** as well as the construction of a Comet assay slide diagrammed in Fig. 2).
10. Once the agarose is set, remove coverslips and incubate the slides in a vertical position in a Coplin jar filled with ice-cold alkaline lysis solution, at 4 °C for 1 h (*see* **Note 14**). Remove the alkaline lysis solution carefully and wash the slides without shaking three times for 15 min per wash in Coplin jars filled with ice-cold ddH_2O at 4 °C. Ensure that the agarose layer is not damaged during these washing steps (*see* **Note 15**).
11. Carefully transfer all slides from the Coplin jar to the electrophoresis tank and position them as close as possible to the anode to ensure all the slides are positioned in the same way relative to the electrodes (*see* Fig. 3). When there are too few slides to fill the entire tank, fill gaps with empty slides to allow a homogenous current during electrophoresis.

3.4 Electrophoresis

1. Pour ice-cold alkaline electrophoresis buffer into the tank until all slides are covered with a thin layer (~4 mm) of buffer. Close the tank. Avoid bubbles on top of the slides. Place the tank into a cold room at 4 °C to prevent melting of LMP agarose during electrophoresis. Alternatively the tank can be placed on ice or in a closed fridge at 4 °C during the electrophoresis (*see* **Note 16**).

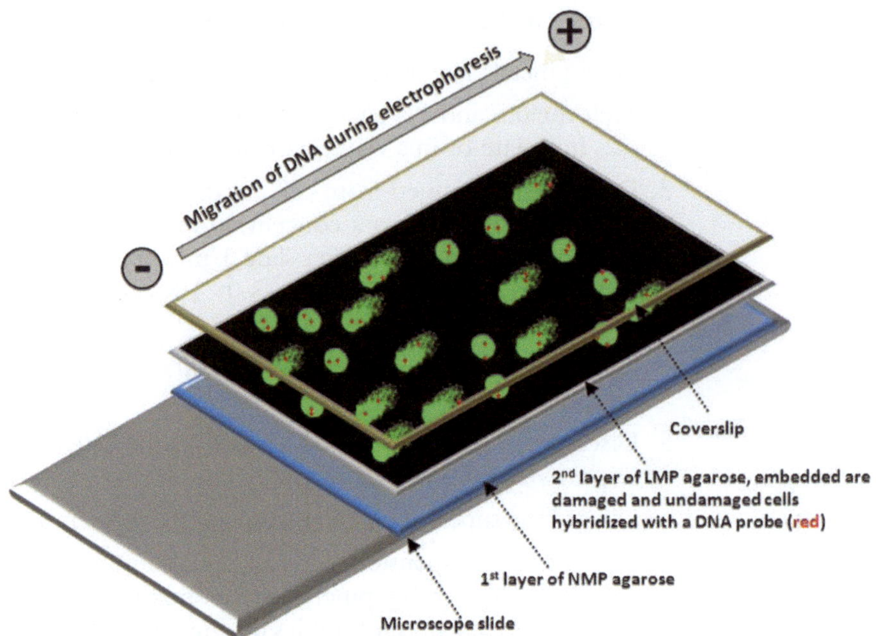

Fig. 2 Schematic construction of a Comet-FISH slide, showing two different agarose layers as well as cells with damaged and undamaged DNA hybridized to a DNA probe (*red signals*)

2. Pre-incubate slides for 30 min in alkaline electrophoresis buffer to unwind DNA and induce DNA breaks in alkali-labile sites (*see* **Note 17**).
3. Electrophorese at a constant voltage 0.75 V/cm for 30 min. The current should be around 300 mA and if it is too high or too low, remove or add, respectively, the alkaline electrophoresis buffer using a 50 mL syringe (*see* **Note 18**).
4. Remove all the slides from the electrophoresis tank and drain the edges of the slides using a paper towel.
5. Wash slides three times with neutralization buffer, 5 min each wash. Rinse the slides carefully with ddH_2O.

3.5 Denaturation and Hybridization

1. Immerse slides in 100 % ethanol at 4 °C for at least 2 h (at this step slides can be stored for up to several weeks).
2. Remove slides from ethanol and let ethanol fully evaporate.
3. To denature DNA, incubate slides in 0.5 M NaOH for 30 min at room temperature (*see* **Notes 19** and **20**).
4. Rinse slides with 1× PBS and dehydrate the samples by submerging slides in a 70, 85, 100 % ethanol series for 2 min each at room temperature.
5. Drain excess ethanol from the slides by blotting the edges of each slide to a Kimwipe© and let the gels air dry for 2–4 h until ethanol is evaporated completely.

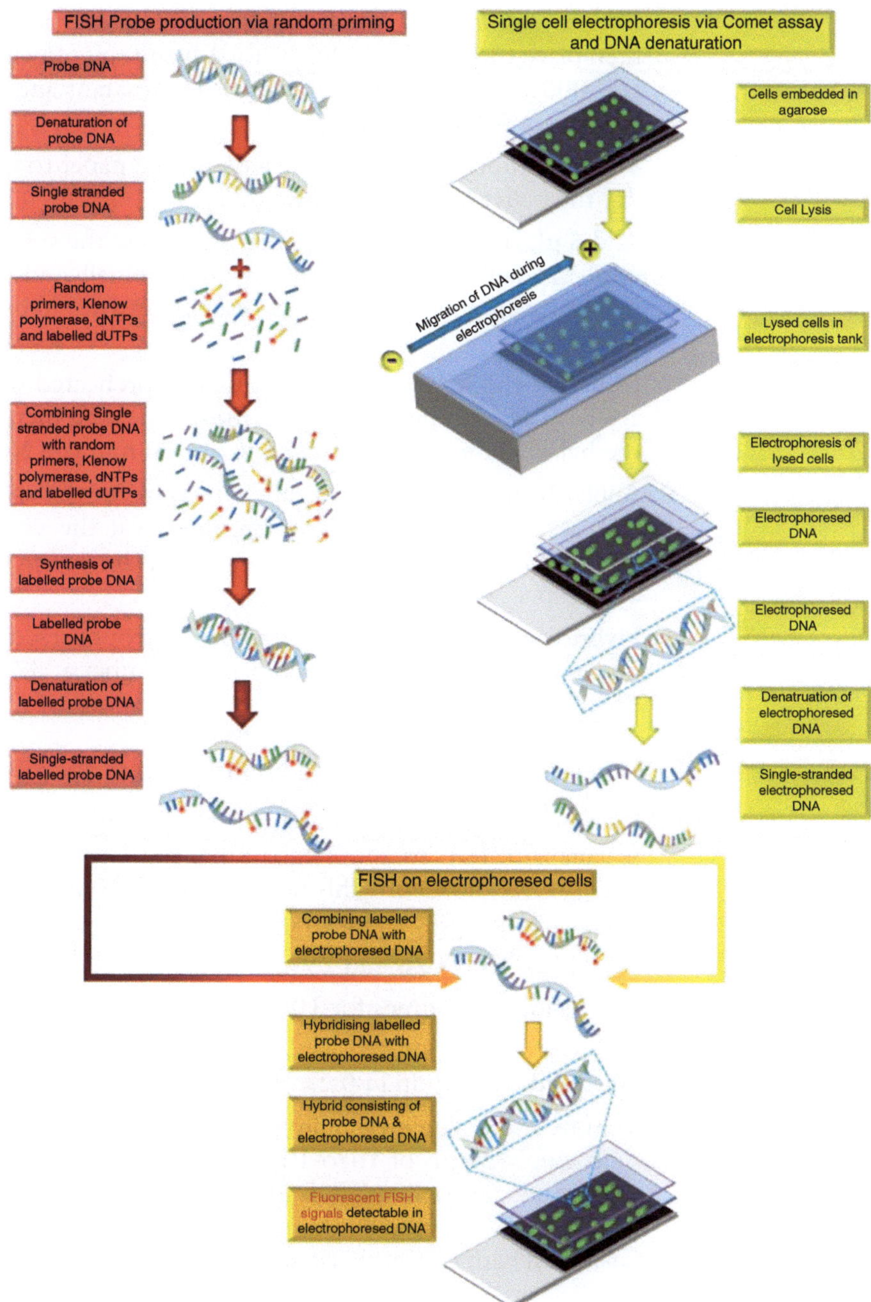

Fig. 3 Schematic illustration of the FISH on Comet method

6. In the meantime, set up the hybridization mixture by combining 10 μL of labelled probe DNA, 10 μL of 1 mg/mL Cot-1 DNA, and 10 μL of 10 mg/mL salmon sperm DNA (We only use premade 1 mg/mL Cot-1 DNA and 10 mg/mL salmon sperm DNA solutions, as preparation of dry DNA solution usually results in higher background levels). Precipitate the

DNA by adding 1 μL of glycogen and 90 μL of 100 % ethanol, then place the sample at −80 °C for 1 h and spin down the tube at 9,000 × *g* for 30 min. Remove all supernatant and wash the pellet with 70 % ethanol.

7. Place the tube upside down on a tissue or a paper towel and air dry the pellet (*see* **Notes 21** and **22**).
8. Add 3 μL of ddH_2O to the pellet and vortex the solution hard until the pellet is completely dissolved. Finally, add 7 μL of hybridization master mix to complete the hybridization mixture.
9. Denature the hybridization mixture in a preheated water bath at 75 °C for 10 min, followed by pre-annealing incubation of the hybridization mixture at 37 °C for 30 min (*see* **Note 22**).
10. Apply 20 μL of the hybridization mixture to the slide, cover with a 22 mm × 22 mm coverslip and seal the coverslip by applying Fixogum© rubber cement along the perimeter of the coverslip.
11. Hybridize the DNA probe with the electrophoresed chromatin for 24–48 h in a humidified plastic chamber at 37 °C in the dark (*see* **Notes 23** and **24**).

3.6 Washing and Staining

See **Notes 25–27** for general requirements for this section.

1. Remove Fixogum© rubber cement by carefully pulling up the sealant with forceps and incubate the slide in PN buffer until the coverslip floats off (this takes usually 5–10 min).
2. Wash slide for 30 min in 30 % formamide in 2× SSC buffer at 45 °C (*see* **Notes 28** and **29**).
3. Wash slide three times for 10 min each wash in 2× SSC buffer at 37 °C.
4. Wash slide for 5 min in 0.2× SSC buffer at 45 °C.
5. Remove excess liquid carefully with a paper towel and air dry the slide. Apply 20 μL of YoYo-1 iodide diluted 1:10,000 from the stock in antifade or 20 μL of DAPI diluted 1:1,000 from the stock in antifade to stain the DNA. Cover the slide with a coverslip and finally seal with conventional nail polish (*see* **Note 30**).

3.7 Quantification

1. There is currently no manual or computerized system to quantify Comet-FISH results. However, each single cell can first be analyzed by eye for their amount of DNA in the tail, followed by scoring the FISH domains detected in the tail or head. Usually cells are graded in five categories according to the following amounts of DNA in the tail and each cell is assigned a score on the scale from 0 to 4: 0 = <5 % of the DNA is in the tail (undamaged cell with a faint comet tail); 1 = 5–20 % (low level damage); 2 = 20–40 % (medium level damage); 3 = 40–95 %

(high level damage); 4 = >95 % (total damage). For 50 scored cells the overall score (the sum of scores of the 50 cells) will be between 0 and 200 arbitrary units.

3.8 Variations for the Neutral Comet Assay

There are four variations compared to the alkaline comet assay protocol described in Subheadings 3.3 and 3.4:

1. In **step 10**, Subheading 3.3, use the neutral lysis solution instead of the alkaline lysis solution for the cell lysis.
2. In **steps 1–5**, Subheading 3.4, use the neutral electrophoresis buffer instead of the alkaline electrophoresis buffer for the incubations and electrophoresis.
3. In **step 3**, Subheading 3.4, use the voltage of 1 V/cm during the electrophoresis.
4. Skip the washes in the **step 5**, Subheading 3.4.

3.9 Variations for the Sperm Comet Assay

There are three alterations compared to the alkaline comet assay protocol described in Subheadings 3.3 and 3.4:

1. In **step 7**, Subheading 3.3 mix the sperm cell suspension with 2 % LMP agarose solution instead of a 1 % LMP agarose solution.
2. In **step 10**, Subheading 3.3 supplement the alkaline lysis solution with 10 mM dithiothreitol and incubate slides for 1 h at 4 °C. Then make alkaline lysis solution with 0.05 g/mL proteinase K and incubate slides for another hour at 4 °C.
3. In **step 3**, Subheading 3.4, electrophorese slides for 20 min.

4 Notes

1. Purchasing a random priming kit is the most reliable and cheapest way to obtain the random priming mix and Klenow polymerase with high activity compared to buying these reagents separately. We are using Invitrogen Bioprime Labelling System with highly reliable results.
2. We highly recommend using fluorescently labelled nucleotides rather than detecting nucleotides, such as Biotin-11-dUTP or Digoxigenin-11-dUTP, with antibodies to the nucleotide modifications. The latter usually results in an increased background and decreased fluorescent intensity of the probe, perhaps due to the cells being embedded in agarose.
3. Addition of DMSO scavenges reactive radicals produced by iron which is released during lysis from hemoglobin. When using cells/tissue without hemoglobin, it is not necessary to add DMSO.
4. Earlier experimental protocols used the anionic detergent sodium sarcosinate to in the lysis solution. This is not necessary

for successful lysis and may induce precipitation of the lysis solution at 4 °C [27].

5. It is highly recommended to use the rubber cement Fixogum© manufactured by Marabu; it comes as a highly viscous liquid. Our laboratory and many of our colleagues use only this brand of rubber cement, as the rubber cement of other providers usually has significantly more liquid form, and therefore may leak under the coverslip and interfere with the hybridization mixture and tissue.
6. Multiple DNA sources can be uses for making a probe DNA by random priming: DNA from BAC/PAC clones, extracted via standard alkali lysis protocols; PCR products of amplified DNA sequences or whole chromosomes, derived from flow-sorted chromosomes, to name a few.
7. Difficulties during the random priming reaction can occur when the probe DNA solution is contaminated with remaining RNA molecules. RNA must be eliminated by purification and enzymatic digestion procedures before the DNA sample can be used for random priming.
8. Fluorochrome-labelled dUTP always need to be at a 1:2 ratio with the dTTP so that steric hindrance of the Klenow polymerase can be avoided.
9. Longer storage of agarose pre-coated slides can increase the chance of detaching agarose layers during lysis or electrophoresis.
10. This protocol works equally well on any eukaryotic cells except spermatozoa (*see* Subheading 3.9).
11. It is very important that the entire experiment (Subheadings 3.3–3.4) is performed on ice, with cold (4 °C) solutions throughout the experiment to avoid DNA repair taking place; in the dark or in a room with an orange filter to prevent UV light-induced DNA damage.
12. It is imperative that temperature, settings for voltage, incubation, lysis and electrophoresis time, electrophoresis buffer volume, and concentration/type of the nucleic acid dyes are not changed within matching experiments where more than one electrophoresis run have to be performed.
13. The dimensions of the agarose gel layer should match the dimensions of the respective coverslip used. The gel should not fill the entire slide or contain any air bubbles.
14. Extending the lysis time for up to 8 h does not influence the success of the experiment. Longer lysis time has been reported in the literature but, in our opinion, it significantly increases the chance of losing cells. The pH should not be greater than 10.5 when increasing the lysis time [8].

15. Remaining salt traces from the lysis buffer can lead to an artificially expressed Comet tail during electrophoresis. Hence, rinsing well with ddH_2O is essential.
16. An even, well controlled layer of electrophoresis buffer covering the slides is essential for reproducible results, as even small differences in electrophoresis buffer volumes result in variable electric fields strength, subsequently producing inconsistent Comet tail lengths.
17. Incubation in electrophoresis buffer prior to electrophoresis depends on the type of cell suspension used, and incubation time can vary between 20 and 70 min. A 30 min incubation for cultured cells and blood samples is generally recommended and is the standard in our laboratory. Shorter incubation periods, however, can reduce the background DNA damage level found in control cells, but at the same time will decrease the sensitivity of DNA damage detection in exposed cells.
18. Changing the current of the power supply during electrophoresis results in dramatic increases or decreases of Comet tails. The settings described here are the international standards and yield the most reproducible results. As a guide, cells not being exposed to DNA damaging agents (control cells) should have around 5–12 % of their DNA in the tail and the tail lengths of no more than 15–25 μm [28].
19. Conventional FISH protocols use heat to denature DNA in cells attached to glass slides. As the electrophoresed cells are embedded in low melting point agarose, which would melt at the temperatures used for DNA denaturation in the conventional FISH experiments (e.g., 73 °C in 75 % formamide), chemical denaturation is used instead.
20. The level of DNA denaturation is dependent on the concentrations of GC nucleobases, monovalent cations, and the length of all DNA double-strand stretches as well as on time the DNA has been exposed to alkaline conditions. In this context, denaturation of the electro-stretched DNA with 0.5 M NaOH for 30 min at RT became the standard in most laboratories. Exceeding these denaturation conditions significantly can easily result in a dramatically increased hybridization background caused by uncoiling of the highly compacted bulk DNA structure. However, if the denaturation is not sufficient, the DNA may not be denatured enough, resulting in decreased or even no hybridization signals.
21. DNA probes made from highly repetitive sequences, such as centromeric DNA, are hybridized in the presence of a 50- to 100-fold excess of total genomic DNA. For DNA probes, which contain unique sequences to a specific region of the genome, hybridization experiments are performed with a similar

mass excess of Cot-1 DNA. Cot-1 DNA consists of DNA fragments, predominantly 50–300 bp in size and highly enriched for repetitive alphoid sequences and intermediate repetitive sequences such as the Kpn and Alu families. These highly repetitive stretches are present in virtually all genomic DNA fragments, from which DNA probes are constructed. During hybridization those highly repetitive sequences anneal to the complementary genomic DNA distributed all over the genome. This may result in a background which can have the same intensity as the hybridized DNA probe without the addition of Cot-1 DNA. The Cot-1 value of a particular DNA sequence is defined as the product of the nucleotide concentration, its reassociation time, and an appropriate buffer factor based on the cation concentration [29].

22. For centromere and telomere FISH-probes, which usually consist of only α-repeat minisatellites, do not add Cot-1 DNA and do not preanneal the probe DNA, if the relative amount of the repetitive sequences within the probe is limited. Otherwise the repetitive sequences of the probe DNA will be blocked by Cot-1 DNA.

23. The time of hybridization time depends on the sequence complexity of the FISH-probe. When genes or entire chromosomes are detected, the FISH-probe contains a significant number of unique sequences, which need an incubation of 24 h (for genes) to 72 h (for entire chromosomes). Telomeric, centromeric, or pericentromeric satellite probes, however, require an incubation time of only 4–6 h.

24. Altering the hybridization temperature (37–41 °C) can be used to differentiate between directly related repetitive sequences in FISH-probes.

25. Compared to the conventional FISH experiments, applying FISH to electrophoresed cells embedded in agarose requires significantly more time for all washing solutions to penetrate the agarose layers. Hence, longer incubation time is.

26. Agarose gels can melt or detach from slides during electrophoresis and washes. Therefore (1) work during all washing steps very gently, i.e. only move the slides slowly from one washing solution to another, as the gels are very fragile and detach from slides easily. Histological staining jars and racks can be used as alternatives to Coplin jars. However, the slides should not placed in the conventional rack positions, instead, place them on top of the rack. Consequently, the gels are not exposed to gravity and therefore remain more strongly attached to the slide. Unfortunately, this modification usually results in higher levels of hybridization background, due to less efficient washing. (2) All steps should be performed in the smallest possible volumes of solutions, to minimize the risk of agarose being removed.

Table 1
Washing conditions and the respective stringency values, assuming a hypothetical probe length of 600 bp, no formamide addition and a GC content of 55 %

SSC buffer	T_w	T_m	ΔT
2×	30	95	65
2×	37	95	58
2×	45	95	50
2×	50	95	45
0.2×	30	58	28
0.2×	37	58	21
0.2×	45	58	13
0.2×	50	58	8

$\Delta T = T_m - T_w$
$T_m = 81.5\ °C + 16.6\ \log[Na^+] + 0.41\ (\%GC) - 0.63\ (\%formamide) - (600/length)$
T_w washing temperature, T_m melting temperature, ΔT stringency

27. The washing stringency is inversely proportional to the concentration of SSC. The lower the concentration of SSC, the higher the stringency is. Do not go lower than 0.1× SSC, because successfully hybridized DNA probes may be removed. For an accurate calculation of washing stringency, one can subtract the wash temperature (T_w) from the DNA melting temperature (T_m), as follows $\Delta T = T_m - T_w$. A decrease in ΔT is proportional to an increase in stringency [30]. To calculate the melting temperature (T_m), the following formula can be used: $T_m = 81.5\ °C + 16.6\ \log[Na^+] + 0.41\ (\%\ GC) - 0.63\ (\%\ formamide) - (600/length)$,

    where $[Na^+]$ represents the sodium concentration); % GC represents the percentage of guanine and cytosine nucleobases in the probe DNA sequence; % formamide represents the percentage (volume) of formamide in the solution; and length represents the length of the DNA probe that is being hybridized. Table 1 gives an example of stringencies using different washing temperatures for the 2× SSC and 0.2× SSC solutions used in this section, with an assumed GC content of 55 %, a 600 bp DNA probe in the absence of formamide.

28. After the purchase of formamide, immediately aliquot and store 50 mL aliquots at −20 °C to avoid decay.

29. The use of formamide can cause the gels to slip from the slides. Longer incubation periods in washing buffer with lower (or even without) formamide concentrations can prevent this problem.

30. Fading of the fluorescent signals of the FISH-probes as well as the DNA staining after a few minutes of UV light exposure are caused by oxidation of the antifade solution. The oxidized antifade solution can be identified by the color change of the solution from clear to brownish and is mostly caused by inappropriate storage of antifade solution at RT for too long. Hence, always store this solution at −20 °C and make it fresh frequently to obtain optimal signal intensities.

References

1. Pardue ML, Gall JG (1970) Chromosomal localization of mouse satellite DNA. Science 168:1356–1358
2. Pinkel D, Straume T, Gray JW (1986) Cytogenetic analysis using quantitative, high-sensitivity, fluorescence hybridization. Proc Natl Acad Sci U S A 83:2934–2938
3. Baumgartner A, Schmid TE, Cemeli E, Anderson D (2004) Parallel evaluation of doxorubicin-induced genetic damage in human lymphocytes and sperm using the comet assay and spectral karyotyping. Mutagenesis 19:313–318
4. Swiger RR, Tucker JD (1996) Fluorescence in situ hybridization: a brief review. Environ Mol Mutagen 27:245–254
5. Anderson D, Dhawan A (2009) The comet assay. Royal Society of Chemistry, Cambridge
6. Cook PR, Brazell IA (1976) Conformational constraints in nuclear DNA. J Cell Sci 22: 287–302
7. Ostling O, Johanson KJ (1984) Microelectrophoretic study of radiation-induced DNA damages in individual mammalian cells. Biochem Biophys Res Commun 123:291–298
8. Collins AR, Oscoz AA, Brunborg G, Gaivao I, Giovannelli L, Kruszewski M, Smith CC, Stetina R (2008) The comet assay: topical issues. Mutagenesis 23:143–151
9. Klaude M, Eriksson S, Nygren J, Ahnstrom G (1996) The comet assay: mechanisms and technical considerations. Mutat Res 363: 89–96
10. Sipinen V, Laubenthal J, Baumgartner A, Cemeli E, Linschooten JO, Godschalk RW, Van Schooten FJ, Anderson D, Brunborg G (2010) In vitro evaluation of baseline and induced DNA damage in human sperm exposed to benzo[a]pyrene or its metabolite benzo[a]pyrene-7,8-diol-9,10-epoxide, using the comet assay. Mutagenesis 25:417–425
11. Albertini RJ, Anderson D, Douglas GR, Hagmar L, Hemminki K, Merlo F, Natarajan AT, Norppa H, Shuker DE, Tice R et al (2000) IPCS guidelines for the monitoring of genotoxic effects of carcinogens in humans. International Programme on Chemical Safety. Mutat Res 463:111–172
12. Santos SJ, Singh NP, Natarajan AT (1997) Fluorescence in situ hybridization with comets. Exp Cell Res 232:407–411
13. Rapp A, Bock C, Dittmar H, Greulich KO (2000) UV-A breakage sensitivity of human chromosomes as measured by Comet-FISH depends on gene density and not on the chromosome size. J Photochem Photobiol B 56:109–117
14. Kumaravel TS, Bristow RG (2005) Detection of genetic instability at HER-2/neu and p53 loci in breast cancer cells sing Comet-FISH. Breast Cancer Res Treat 91:89–93
15. McKelvey-Martin VJ, Ho ET, McKeown SR, Johnston SR, McCarthy PJ, Rajab NF, Downes CS (1998) Emerging applications of the single cell gel electrophoresis (comet) assay. I. Management of invasive transitional cell human bladder carcinoma. II. Fluorescent in situ hybridization comets for the identification of damaged and repaired DNA sequences in individual cells. Mutagenesis 13:1–8
16. Mosesso P, Palitti F, Pepe G, Pinero J, Bellacima R, Ahnstrom G, Natarajan AT (2010) Relationship between chromatin structure, DNA damage and repair following X-irradiation of human lymphocytes. Mutat Res 701:86–91
17. Amendola R, Basso E, Pacifici PG, Piras E, Giovanetti A, Volpato C, Romeo G (2006) Ret, Abl1 (cAbl) and Trp53 gene fragmentations in Comet-FISH assay act as in vivo biomarkers of radiation exposure in C57BL/6 and CBA/J mice. Radiat Res 165:553–561
18. Glei M, Schaeferhenrich A, Claussen U, Kuechler A, Liehr T, Weise A, Marian B, Sendt W, Pool-Zobel BL (2007) Comet fluorescence in situ hybridization analysis for oxidative stress-induced DNA damage in colon cancer relevant genes. Toxicol Sci 96:279–284
19. Park E, Glei M, Knobel Y, Pool-Zobel BL (2007) Blood mononucleocytes are sensitive to the DNA damaging effects of iron overload – in

vitro and ex vivo results with human and rat cells. Mutat Res 619:59–67
20. Arutyunyan R, Rapp A, Greulich KO, Hovhannisyan G, Haroutiunian S, Gebhart E (2005) Fragility of telomeres after bleomycin and cisplatin combined treatment measured in human leukocytes with the Comet-FISH technique. Exp Oncol 27:38–42
21. Escobar PA, Smith MT, Vasishta A, Hubbard AE, Zhang L (2007) Leukaemia-specific chromosome damage detected by comet with fluorescence in situ hybridization (Comet-FISH). Mutagenesis 22:321–327
22. Fernandez JL, Vazquez-Gundin F, Rivero MT, Genesca A, Gosalvez J, Goyanes V (2001) DBD-fish on neutral comets: simultaneous analysis of DNA single- and double-strand breaks in individual cells. Exp Cell Res 270: 102–109
23. Menke M, Angelis KJ, Schubert I (2000) Detection of specific DNA lesions by a combination of comet assay and FISH in plants. Environ Mol Mutagen 35:132–138
24. Glei M, Hovhannisyan G, Pool-Zobel BL (2009) Use of Comet-FISH in the study of DNA damage and repair: review. Mutat Res 681:33–43
25. Spivak G, Cox RA, Hanawalt PC (2009) New applications of the comet assay: Comet-FISH and transcription-coupled DNA repair. Mutat Res 681:44–50
26. Baumgartner A, Cemeli E, Laubenthal J, Anderson A (2009) The comet assay in sperm-assessing genotoxins in male germ cells. Elsevier. RSC Publishing in London, UK. DOI: 10.1039/9781847559746-00331
27. Singh NP, McCoy MT, Tice RR, Schneider EL (1988) A simple technique for quantitation of low levels of DNA damage in individual cells. Exp Cell Res 175:184–191
28. Tice RR, Agurell E, Anderson D, Burlinson B, Hartmann A, Kobayashi H, Miyamae Y, Rojas E, Ryu JC, Sasaki YF (2000) Single cell gel/comet assay: guidelines for in vitro and in vivo genetic toxicology testing. Environ Mol Mutagen 35:206–221
29. Wienberg J, Stanyon R (1997) Comparative painting of mammalian chromosomes. Curr Opin Genet Dev 7:784–791
30. Maniatis T, Fritsch EF, Sambrook J (1982) Molecular cloning : a laboratory manual. Cold Spring Harbor Laboratory, Cold Spring Harbor, NY

Chapter 16

Analysis of DNA-Protein Interactions Using PAGE: Band-Shift Assays

Lynn Powell

Abstract

The band-shift assay using polyacrylamide gel electrophoresis is a powerful technique used to investigate DNA–protein interactions. The basis of the method is the separation of free DNA from DNA–protein complexes by virtue of differences in charge, size, and shape. The band-shift assay can be used to determine thermodynamic and kinetic binding constants and also to analyze the composition and stoichiometries of DNA–protein complexes.

Key words Band shift, DNA–protein interaction, DNA binding constant, EMSA, PAGE, Gel retardation

1 Introduction

The band-shift assay also known as gel retardation or electrophoretic mobility shift assay (EMSA) is a simple yet very effective method for the investigation of many aspects of DNA–protein interactions. It is a very sensitive technique for the detection of DNA–protein complexes and is relatively quick to perform. The band-shift assay is based on the fact that association of protein with a binding site within a fragment of DNA results in decreased mobility of the DNA within a non-denaturing polyacrylamide gel. The degree of retardation of the DNA is related to the charge, size, and shape of the protein bound. Even DNA–protein complexes of low kinetic stability may be detected in the band-shift assay due to the caging effect of the polyacrylamide which promotes reassociation of rapidly dissociating complexes by keeping them in close proximity to one another [1, 2]. Band-shift assays can be used to determine binding affinities of proteins for specific DNA binding sites [3, 4] or to determine kinetic parameters [5, 6]. It can be used to assess the effects of protein cofactors on specific binding affinities [3] and can also indicate conformational changes either in the protein or via DNA bending [3, 7, 8]. A further refinement to the band-shift

Svetlana Makovets (ed.), *DNA Electrophoresis: Methods and Protocols*, Methods in Molecular Biology, vol. 1054,
DOI 10.1007/978-1-62703-565-1_16,

assay is to perform a super-shift assay where antibody to a particular protein is added to the binding reaction in order to determine whether this results in an alteration in the mobility of the complex of interest [9, 10]. Preparative band-shift gels are used for footprinting methods such as methylation interference [11], DNaseI footprinting and copper orthophenanthroline footprinting [8] and in combination with SDS-PAGE and Western blotting can also be used to determine protein complex composition and stoichiometries [8, 12].

Optimal binding conditions for the DNA–protein interaction under observation must be determined empirically. Factors to be optimized include pH, ionic strength, temperature, and incubation time. Typically binding buffers contain between 50 and 150 mM NaCl or KCl, 10–20 mM Tris–HCl or HEPES, and have a pH between 6 and 8 [7, 8, 13]. However, lower and higher pH, and ionic strength, may be advantageous depending upon the system [13]. Reducing agents, for example, 1 mM DTT or β-mercaptoethanol, and EDTA are also often included, although for certain proteins, for example, Zn finger transcription factors, the EDTA must be excluded and inclusion of Zn^{2+} in the form of $ZnCl_2$ or $ZnSO_4$ in the binding buffer is necessary [14]. Other additives include low concentrations of BSA (≤0.1 mg/ml) to prevent nonspecific loss of protein on surfaces during processing, nonionic detergents to improve protein solubility, and specific cofactors, for example, ATP for Rad51 recombinase [15] or AdoMet for *Eco*KI methyltransferase [3]. Competitor DNA, for example, poly d(I-C), can also be included to prevent nonspecific binding by other factors if a crude extract, rather than purified protein, is used [16].

Here example methods are described for band-shift assays using purified protein to measure dissociation constants by protein titration and competition assays (Fig. 2) and illustrations of analysis of multiprotein complexes are also given (Figs. 3 and 4).

2 Materials

All solutions should be prepared with ultrapure nuclease-free water and analytical grade reagents. All plastics used must be sterile and nuclease-free. Polyacrylamide gels should be made fresh and binding buffer may be frozen in small working aliquots for use in batches. Please note acrylamide is a potent neurotoxin and appropriate safety measures must be taken when handling it. Similarly, the radioisotopes used to label the DNA are hazardous and procedures for their safe handling must be followed.

2.1 Polyacrylamide Gel Electrophoresis Components

1. 10×TBE (Tris–Borate–EDTA) running buffer: 0.89 M Tris, 0.89 M Borate, 0.02 M EDTA, pH 8.3. Weigh 108 g Tris base and 55 g Boric acid and add water to 900 mL. Mix to dissolve,

then add 40 mL of 0.5 M EDTA (pH 8.0) before adding water to a final volume of 1 L. Dilute this stock 1 in 20 to give 0.5× TBE running buffer (*see* **Note 1**).

2. 30 % acrylamide/bisacrylamide solution (37.5:1 acrylamide: bisacrylamide) (Severn Biotech or Sigma). Store at 4 °C (*see* **Note 2**).
3. Ammonium persulfate: 10 % v/v in water (*see* **Note 3**).
4. *N*, *N*, *N*, *N′*-tetramethylethylenediamine (TEMED, Sigma). Store at 4 °C (*see* **Note 4**).
5. Glass plates of medium size (~15 cm × 15 cm) for vertical gel electrophoresis: one notched and one plain with 2 mm spacers and a 12 well comb. A matching electrophoresis tank (Atto or similar).
6. Bulldog clips (40 mm).
7. Wash bottle.
8. Power supply (Bio-Rad PowerPac 300 or similar).
9. PAGE sample buffer: 2 mL glycerol, 0.5 mL 10× TBE, 100 μL 0.1 % bromophenol blue, 7.4 mL water (*see* **Note 5**).
10. Whatman 3MM paper (Sigma).
11. DE81 paper (Sigma).
12. Gel dryer with a vacuum pump.

2.2 DNA-Labelling Components

1. Upper and lower strand unphosphorylated synthetic oligonucleotides containing the binding site of interest made up to 100 μM (100 pmol/μL) for stock solution and diluted to 1 μM (1 pmol/μL) for use in labelling reactions (*see* **Note 6**).
2. γ^{33}P-ATP (3,000 Ci/mmol; 10 μCi/μl) (GE Biosciences) (*see* **Note 7**).
3. Polynucleotide kinase with 10× reaction buffer (New England Biolabs).
4. ProbeQuant G-50 micro columns (GE Healthcare).
5. STE (Sodium chloride–Tris–EDTA) buffer: 10 mM Tris–HCl, pH 8.0, 150 mM NaCl, 1 mM EDTA (pH 8.0).
6. 37 and 65 °C water baths or heat blocks.
7. PCR machine.
8. Microcentrifuge.

2.3 DNA Binding Components

1. Labelled double-stranded oligonucleotide containing the binding site of interest prepared as described in Subheadings 2.2 and 3.1.
2. Purified DNA binding protein of interest (e.g., protein purification protocols can be found in ref. 17).

3. 5×DNA binding buffer: 50 mM Tris–HCl, pH 7.5, 5 mM DTT, 5 mM EDTA, 500 mM NaCl. Aliquot and store at −80 °C. Dilute 1 in 5 for use in binding reactions (*see* **Note 8**).
4. Ultrapure glycerol (Life Technologies).
5. Presiliconized tubes (Bioquote UK or Sigma-Aldrich) (*see* **Note 9**).

2.4 Components for DNA–Protein Complex Visualization and Extraction

1. Gel dryer.
2. Saran wrap.
3. Phosphor storage screen and cassette (GE Healthcare) or autoradiography cassette and X-ray film (Kodak).
4. PhosphorImager (GE Healthcare) or X-ray film developer.
5. Phosphorescent tape (e.g., Trackertape from GE Healthcare).
6. Scalpel.
7. Protease inhibitors (complete Protease Inhibitor Cocktail Tablets, Roche).
8. SDS-PAGE loading buffer: 50 mM Tris–HCl pH 6.8, 2 % w/v SDS, 0.1 % bromophenol blue, 10 % glycerol, 0.1 M DTT.

3 Methods

3.1 Labelling of Oligonucleotides

1. Set up 20 μL oligonucleotide end-labelling reaction in a microcentrifuge tube as follows. Add 10.5 μL of sterile nuclease-free water, then 2 μL of 10×polynucleotide kinase reaction buffer, 2 μL of 1 μM top-strand oligonucleotide, and 4.5 μL γ^{33}P-ATP (3,000 Ci/mmol; 10 μCi/μl) (*see* **Note 10**).
2. To start the reaction add 1 μL (10 units) of PNK, mix well, and incubate at 37 °C for 1 h before heat inactivation of the kinase at 65 °C for 10 min.
3. Prepare ProbeQuant G-50 column for use (as described in the manufacturer's protocol, i.e., vortex to resuspend the resin, loosen cap one-quarter turn and twist off the bottom closure. Place the column in a collection tube and spin for 1 min at 735×*g* to remove the column storage buffer).
4. Place the column in a fresh 1.5 mL collection tube.
5. Add 30 μL of STE to the tube with radiolabelled oligonucleotide, spin down mixture in a microcentrifuge to mix, then slowly and carefully load the whole 50 μL onto the prepared G-50 column taking care not to disturb the resin surface.
6. Spin the column for 2 min at 735×*g* to collect the labelled oligonucleotide in the collection tube. The free γ^{33}-P ATP will be retained in the column, which can now be discarded. The labelled top-strand DNA is at the bottom of the collection tube and can now be hybridized to its complementary strand.

Table 1
Recipes for polyacrylamide gels (15 cm × 15 cm × 2mm)

Solution	4 % Gel	5 % Gel	6 % Gel	7.5 % Gel	10 % Gel
Water	40.2 mL	38.6 mL	36.9 mL	34.4 mL	30.2 mL
10 × TBE	2.5 mL	2.5 mL	2.5 mL	2.5 mL	2.5 mL
30 % acrylamide	6.7 mL	8.3 mL	10 mL	12.5 mL	16.7 mL
10 % ammonium persulfate	500 μL	500 μL	500 μL	500 μL	500 μL
TEMED	50 μL	50 μL	50 μL	50 μL	50 μL
Total volume	50 mL	50 mL	50 mL	50 mL	50 mL

7. Add 2.1 μL of 1 μM unlabelled complementary strand oligonucleotide and 7.9 μL dH_2O to the 50 μL of labelled top-strand oligonucleotide in a 0.5 mL tube and hybridize by heating the tube to 95 °C for 5 min, followed by stepwise cooling (68 °C, 42 °C then 37 °C for 10 min each) using a PCR machine before the final cooling step on ice for 10 min (*see* **Note 11**). This gives a 33 nM stock of double-stranded DNA containing the protein binding site of interest.
8. Store labelled double-stranded DNA at −20 °C before use.

3.2 Casting Polyacrylamide Gels

1. Assemble a gel casting cassette using two clean 15 cm × 15 cm glass plates, one flat and one notched (*see* **Note 12**), and 2 mm spacers. Use bulldog clips to hold the plates together.
2. Mix 37 mL dH_2O, 2.5 mL of 10× TBE, 10 mL of 30 % acrylamide, 500 μL of 10 % ammonium persulfate, and 50 μL TEMED in a 250 mL conical flask to make a 6 % polyacrylamide gel (*see* Table 1 for mixes for alternative percentages; *see* **Note 13**). Pour into the gel cassette allowing space for the comb. Insert a 12 well comb immediately, ensuring no air bubbles are trapped beneath it (*see* **Note 14**). Retain excess mix in the conical flask to check that the gel mix has set. Leave to set for at least 1 h (*see* **Note 15**).

3.3 DNA–Protein Binding Reactions for Determination of Apparent Dissociation Constants (Kds)

To determine the apparent Kd of a DNA–protein interaction, the binding reactions are made up with constant limiting radiolabelled DNA (0.1 nM) and increasing amounts of protein. Protein concentrations used are designed to cover the range between 0 and 100 % binding of the DNA target. Pilot experiments are normally required to determine the range of protein concentrations to be used in a follow-up experiment for accurate measurements of Kd (*see* **Note 16**).

1. Set up DNA–protein binding reactions containing 5 % glycerol, 0.1 nM DNA, and purified protein at a range of concentrations (*see* **Note 16**) in a total volume of 20 μL in the appropriate 1 × binding buffer (*see* Subheading 2.3 for an example and *see* **Note 8**).
2. Incubate reactions for 10 min between 4 and 37 °C (*see* **Note 17**).
3. Load 18 μL of each binding reaction onto the gel (*see* **Note 5)** and electrophorese, visualize, and analyze the gel as described in Subheadings 3.5, 3.6, and 3.7 (*see* Fig. 2a–d for examples of titrations and binding curves).

3.4 Binding Reactions for Competition Assays

A typical competition assay includes both specific and nonspecific competitors (Fig. 2). Firstly competition is tested by the use of cold competitor DNA identical to the labelled specific DNA. This DNA competes very effectively (data not shown) while in contrast, DNA with a mutated binding site fails to compete (Fig. 2f) showing its reduced binding affinity. Using this method alternative binding sites can be quickly compared. Examples shown in Fig. 2e, g and h show alternative binding sites that compete as efficiently as the labelled specific site, consistent with these sites having similarly high affinity for the protein of interest.

1. Mix 0.1 nM ^{33}P-labelled DNA in 1 × binding buffer either with no competitor or with a range of cold (unlabelled) competitor DNA at 5–100 nM giving 50–1,000-fold competition (*see* Fig. 2e–h).
2. Add protein to a final concentration appropriate to just give 100 % binding in the absence of competitor (*see* **Note 18**), and incubate for the appropriate time and at the appropriate temperature as determined in the titration assay.
3. Load 18 μL of each 20 μL sample onto the gel, then run and dry the gel before PhosphorImager analysis as described in Subheadings 3.5 and 3.6.

3.5 Polyacrylamide Gel Electrophoresis

1. Once the gel has set, remove the comb and wash out the wells with deionized water using a wash bottle (*see* **Note 19**).
2. Assemble the gel cassette in the electrophoresis tank and add running buffer to the lower chamber first in order to allow removal of air bubbles trapped below the gel which would otherwise cause disruption of the gel front (*see* **Note 20**). After removal of air bubbles add running buffer to fill the top gel chamber.
3. Load 18 μL from the 20 μL binding reactions prepared as described in Subheading 3.3 (*see* **Note 21**) and load 5 μL sample buffer to the end lane to visualize the progress of the electrophoresis using the bromophenol blue marker (*see* **Notes 5** and **22**).

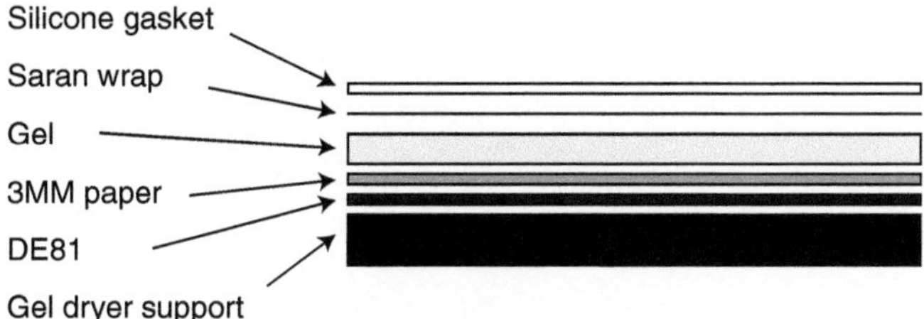

Fig. 1 Assembly of gel on the vacuum gel dryer. Following polyacrylamide gel electrophoresis to separate bound and free DNA, the gel is dried onto Whatman 3MM paper, underneath which is placed a layer of DE81 paper to prevent contamination of the gel dryer. 2 mm gels are dried under vacuum for 2 h at 80 °C

4. Run the gel at 40 mA constant current for a 2 mm gel (*see* **Note 23**) until the bromophenol blue of the sample buffer has migrated to an appropriate position to give good separation of DNA–protein complex from the free DNA (*see* **Note 22**).
5. During the gel run prepare a square of DE81 paper and a square of Whatman 3MM paper, each 20 cm × 20 cm. Place the DE81 paper on the gel drier surface and on top of this place the square of Whatman paper (*see* **Note 24** and Fig. 1).
6. At the end of the run prise the gel plates open using the "tail" of a plastic pen lid (*see* **Note 25**). The gel will remain on the notched plate. Use the pen lid to score along the edges of the gel abutting the spacers. Invert the plate onto the Whatman paper on the gel drier prepared as above, and transfer the gel to the paper by pressing the back of the gel plate. Cover the gel with Saran wrap, cover with the silicone seal of the gel dryer and apply vacuum to dry (*see* Fig. 1). Dry the gel at 80 °C for 2 h.

3.6 PhosphorImage Analysis

1. At the end of the drying period (*see* **Note 26**), place the gel in a cassette with a phosphor storage screen and expose the gel to the screen for 3–6 h depending on the intensity of the bands (*see* **Note 27**).
2. At the end of the exposure time scan the screen using a PhosphorImager and assess the resultant image to determine whether alternative exposure times are necessary. Typical titrations are shown in Fig. 2a, b.

3.7 Analysis of Binding Data to Determine Dissociation Constants Kds

1. Binding constants are determined by titrating limited DNA with protein at a range of concentrations (as described in Subheading 3.3, **step 1**) such that the DNA is partly or wholly shifted in order to generate data for plotting.
2. The percentage of bound DNA is assessed by measuring the depletion of free DNA using ImageQuant software to analyze the PhosphorImager results (*see* **Note 28**).

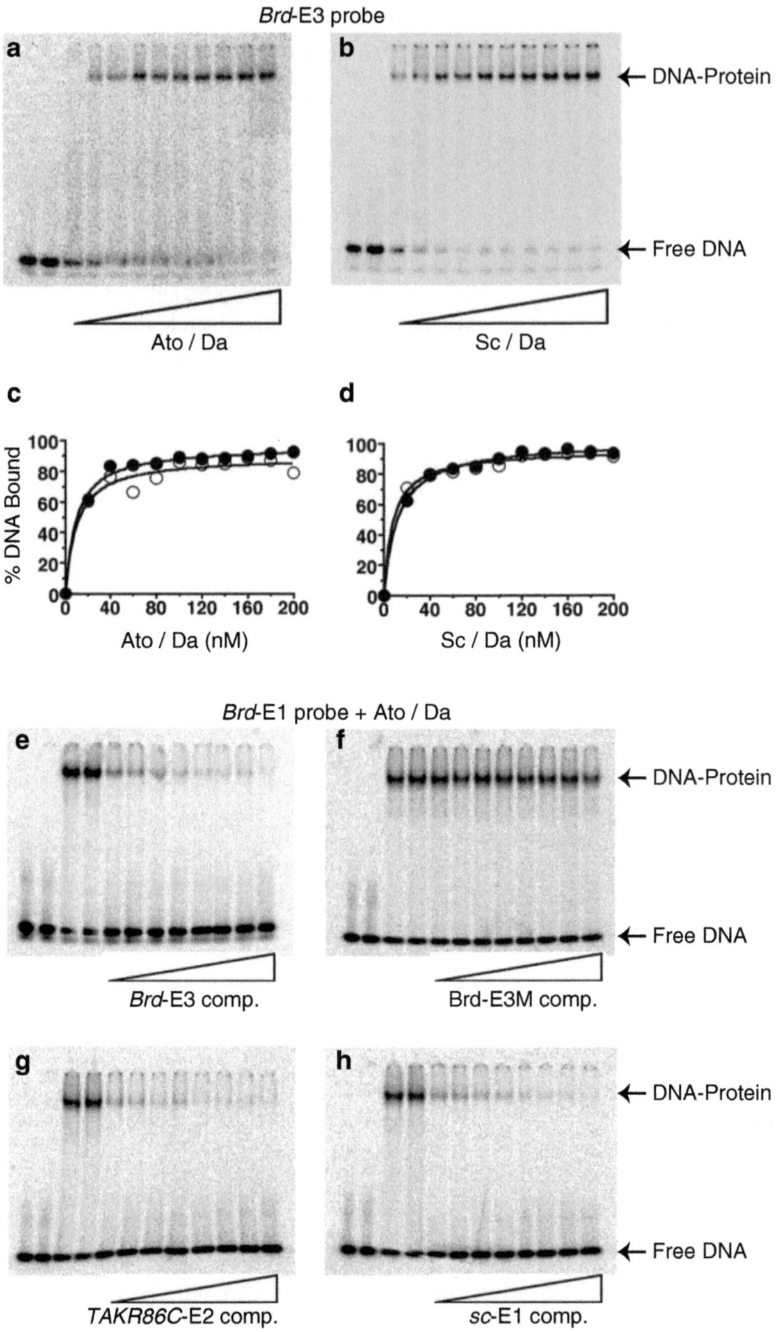

Fig. 2 Use of the band-shift assay to assess relative DNA binding affinities of *Drosophila* proneural proteins for different E-box binding sites using titration and competition assays [4]. In vitro DNA binding analysis for the proneural protein—Daughterless (Da) heterodimers Atonal/Da (Ato/Da) and Scute/Da (Sc/Da) binding to E-box recognition sites using band-shift assays [4]. (**a**–**d**) Relative DNA binding affinities for the DNA binding sequences Brd-E1 (Sc-specific in vivo) and Brd-E3 (Ato-specific in vivo) were assessed by titration. (**a**) Representative titration gel for Brd-E3 with Ato/Da. (**b**) Representative titration gel for Brd-E3 with Sc/Da.

3. A plot of percentage bound DNA versus protein concentration is fitted using a single site binding equation using a package such as Prism 4 (GraphPad) to give apparent dissociation constants (Kds) (*see* **Note 29**). Examples of fitted plots are shown in Fig. 2c, d.

3.8 Protocol for Excision of Bands from Preparative PAGE Gels for Use in SDS-PAGE for the Analysis of Protein Composition in Complexes of Varying Electrophoretic Mobility

1. Prepare binding reactions resulting in multiple bound species under conditions determined as described in Subheadings 3.3 and 3.7 (e.g., *see* Fig. 3a).
2. After electrophoresis retain the wet gel on the notched plate and wrap it in cling film.
3. Affix fluorescent marker tape onto the cling film to allow later alignment of the gel with X-ray film.
4. Carry out autoradiography of the wet gel to determine the positions of the bound complexes within the gel. Make two exposures—one to retain and one to make a template for excision of the wet band from the gel (*see* **Note 30**).
5. Using a scalpel, remove a square of the film corresponding to the band of interest to make a template for alignment with the preparative PAGE gel.
6. Align the film with the gel by using the fluorescent marker tape, and excise a gel slice corresponding to the appropriate band.
7. Elute the DNA–protein complexes from the gel slice by incubating the slice in 300 μL SDS-PAGE sample buffer containing protease inhibitors on a rotating wheel overnight at 4 °C.

Fig. 2 (continued) (**c**) Binding curves of 0.1 nM Brd-E1 DNA probe (*open circles*) and 0.1 nM Brd-E3 probe (*closed circles*) with 0–200 nM Ato/Da (in 20 nM increments) in 1× binding buffer described in Subheading 2.3, **item 3**. Percentage DNA bound was deduced from measurement of free DNA as described in Subheading 3.7, **step 2**. (**d**) Similar plot for Sc/Da. Apparent Kds were determined by fitting the data to a one-site binding equation using Prism 4 (GraphPad). The titration assays showed that the two proneural proteins made no distinction between the DNA sequences analyzed. The 4 combinations tested yielded Kds of ~10 nM in each case. (**e–h**) Competition band-shift assays with Ato/Da binding to radiolabelled 0.1 nM Brd-E1 competed with 0, 5, 10, 15, 20, 30, 40, 50, and 100 nM cold competitor DNA (50–1,000-fold competition). In each case, lanes 1 and 2 include no protein, lanes 3 and 4 include 150 nM Ato/Da and no competitor, and lanes 5–12 include 150 nM Ato/Da and increasing amounts of competitor DNA as indicated. The mutated binding site Brd-E3 M failed to compete for protein binding (**f**), but the other E-boxes (**g**) and (**h**) competed with equal efficiency to BrdE3 (**e**) suggesting similar binding affinities. For additional details *see* [4] (reproduced from ref. 4 with permission from the American Society for Microbiology)

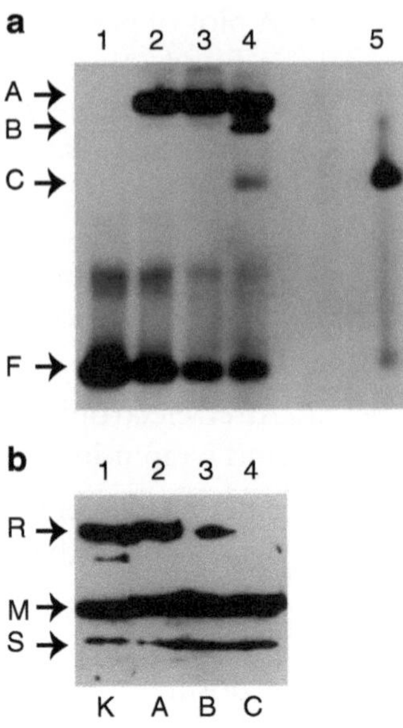

Fig. 3 Identification of protein subunits in multiple complexes found by band shifts of unmethylated DNA by *Eco*KI type I restriction endonuclease [12]. The *Eco*KI restriction enzyme has two restriction (R) subunits, two modification (M) subunits and one specificity (S) subunit ($R_2M_2S_1$), whereas, the *Eco*KI methyltransferase lacks the R subunits (M_2S_1). Both the restriction enzyme and the methyltransferase as well as an $R_1M_2S_1$ intermediate bind to 45 bp duplexes containing the *Eco*KI recognition site AAC(N_6)GTGC [12]. (**a**) Binding of unmethylated DNA by *Eco*KI restriction endonuclease and methyltransferase. Unmethylated specific 45 bp DNA (315 nM) and *Eco*KI nuclease (342 nM; lanes *2–4*) or methyltransferase (2M and 1S subunit) (342 nM; lane *5*) were incubated in 20 mM Tris–HCl (pH 8.0), 100 mM NaCl, 6 mM $MgCl_2$, 7 mM β-mercaptoethanol, 5 % glycerol for 20 min at 22 °C prior to running for 2 h on a 5 % non-denaturing polyacrylamide gel. AdoMet was at 100 mM and ATP at 2 mM, if present. Lane *1*: free DNA; lane *2*: DNA + *Eco*KI, no cofactors; lane *3*: DNA + *Eco*KI + AdoMet; lane *4*: DNA + *Eco*KI + AdoMet + ATP; lane *5*: DNA + methyltransferase + AdoMet + ATP. (**b**) Western blot of SDS-PAGE of protein eluted from DNA–protein complexes in non-denaturing gel. Protein was eluted from complexes of the type labelled A, B, and C in the non-denaturing gel shown in (**a**) and run on a 7.5 % SDS-PAGE gel. This gel was blotted onto a PVDF membrane and the blot developed using a chemiluminescence horse radish peroxidase system and antiserum raised against purified *Eco*KI protein. Lane *1*: purified *Eco*KI; lane *2*: protein eluted from complex A; lane *3*: protein eluted from complex B; lane *4*: protein eluted from complex C. Complex C lacks R subunits and complex B contains significantly less R subunit than complex A, consistent with complexes A, B, and C corresponding to $R_2M_2S_1$, $R_1M_2S_1$ and M_2S_1 as shown in Fig. 4 (reproduced from ref. 12 with permission from Elsevier)

8. Pipette eluates into fresh microcentrifuge tubes and boil the samples for 5 min. Run on an SDS-PAGE gel (as described in ref. 18) and subject to Western blotting (as described in ref. 19) with appropriate antibodies to detect the proteins of interest. An example of such an experiment is shown in Fig. 3.

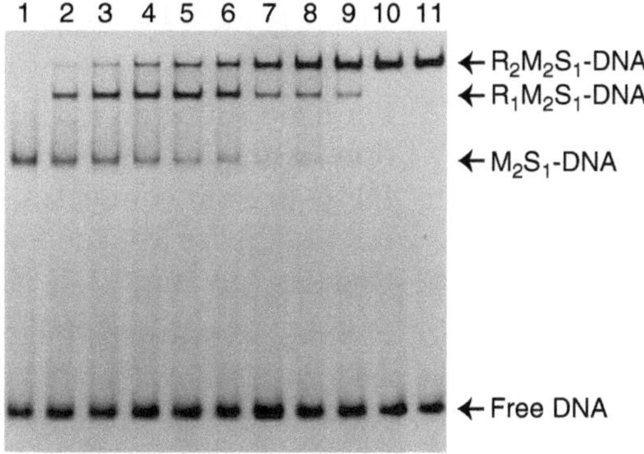

Fig. 4 Use of the band-shift assay to assess reconstitution of a multisubunit DNA–binding protein complex. The band-shift assay can be used to monitor the assembly of multisubunit DNA–binding protein complexes. For example, the *Eco*KI restriction enzyme ($R_2M_2S_1$) can be assembled in vitro by mixing purified R subunit with the *Eco*KI methyltransferase (M_2S_1 [12]). R subunit and methyltransferase (300 nM) were mixed in varying ratios and incubated for 10 min on ice prior to the addition of DNA. Then, unmethylated *Eco*KI-specific 45 bp DNA was added and incubated for 10 min at 22 °C before loading onto a 5 % polyacrylamide non-denaturing gel and running for 2 h at 40 mA. Lanes *1–11*: methyltransferase (300 nM), DNA (300 nM) and increasing amounts of R subunit to give R:methyltransferase ratios of 0:1, 0.25:1, 0.5:1, 0.75:1, 1.0:1, 1.25:1, 1.5:1, 1.75:1, 2.0:1, 5:1, and 10:1. The DNA complexes with M_2S_1, $R_1M_2S_1$ and $R_2M_2S_1$ proteins, which migrate progressively more slowly with size, are indicated (reproduced from ref. 12 with permission from Elsevier)

3.9 Use of the Band-Shift Assay to Monitor the Assembly of Multisubunit Protein Complexes

1. To assemble protein complexes, combine the relevant subunits in appropriate binding buffer and allow the reactions to equilibrate on ice.
2. Add radiolabelled DNA and allow it to bind to the protein complexes.
3. Load the samples onto a non-denaturing gel. Run and process the gel as described in Subheadings 3.4 and 3.5 to detect the various complexes, as shown in the example in Fig. 4.

4 Notes

1. 1×TBE or TAE may also be used. The low ionic strength of the binding buffer helps to stabilize transient interactions. The buffer used in the gel must be the same concentration and composition as the running buffer.
2. Unpolymerized acrylamide is a potent neurotoxin so appropriate protective measures must be taken when handling solutions containing it.
3. 10 % ammonium persulfate is best prepared fresh each time.

4. TEMED must be dispensed in a fume hood due to its noxious fumes.
5. For band shifts, do not add sample buffer to the binding reaction, as the bromophenol blue can result in the disruption of DNA–protein complexes. Include glycerol in incubation mixture to level of 5 % and load sample buffer alone in an end well to assess run length.
6. Hybridized synthetic oligonucleotides (20–45 bp) focusing on a single binding site are generally used for ease of analysis of a particular DNA–protein interaction. (Minimal binding sites can be ascertained by trying a range of fragment sizes in the band-shift assay [3].) However, restriction or PCR DNA fragments up to ~300 bp can also be used and end-labelling can either be as described here, using polynucleotide kinase, or *E. coli* DNA polymerase Klenow fragment can be used for filling in of a 5′ overhang of a restriction fragment [18]. Purity of oligonucleotides can be an issue as failure sequences can be present. HPLC purification is recommended to remove such impurities.
7. $\gamma^{33}P$ is preferred over $\gamma^{32}P$ for two reasons. Firstly, it is lower energy than ^{32}P; hence it is less hazardous, and secondly it produces sharper bands on X-ray film or in PhosphorImager results. (However, ^{32}P can be advantageous if sensitivity is paramount.) Due to their hazardous nature, appropriate procedures for safe handling of radioisotopes must be followed. Alternative nonradioactive labelling methods including the use of fluorescent tags on both DNA and protein have also been described [20, 21].
8. Optimal pH and ionic strength for the DNA–protein interaction under observation must be determined empirically (*see* Subheading 1).
9. Presiliconized tubes are used for the DNA binding reactions to avoid loss of protein by adherence to tube surfaces.
10. Labelling of one strand of the oligonucleotide duplex prior to hybridization to a slight molar excess of the complementary strand is recommended to prevent the appearance of labelled single-stranded DNA on the gel which may impair quantification of the free duplex DNA for determination of binding constants as described in Subheading 3.7, **step 2** (*see* **Note 28**).
11. More efficient hybridization of the two DNA strands can be achieved by stepwise cooling than by the alternative method of boiling in a large volume of water and leaving to cool.
12. The glass plates are cleaned thoroughly with water followed by ethanol. The use of detergent is not recommended, but if used must be rinsed thoroughly to prevent disruption of DNA binding.

13. Polyacrylamide concentrations between 4 and 8 % are typically used. Lower percentage gels with a lower acrylamide to bisacrylamide ratio may be needed for larger complexes. Alternatively, for very large complexes agarose gels may be used (*see* Chapter 17).
14. Insertion of the comb at an angle, e.g., left side first before pushing into the gel mix helps to avoid air bubbles.
15. Gels may be stored at 4 °C overnight if wrapped in tissue dampened with 0.5× TBE and cling film to prevent drying out.
16. Pilot experiments including 10 nM, 100 nM, and 1 μM protein provide a good starting point to determine the range of protein concentrations needed to bind to the DNA of interest. Typical dissociation constants for a specific DNA–protein interaction are generally in the low nM region; however, the fraction of active protein within the purified pool may be relatively low necessitating the use of higher concentrations of protein. The contribution of the protein storage buffer to the glycerol concentration and ionic strength must be considered. For example, many purified proteins are stored in buffers with a high concentration of glycerol, so the glycerol in the binding buffer must be reduced so that the final glycerol concentration is 5 %, as DNA–protein interactions can be sensitive to increased glycerol concentration. (Conversely glycerol has been found to stabilize some DNA–protein complexes [13].)
17. Optimal binding temperature and incubation times vary between different DNA–protein complexes and must be determined empirically.
18. Prior titration analysis to determine the minimum protein concentration necessary to give 100 % binding is essential to prevent oversaturation of the complex with excess protein, which would reduce the efficiency of the competitor.
19. Washing the wells removes residual unpolymerized acrylamide which would otherwise impede the progress of the DNA–protein complexes into the gel.
20. Removal of air bubbles from beneath the gel can be done by tilting the gel tank or alternatively by using a syringe and bent needle.
21. Loading of the samples onto the gel must be done with care to avoid sample dilution by mixing with the running buffer in the wells. The glycerol within the binding buffer should ensure that the samples settle down to the bottom of the well if loaded with care.
22. The length of the gel run needed for separation of DNA–protein complexes from free DNA will vary depending upon the system.

It should be noted that bromophenol blue co-migrates with DNA of approximately 65 base pairs in a 5 % gel. If larger fragments of DNA are to be tested, it may be useful to include xanthene cyanol in the PAGE sample buffer as this co-migrates with DNA of 260 bp in the same conditions [8].

23. Gels can be run singly or in pairs. Use 40 mA constant current for one and 80 mA for two 2 mm gels. The bromophenol blue will reach the bottom of the gel within 2 h at this current. Shorter runs may be sufficient for adequate separation of free and bound DNA. The optimum length of migration must be determined empirically for the DNA–protein interaction of interest. To ensure the free DNA remains in the gel use bromophenol blue to estimate the relative migration of free DNA (*see* **Note 22**).
24. DE81 paper is positively charged and placed under the Whatman paper onto which the gel is dried. It prevents the gel dryer from becoming radioactive by trapping any DNA (negatively charged) which might leach from the gel during drying (*see* Fig. 1).
25. Use of a plastic pen lid (specially designated for this purpose) avoids damage to the glass plates that could be inflicted by the use of a metal spatula.
26. It is essential to check that the gel is completely dry. Press the edge of the gel with a gloved hand to ensure it is not tacky.
27. Alternatively the gel is placed in an autoradiography cassette and exposed to X-ray film for 12–24 h before developing.
28. Measurements of depletion of free DNA rather than the increase in DNA–protein complex give a more accurate indication of binding as the former is unaffected by dissociation during the gel run [22].
29. Kds determined are referred to as "apparent dissociation constants" as their values depend on the mode of determination of the protein concentration and the fraction of active protein (*see* also **Note 16**).
30. Two pieces of film can be exposed simultaneously if using ^{32}P.

Acknowledgements

I would like to acknowledge the help of Professors Noreen Murray and Andrew Jarman in whose laboratories this work took place. Thanks are also due to my colleagues in the Murray and Jarman laboratories, in particular David Dryden, Laurie Cooper, and Petra zur Lage for their help and advice with protein purification.

References

1. Garner MM, Revzin A (1981) A gel electrophoresis method for quantifying the binding of proteins to specific DNA regions: application to components of the *Escherichia coli* lactose operon regulatory system. Nucleic Acids Res 9:3047–3060
2. Fried M, Crothers DM (1981) Equilibria and kinetics of lac repressor-operator interactions by polyacrylamide gel electrophoresis. Nucleic Acids Res 9:6505–6525
3. Powell LM, Dryden DT, Willcock DF, Pain RH, Murray NE (1993) DNA recognition by the EcoK methyltransferase. The influence of DNA methylation and the cofactor S-adenosyl-l-methionine. J Mol Biol 234:60–71
4. Powell LM, Zur Lage PI, Prentice DR, Senthinathan B, Jarman AP (2004) The proneural proteins Atonal and Scute regulate neural target genes through different E-box binding sites. Mol Cell Biol 24:9517–9526
5. Dey B, Thukral S, Krishnan S, Chakrobarty M, Gupta S, Manghani C, Rani V (2012) DNA–protein interactions: methods for detection and analysis. Mol Cell Biochem 365:279–299
6. Henriksson-Peltola P, Sehlen W, Haggard-Ljungquist E (2007) Determination of the DNA-binding kinetics of three related but heteroimmune bacteriophage repressors using EMSA and SPR analysis. Nucleic Acids Res 35:3181–3191
7. Buratowski S, Chodosh LA (2001) Mobility shift DNA-binding assay using gel electrophoresis. In: Ausubel FM et al (eds) Current protocols in molecular biology, 36: 12.2.1–12.2.11, John Wiley and Sons inc.
8. Lane D, Prentki P, Chandler M (1992) Use of gel retardation to analyze protein–nucleic acid interactions. Microbiol Rev 56:509–528
9. Jiang D, Jarrett HW, Haskins WE (2009) Methods for proteomic analysis of transcription factors. J Chromatogr 1216:6881–6889
10. zur Lage PI, Powell LM, Prentice DR, McLaughlin P, Jarman AP (2004) EGF receptor signaling triggers recruitment of Drosophila sense organ precursors by stimulating proneural gene autoregulation. Dev Cell 7:687–696
11. Powell LM, Murray NE (1995) S-adenosyl methionine alters the DNA contacts of the EcoKI methyltransferase. Nucleic Acids Res 23:967–974
12. Powell LM, Dryden DT, Murray NE (1998) Sequence-specific DNA binding by EcoKI, a type IA DNA restriction enzyme. J Mol Biol 283:963–976
13. Hellman LM, Fried MG (2007) Electrophoretic mobility shift assay (EMSA) for detecting protein–nucleic acid interactions. Nat Protoc 2:1849–1861
14. Kothinti R, Tabatabai NM, Petering DH (2011) Electrophoretic mobility shift assay of zinc finger proteins: competition for Zn(2+) bound to Sp1 in protocols including EDTA. J Inorg Biochem 105:569–576
15. Chen J, Villanueva N, Rould MA, Morrical SW (2010) Insights into the mechanism of Rad51 recombinase from the structure and properties of a filament interface mutant. Nucleic Acids Res 38:4889–4906
16. Carey MF, Peterson CL, Smale ST (2012) Experimental strategies for the identification of DNA-binding proteins. Cold Spring Harb Protoc 2012:18–33
17. Cutler PE (2004) Protein purification protocols. Methods Mol Biol (Clifton, NJ) 244: 1–496
18. Sambrook J, Russell DW (2001) Molecular cloning: a laboratory manual. Cold Spring Harb Laboratory Press, NewYork
19. Kurien BT, Scofield RH (2009) Introduction to protein blotting. Methods Mol Biol (Clifton, NJ) 536:9–22
20. Forwood JK, Jans DA (2006) Quantitative analysis of DNA-protein interactions using double-labeled native gel electrophoresis and fluorescence-based imaging. Electrophoresis 27:3166–3170
21. Pagano JM, Clingman CC, Ryder SP (2011) Quantitative approaches to monitor protein–nucleic acid interactions using fluorescent probes. RNA 17:14–20
22. Revzin A (1989) Gel electrophoresis assays for DNA–protein interactions. Biotechniques 7: 346–355

Chapter 17

Assaying Cooperativity of Protein–DNA Interactions Using Agarose Gel Electrophoresis

Tanya L. Williams and Daniel L. Levy

Abstract

DNA-binding proteins play essential roles in many cellular processes. Understanding on a molecular level how these proteins interact with their cognate sequences can provide important functional insights. Here, we describe a band shift assay in agarose gel to assess the mode of protein binding to a DNA molecule containing multiple protein-binding sites. The basis for the assay is that protein–DNA complexes display retarded gel electrophoresis mobility, due to their increased molecular weight relative to free DNA. The degree of retardation is higher with increasing numbers of bound protein molecules, thereby allowing resolution of complexes with differing protein–DNA stoichiometries. The DNA is radiolabeled to allow for visualization of both unbound DNA and all the different DNA–protein complexes. We present a quantitative analysis to determine whether protein binding to multiple sites within the same DNA molecule is independent or cooperative.

Key words DNA-binding proteins, Agarose gel electrophoresis, Band shift, Cooperative binding, Independent binding, DNA-binding site arrays, Dot blot, Dissociation constant, Radiolabeled DNA, DEAE cellulose paper

1 Introduction

DNA-binding proteins are essential for many aspects of cell biology, including chromatin structure [1], DNA replication and repair [2–4], and transcription [5]. Molecular mechanisms and regulation of these processes are often determined by the affinity of specific DNA-binding proteins for their cognate sequences and by their mode of binding. In this chapter, we describe a band shift assay in agarose gels to assess cooperative binding of DNA-binding proteins to tandem arrays of binding sites.

Cooperative protein binding can occur when multiple binding sites are present within the same stretch of DNA. Cooperativity is indicated when binding of one protein promotes the binding of proteins at neighboring sites, decreasing their apparent K_d.

Svetlana Makovets (ed.), *DNA Electrophoresis: Methods and Protocols*, Methods in Molecular Biology, vol. 1054, DOI 10.1007/978-1-62703-565-1_17, © Springer Science+Business Media New York 2013

Binding of the same protein [2, 6] or different proteins [5] can exhibit cooperativity. Independent binding occurs when protein binding at each individual site is dictated only by its affinity for the DNA-binding site and not by proteins bound at neighboring sites [3]. Non-independent binding occurs when protein bound at one site influences protein binding at neighboring sites [7], increasing the apparent K_d in the case of competitive binding or decreasing the apparent K_d in the case of cooperative binding. Such binding modes can be critically important for cellular function. Band shift assays provide a powerful in vitro approach to assess cooperativity.

In order to perform quantitative band shift experiments, one must first determine the affinity of the DNA-binding protein of interest for its cognate site. We describe a filter binding assay in Subheading 3.2 for measuring this K_d, using radiolabeled DNA (Subheading 3.1 describes how to end-label DNA with ^{32}P). Protein–DNA complexes are assembled in vitro with labeled DNA and varying concentrations of protein. These complexes are applied to nitrocellulose, which strongly binds protein and protein–DNA complexes but not free DNA. Protein bound to DNA is visualized as a ^{32}P signal using x-ray film or a phosphorimager. Under the appropriate conditions where the DNA concentration is kept low, the K_d is equal to the protein concentration where DNA binding is half maximal [8].

To assess cooperativity, band shift assays in agarose are performed as described in Subheading 3.3. Again, radiolabeled DNA is utilized (Subheading 3.1), in this case DNA containing tandem arrays of protein-binding sites. Protein–DNA complexes are assembled in vitro and separated by agarose gel electrophoresis. Free DNA migrates rapidly in the gel while protein–DNA complexes are retarded due to their increased molecular weight. When it is necessary to resolve multiple high molecular weight protein–DNA complexes, agarose gels are preferred to polyacrylamide gels, which are generally only useful for assaying single protein-binding events to DNA. Agarose gel electrophoresis conditions are established such that complexes containing variable numbers of bound proteins can be resolved and relative amounts of each bound state can be quantified upon drying the gel onto ion-exchange paper. A sufficiently high DNA concentration is used to ensure that all sites compete for protein binding, and the protein concentration is varied but kept below or equal to the total number of DNA recognition sites in the reaction. Assessment of cooperative versus independent binding comes from a quantitative comparison of the observed distribution of protein among the DNA molecules to the distribution predicted by random assortment (*see* **Note 1**).

2 Materials

2.1 DNA Radiolabeling

1. Plasmids containing single-binding sites and tandem arrays of sites: these plasmids can be generated through standard molecular biology approaches [9] and/or using a reiterative cloning strategy [10].
2. Restriction enzymes to cut out DNA-binding arrays from the plasmids.
3. Qiagen Gel Extraction Kit or equivalent.
4. Shrimp alkaline phosphatase (Roche).
5. T4 polynucleotide kinase (New England Biolabs).
6. (γ)-^{32}P-ATP (*see* **Note 2**).
7. ProbeQuant G-50 micro spin columns (GE Healthcare).
8. Thermal cycler.
9. 5 M NaCl: Dissolve 292 g of NaCl in water to a final volume of 1 L, autoclave, and store at room temperature.

2.2 Determining the K_d of Protein–DNA Binding

1. Purified protein of interest (*see* **Note 3**).
2. Radiolabeled DNA containing a single-binding site (*see* Subheading 2.1).
3. Buffer A (5×): 100 mM HEPES pH7.9, 750 mM NaCl, 5 mM dithiothreitol (DTT), 50 % glycerol, 25 mM $MgCl_2$. Dissolve 11.9 g HEPES, 21.9 g NaCl, and 1.2 g $MgCl_2$ in 200 mL of water. Adjust pH to 7.9 with 10 N NaOH. Adjust volume to 250 mL. Add 250 mL of glycerol. Filter sterilize and store at 4 °C. Add DTT fresh from 1 M DTT stock on day of use. For instance, to 25 mL of 5 × buffer A, add 125 μL of 1 M DTT.
4. 1 M DTT: Dissolve 1.54 g DTT in enough water to make 10 mL of solution. Filter sterilize and store 1 mL aliquots at −20 °C.
5. 10 mg/mL bovine serum albumin (BSA)—New England Biolabs. The stock solution can also be prepared in water from fraction V protease-free BSA (Sigma A3059) and stored in aliquots at −20 °C.
6. 1 mg/mL poly(dI-dC): Dissolve poly(dI-dC) sodium salt (Fluka/Biochemika) in water at 1 mg/mL and store aliquots at −20 °C.
7. 25 °C water bath.
8. Nitrocellulose membrane (Schleicher & Schuell).
9. Minifold I dot blot apparatus (Schleicher & Schuell).
10. Whatman 3MM filter paper.
11. Saran wrap.
12. Autoradiography detection equipment: storage phosphor screen, Typhoon or Storm Biomolecular Imager, and ImageQuant software from GE Healthcare (*see* **Note 4**).

2.3 Analysis of Protein–DNA Interactions by Agarose Gel Mobility Shift Assays

1. Buffer A, BSA, poly(dI-dC), and purified protein of interest (*see* Subheading 2.2).
2. Radiolabeled and unlabeled DNA containing tandem arrays of binding sites (*see* Subheading 2.1).
3. 25 °C water bath.
4. Agarose (we use SeaKem LE agarose).
5. TBE (5×): Dissolve 54 g Tris base, 27.5 g boric acid, and 20 mL 0.5 M EDTA (pH 8.0) in 900 mL water. Take volume up to 1 L. The pH should be ~8.3. Filter sterilize and store at room temperature.
6. 0.5 M EDTA (pH 8.0): Dissolve 186.1 g of disodium EDTA·2H_2O in 800 mL water. Adjust the pH to 8.0 with NaOH (~20 g NaOH pellets). Adjust the volume to 1 L. Autoclave and store at room temperature.
7. Horizontal agarose gel electrophoresis box containing a self-recirculating buffer system, with a casting tray for a 20 cm ×25 cm gel (Thermo Scientific) (*see* **Note 5**).
8. Electrophoresis power source.
9. Gel-loading dye: 10 % glycerol, 0.1 % bromophenol blue (BPB). Dissolve 0.1 g BPB in 90 mL water and 10 mL glycerol.
10. DE81 DEAE cellulose chromatography paper (Whatman).
11. Paper towels.
12. Filter paper (Whatman).
13. Saran wrap.
14. Cover plate—a flat glass or metal plate approximately the size of the gel.
15. Weights.
16. Vacuum gel dryer.
17. Autoradiography detection equipment: storage phosphor screen, Typhoon or Storm Biomolecular Imager, and ImageQuant software from GE Healthcare (*see* **Note 4**).
18. Data analysis software (we use Microsoft Excel).

3 Methods

Radiolabeled DNA is required both for measuring the protein affinity for a single DNA-binding site and for determining cooperative binding to multiple sites.

3.1 DNA Radiolabeling

1. Isolate DNA fragments containing binding sites of interest. Digest plasmids with appropriate restriction enzymes and separate the fragments by standard agarose gel electrophoresis.

Excise and purify the DNA fragments using Qiagen Gel Extraction Kit or similar method.

2. Dephosphorylate the purified DNA fragments with shrimp alkaline phosphatase (SAP) at 37 °C for 15 min and then inactivate the enzyme at 65 °C for 15 min. A sample reaction might contain 50 ng DNA fragment, 1 μL 10× dephosphorylation buffer, 1 unit SAP, and water to a final volume of 10 μL (*see* **Note 6**).
3. End-label the DNA fragments with ^{32}P using T4 polynucleotide kinase (PNK) at 37 °C for 30–60 min (*see* **Note 7**). A sample reaction might contain 10 μL dephosphorylated DNA fragment from **step 2**, 2 μL 10×T4 PNK buffer, 20 pmol (γ)-^{32}P-ATP, 10 units T4 PNK, and water to a final volume of 20 μL (*see* **Note 8**).
4. Remove unincorporated nucleotides using a ProbeQuant G-50 micro spin column. Briefly, centrifuge the column at 735 ×*g* for 1 min, place the column in a clean 1.5 mL centrifuge tube, adjust DNA sample volume to 50 μL with probe buffer type 1 provided with the spin columns (10 mM Tris–HCl, 1 mM EDTA, 150 mM NaCl, pH 8.0), apply the sample to the column, and collect the purified DNA by centrifuging at 735 ×*g* for 2 min.
5. To ensure proper annealing of the labeled DNA fragments, heat samples in a PCR block at 95 °C for 5 min, then drop the temperature 1 °C/min to 25 °C (*see* **Note 9**).

3.2 Determining the K_d of Protein–DNA Binding

To determine if protein binding to an array of tandem DNA sites is cooperative, it is essential to determine the affinity of the protein for a single site.

1. Binding reactions are assembled in 1× buffer A plus 0.1 mg/mL BSA and 5 μg/mL dI-dC (*see* **Note 10**). For each binding reaction, prepare a 500 μL aliquot of ^{32}P-labeled DNA with a single-binding site at a concentration at least tenfold below the K_d for interaction (*see* **Note 11**). Also, prepare 250 μL of different protein dilutions on ice (*see* **Note 12**).
2. Initiate binding reactions by mixing DNA and protein aliquots and incubating at 25 °C for 30 min (*see* **Note 13**). Include a control reaction lacking protein.
3. Pre-wet nitrocellulose membrane in water and then equilibrate in buffer A. To assemble the dot blot apparatus, first layer a piece of Whatman 3MM filter paper soaked in buffer A on the base of the apparatus. Then layer the equilibrated nitrocellulose membrane on the filter paper and top with the cover plate.
4. Apply each binding reaction to a separate well under vacuum. Once the buffer has run through, wash each well twice with 500 μL buffer A (*see* **Note 14**).

5. Disassemble the dot blot apparatus, wrap the membrane in Saran wrap, expose it to a storage phosphor screen, and scan the screen using a Typhoon scanner. Quantify the intensity of each dot using ImageQuant (*see* **Note 4**). To correct for background binding of DNA to nitrocellulose, subtract the intensity of the control dot lacking protein from the intensity measurements for all other binding reactions.
6. Plot dot intensity versus input protein concentration. The K_d is estimated from the curve as the protein concentration where DNA binding is half maximal. Multiple experiments should be performed to obtain a standard deviation for each measured K_d.

3.3 Analysis of Protein–DNA Interactions by Agarose Gel Mobility Shift Assays

1. Prepare binding reactions (final volume 15 μL): 1× buffer A, 0.3 mg/mL BSA, 5–10 μg/mL dI-dC, and appropriate concentrations of protein and DNA array fragment (*see* **Note 10**).

   For each reaction, prepare 10 μL aliquots of DNA that contain cold, unlabeled DNA fragment at a concentration at least twofold greater than the K_d for binding a single site multiplied by the number of binding sites (*see* **Note 15**). Add ^{32}P-labeled DNA at 1/100th the concentration of cold DNA to allow for visualization by autoradiography.

   Prepare various dilutions of protein in 5 μL aliquots. Final protein concentrations should vary to accommodate single to full binding site occupancy.
2. Initiate reactions by mixing 10 μL DNA aliquots with 5 μL protein aliquots. Incubate reactions at 25 °C for 30–60 min (*see* **Note 13**).
3. Prepare a 1 % solution of melted agarose in 0.5× TBE and cast a 20 cm by 25 cm gel, using a comb that will form wells sufficient to hold each 15 μL binding reaction. Fill the gel box with 0.5× TBE, and prechill the gel and buffer at 4 °C before loading. Ensure that buffer recirculates between the anode and cathode chambers (*see* **Note 5**).
4. No loading dye is added to the samples. Glycerol in the buffer will ensure the samples settle to the bottom of the well (*see* **Note 16**). Carefully load each sample, one reaction per well. It is useful to run one lane that contains BPB dye in order to track the progress of sample migration in the gel.
5. Run the gel at 4 °C at 10 V/cm for 5–6 h, depending on the size of the DNA fragment, until the BPB dye is a few centimeters from the bottom of the gel (*see* **Notes 17** and **18**).
6. Arrange an 8–10 cm stack of paper towels on a bench top. Place a few sheets of filter paper on top of the paper towels, followed by a sheet of DE81 paper. Carefully slide the gel onto the DE81 paper (*see* **Note 19**). Cover the gel with Saran wrap and a flat

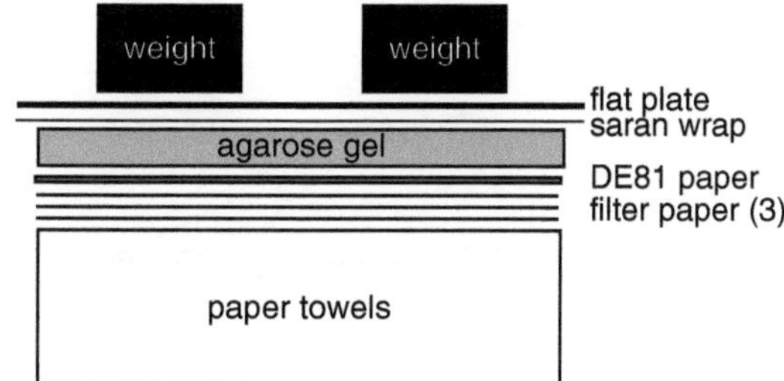

Fig. 1 Assembly for drying agarose gels to DE81 paper. Three sheets of filter paper are layered on top of an 8–10 cm stack of paper towels. A sheet of DE81 paper is placed on the filter paper, and the gel is sandwiched between the DE81 paper and weights. The gel is allowed to adhere and dry to the DE81 paper overnight at room temperature. Subsequently, the gel and DE81 paper are dried under vacuum

plate, such as the tray used to cast the gel. Place weights on top of the tray, such as several 1 L glass bottles with water (Fig. 1).

7. Leave the gel overnight at room temperature. Then dry the gel, adhered to the DE81 paper, under vacuum at 80 °C for 1 h.
8. Expose the dried gel to a storage phosphor screen, and scan the screen on a Typhoon scanner (*see* **Note 20** and Fig. 2b).
9. Quantify band intensities using ImageQuant by drawing a rectangle around each band. Correct for background signal by subtracting the intensity measured for a rectangle of the same size positioned over a blank area of the blot (*see* **Note 4**).
10. Analyze gel shifts to determine if binding is cooperative or independent (*see* **Note 15** and Fig. 2). Binding site sequence, spacing, numbers, and affinity can be varied to examine how such changes influence cooperativity.
11. Modifications of this procedure include protein super shifts (*see* **Note 21**) and competitive release of bound protein (*see* **Note 22**).

4 Notes

1. A thorough quantitative discussion of cooperativity in the assembly of protein–DNA complexes is presented elsewhere [11].
2. Be sure the gamma phosphate is ^{32}P labeled as this is the phosphate that will be transferred to the DNA fragment upon end labeling.
3. Details of protein preparation will depend on the specific protein under study. Factors to consider are that the protein is sufficiently pure to avoid DNA binding by contaminant proteins and that the protein is stable and active in the buffer used for storage and for assaying binding.

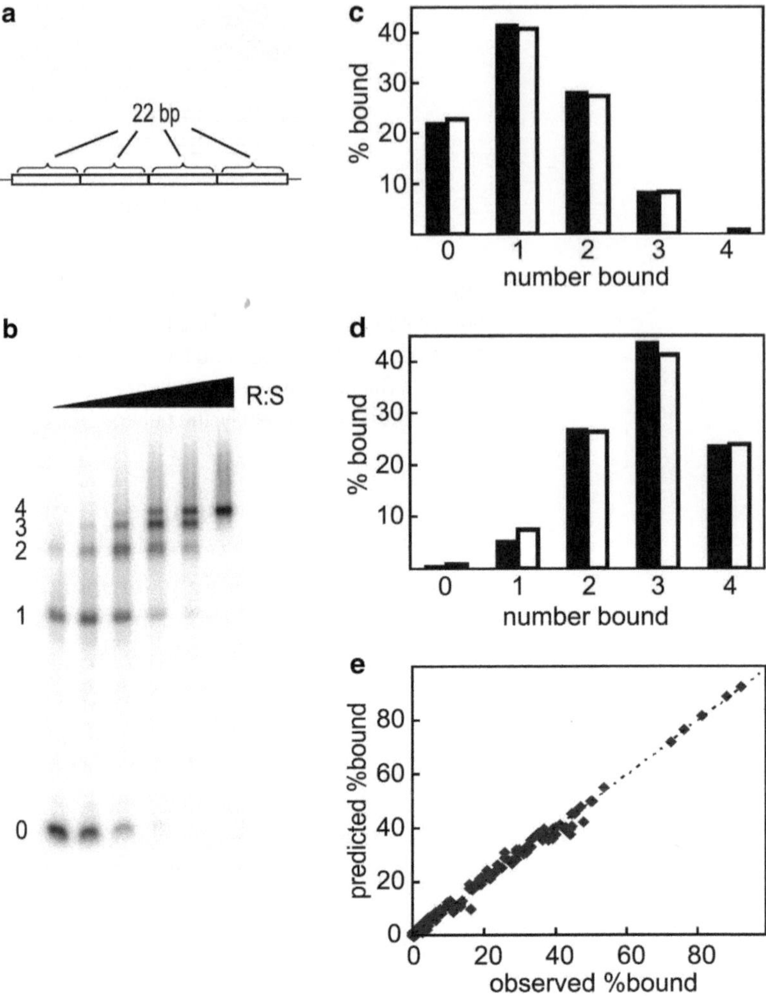

Fig. 2 Example band shift assay in agarose and associated analysis. Here Rap1 protein binds independently to each site within an array of four recognition sites. (**a**) Diagram of the four-site DNA fragment. Each rectangle contains an identical high-affinity Rap1 recognition site; the sequence repeats every 22 bp. (**b**) Gel shifts of Rap1 bound to the DNA fragment depicted in (**a**). The expected ratio of Rap1 to DNA sites (R:S) denoted by the *triangle* is 0.1075, 0.215, 0.43, 0.645, 0.75, and 0.86 from left to right. The numbers to the left of the gel indicate the number of bound Rap1 molecules/DNA molecule (the bound state, i of DR_i). The [DNA] is 50 nM. (**c**, **d**) Examples of the predicted distribution of bound states based on random assortment that best fit the observed distribution of bound states for a particular expected R:S. Shown are best fit predicted (*white bars*) and observed (*black bars*) distributions for the four-site array with 22 bp spacing; distribution comparison for an expected R:S of 0.425 in (**c**) and 0.645 in (**d**) gives best fit R:S values of 0.31 and 0.7, respectively. (**e**) Correlation between the best fit and observed distributions for the four-site array with 22 bp spacing. Each point represents the paired % bound values for each DR_i from the observed (abscissa) and best fit (ordinate) distributions. The data are shown from seven separate experiments with a total of 31 observed distribution sets that range in

4. If a phosphorimager is not available, traditional autoradiography can be performed. The membrane or gel is exposed to autoradiography film in a cassette at −80 °C, using an intensifying screen if necessary. The length of exposure will depend on the strength of the radioactive signal. The film is then developed, a digital version is generated by scanning, and quantitative analysis is performed using free software such as NIH ImageJ.
5. A pH gradient can develop along the length of the gel as it runs. This can lead to melting of the gel and/or "smiling" of the bands. To avoid this problem, buffer is circulated between the anode and cathode chambers. This can be accomplished using a self-recirculating tank with a conduit tube that mixes buffer between the two chambers. Alternatively, buffer can be circulated using a peristaltic pump.
6. Synthesized oligonucleotides can also be used to generate the DNA fragment of interest. To do so, mix equal molar proportions of complementary oligonucleotides encompassing the DNA-binding site in a buffer containing 0.1 M NaCl. Annealing is accomplished by heating the mix in a PCR block at 95 °C for 5 min, then dropping the temperature 1 °C per minute to 25 °C.
7. This reaction should not be overincubated as dephosphorylation will take place. PNK can both phosphorylate and dephosphorylate. Once ATP is exhausted, only dephosphorylation will occur, resulting in loss of the ^{32}P label.
8. Be sure that any users performing this procedure have been trained to work with ^{32}P radioactivity and that careful monitoring is performed throughout the procedure to avoid radioactive contamination.
9. Alternatively, the tube with labeled DNA can be placed in a heat block containing boiling water for 5 min. Then, the block is removed from the heat source and allowed to cool slowly to room temperature. Take care that the tube lid is sealed to avoid

Fig. 2 (continued) R:S from 0.15 to 0.97. Least square analysis indicates close clustering ($R^2 = 0.99$) of the pair-wise DR_i values about the axis of perfect correlation (*dotted line*), which represents independent binding by Rap1 to the sites within the array. Cooperative binding by Rap1 would lead to a spread in the points away from the axis of perfect correlation (and a lower R^2 value), because the DR_i values within the observed and best fit distributions would fail to closely match. This research was originally published in *The Journal of Biological Chemistry*. T. L. Williams, D. L. Levy, S. Maki-Yonekura, K. Yonekura, E. H. Blackburn. Characterization of the yeast telomere nucleoprotein core: Rap1 binds independently to each recognition site. *JBC.* 2010; 285:35814–35824. ©The American Society for Biochemistry and Molecular Biology

contamination of the workspace with radioactivity. We find that this reannealing step helps to generate a more uniform population of dsDNA molecules.

10. When studying a novel protein–DNA interaction, it is important to examine how buffer components affect binding. Factors to consider include the type of buffer and pH, the type and concentration of salt, and the effects of divalent metal cations, reducing agents, detergents, and glycerol. For example, studies of the cooperative binding behavior of single-strand binding protein indicate that the salt and divalent metal ion concentrations are critically important [12]. BSA and dI-dC are included to minimize nonspecific protein–DNA interactions, and it is also important to determine how these components affect binding affinity and behavior.

11. Determining the proper DNA concentration requires some initial exploratory experiments. When necessary, experiments must be repeated with lower and lower DNA concentrations until the condition is met that the concentration of DNA used in the experiment is at least tenfold below the measured K_d. This ensures that binding is driven by the protein concentration and therefore allows an estimation of the K_d from the protein concentration where binding is half maximal. The volumes recommended here are based on the high-affinity interaction of Rap1 binding its cognate DNA sequence, with a measured K_d of 20 pM (3). To measure this interaction, a DNA concentration of ~1 pM was used and the concentration of added Rap1 was varied in twofold steps from 5 to 160 pM. In situations where the K_d is high and/or protein is precious, reaction volumes can be scaled down to conserve protein.

12. This is most easily accomplished through serial dilution of the protein stock. Twofold dilution steps centered around the K_d are ideal. Again, this requires some trial and error as one repeats experiments to approach an accurate measurement of the K_d.

13. When the assay is first being established, perform time courses to ensure that binding has reached equilibrium. It can also be informative to investigate how temperature affects binding affinity and behavior.

14. When setting up the assay, perform trial runs with different numbers of washes to determine how many washes are required to remove most background DNA binding.

15. To determine whether or not the protein of interest binds arrays of sites cooperatively, one first calculates predicted results for independent binding. The frequency (P_i) for obtaining each bound state DP_i (where D is the DNA fragment, P is the bound protein, and i is the number of bound protein molecules)

given random assortment is a function of the relative concentrations of protein and DNA, the number of protein-binding sites per array (n), and the number of ways of obtaining each DP_i. The relationship is shown in Eq. 1.

$$Pi = \left(\left(\frac{[\text{protein}]}{n[\text{DNA}]} \right)^i \left(1 - \frac{[\text{protein}]}{n[\text{DNA}]} \right)^{n-1} \right) \times \left(\frac{n!}{(n-1)!i!} \right) \quad (1)$$

Equation 1 assumes that all protein available in the reaction is bound to DNA. Therefore, binding is ensured by performing the reactions at a DNA concentration well above the K_d for protein binding to the final site in the array. The K_d for binding the final site is equal to the binding affinity for a single site (k) multiplied by the ratio of the number of ways to obtain DP_n and DP_{n-1} [13]. Thus, the affinity of the protein for the last binding event is nk. The concentration of DNA in each binding reaction is therefore kept above the value measured for binding the recognition sequence as a single site (via filter binding) multiplied by the number of sites within the array. Cooperative binding is assessed by comparing the observed distribution of bound states DP_i, for various input ratios of protein to DNA sites (P:S), to the distribution predicted by multisite random assortment calculated from Eq. 1.

Quantifying relative amounts of each bound state is complicated by uncertainties in knowing precisely the protein and DNA concentrations in each binding reaction, and small deviations can significantly affect the results of the band shift assay. To eliminate error due to such uncertainties, a modified approach is suggested to compare the observed frequency of each DP_i with that predicted from Eq. 1. First, the amount of each DP_i species is normalized to the total radiolabeled DNA in each reaction/lane. Then, the frequency of each observed DP_i for a given P:S ratio is systematically compared with the frequency predicted for P:S values varying from 0 to 1. The distribution that best fits the observed data is determined by minimizing the difference between each DP_i in the predicted and observed distributions (i.e., minimize the residual between each DP_i of the observed and predicted distributions by minimizing the sqrt(sum(residual)^2)). The similarity between the best fit and observed distributions is then visualized by plotting the pair-wise values for each DP_i from the best fit and observed distributions. Close clustering of the pair-wise DP_i values about the axis of perfect correlation, determined by least square analysis, indicates independent binding because the observed distribution of each bound state for a given P:S is highly similar to the frequency predicted for each bound state based on random assortment. On the other hand, cooperative binding behavior is expected to increase the frequency of both the higher bound and the unbound states while decreasing the frequency of the lower bound states as compared with random assortment. Such cooperative behavior would lead to a diffusion of the pair-wise DP_i values away from the axis of perfect correlation.

16. If glycerol is absent from the binding buffer used for a specific application, supplement binding reactions with glycerol at a final concentration of 10 % just prior to loading.
17. Run time and temperature can be varied as appropriate for specific protein–DNA complexes.
18. Running buffer may contain radioactivity, so monitor and dispose of buffer accordingly.
19. Avoid trapping air bubbles between the DE81 paper and gel. Carefully smooth out any air bubbles that are present.
20. If bands appear particularly fuzzy, a number of parameters can be optimized to improve band clarity. Components in the binding buffer can be altered; salt concentration in particular can affect the appearance of the bands. The gel can be cast and run in a different concentration of TBE or different kind of buffer such as TAE. Changing the percentage of agarose can also make a difference. Lastly, if conditions that once produced sharp bands begin generating poor-quality gel shifts, preparing fresh DNA and protein will often solve this problem.
21. If it is suspected that additional proteins bind to the DNA–protein complex under study, then super shift experiments can be performed. DNA–protein complexes are assembled in the presence of an additional protein that can be added before or after assembly of the initial protein–DNA complex. Once equilibrium binding is reached, the reaction is loaded onto an agarose gel, as described, and the various bound states are separated by electrophoresis. Binding of the additional protein to the complex is manifested as a super shifted band or bands with retarded mobility relative to the original protein–DNA complex bands.
22. To assay the kinetics of dissociation for a protein–DNA complex, competitive binding can be performed with cold, unlabeled DNA. Protein–DNA complexes are first assembled, as already described. Subsequently, an excess of cold competitor DNA is added and reactions are run on an agarose gel at different time points after addition of the competing DNA. Complex dissociation is manifested by loss of shifted bands, as dissociated protein binds competing DNA that is not visible by autoradiography. The release time is indicative of complex stability [3].

Acknowledgements

We thank David M. Freed and members of the Blackburn and Stroud labs for the helpful advice and discussions. T.L.W. was supported by a postdoctoral National Research Service Award fellowship. D.L.L. was supported by a predoctoral fellowship from the Howard Hughes Medical Institute.

References

1. Zlatanova J, Yaneva J (1991) Histone H1–DNA interactions and their relation to chromatin structure and function. DNA Cell Biol 10:239–248
2. Lohman TM, Ferrari ME (1994) *Escherichia coli* single-stranded DNA-binding protein: multiple DNA-binding modes and cooperativities. Annu Rev Biochem 63:527–570
3. Williams TL, Levy DL, Maki-Yonekura S, Yonekura K, Blackburn EH (2010) Characterization of the yeast telomere nucleoprotein core: Rap1 binds independently to each recognition site. J Biol Chem 285:35814–35824
4. Takahashi M, Norden B (1994) Structure of RecA–DNA complex and mechanism of DNA strand exchange reaction in homologous recombination. Adv Biophys 30:1–35
5. Zaret KS, Carroll JS (2011) Pioneer transcription factors: establishing competence for gene expression. Genes Dev 25:2227–2241
6. Ptashne M (2004) A genetic switch: phage lambda revisited, 3rd edn. Cold Spring Harbor Laboratory Press, Cold Spring Harbor, NY
7. Stauffer ME, Chazin WJ (2004) Structural mechanisms of DNA replication, repair, and recombination. J Biol Chem 279:30915–30918
8. Nelson DL, Cox MM (2004) Protein function. In: Lehninger principles of biochemistry, 4th edn. Worth Publishers, New York, pp 157–189, Chapter 5
9. Sambrook J, Russell DW (2001) Molecular cloning: a laboratory manual, 3rd edn. Cold Spring Harbor Laboratory Press, Cold Spring Harbor, NY
10. Pierce JC (1994) In-frame cloning of large synthetic genes using moderate-size oligonucleotides. Biotechniques 16:708–715
11. Senear DF, Ross JB, Laue TM (1998) Analysis of protein and DNA-mediated contributions to cooperative assembly of protein–DNA complexes. Methods 16:3–20
12. von Hippel PH (2007) From "simple" DNA–protein interactions to the macromolecular machines of gene expression. Annu Rev Biophys Biomol Struct 36:79–105
13. Cantor C, Schimmel P (1980) Biophysical chemistry part III: the behavior of biological macromolecules. W. H. Freeman and Company, New York, pp 849–886

Chapter 18

DNA Bending by Proteins: Utilizing Plasmid pBendAT as a Tool

Fenfei Leng

Abstract

Protein-induced DNA bending plays a key role in many essential biological processes, such as DNA replication, recombination, and transcription, and can be analyzed by a variety of biochemical and biophysical methods, such as electrophoresis mobility shift assay (EMSA), X-ray crystallography, nuclear magnetic resonance (NMR), and DNA ring closure assay. In this chapter, I will provide a detailed protocol for studying protein-induced DNA bending by utilizing the plasmid pBendAT. pBendAT carries a 230 bp DNA segment containing five pairs of restriction–endonuclease recognition sites and can be used to produce a set of five DNA fragments of identical length, with each fragment having a different positioning of a protein-binding site. Binding of a protein to this site will divide a DNA fragment into two DNA segments. If protein binding leads to DNA bending, then the two DNA segments will be at an angle to each other and the distance between the DNA ends will shorten. As a result, the gel mobility of the protein–DNA complex will be affected as the mobility of a rigid DNA fragment is inversely proportional to the end-to-end distance. Therefore, by analyzing how the position of protein binding within a fragment affects the gel mobility of the protein–DNA complexes, we are able to determine the DNA bending angle and the location of the bend. The DNA fragments of identical length can also be conveniently generated by PCR amplification using pBendAT as the DNA template. Since the 230 bp DNA fragment of pBendAT does not contain more than two consecutive AT base pairs, pBendAT is particularly suitable for the assessment of DNA bending induced by proteins recognizing AT-rich DNA sequences cloned in the 230 bp DNA fragment.

Key words DNA bending, DNA-binding proteins, Electrophoresis mobility shift assay (EMSA), Circular permutation assay, Polymerase chain reaction

1 Introduction

Protein-induced DNA bending plays an important role in DNA replication, recombination, and transcription [1–3]. For instance, the lactose repressor (LacI), a transcription factor controlling metabolism of lactose in *E. coli* and other bacteria [4], sharply bends its DNA-binding site, the *lacO1* operator, which plays a key role in the regulation of expression of β-galactosidase (LacZ), β-galactoside permease (LacY), and β-galactoside transacetylase (LacA) in the *lac* operon [5]. Another well-characterized transcription

Svetlana Makovets (ed.), *DNA Electrophoresis: Methods and Protocols*, Methods in Molecular Biology, vol. 1054, DOI 10.1007/978-1-62703-565-1_18, © Springer Science+Business Media New York 2013

factor that sharply bends its binding site is the *E. coli* cAMP receptor protein (CRP; another name is catabolite gene activator protein – CAP). CRP tightly associates with its binding sites in the promoter regions and regulates transcription of many genes in bacteria [6, 7]. The DNA bending angles for both proteins were estimated to be ~90° by using a variety of different methods, such as EMSA, X-ray crystallography, NMR, and DNA ring closure assay. [8, 9]. Among the methods suitable for the analysis of protein-induced DNA bending, EMSA (or circular permutation assay) is the most commonly used technique to identify a DNA bend, determine the bending site, and estimate the bending angle [10, 11]. If a DNA fragment is smaller than the persistence length, the length over which the direction of a DNA segment persists, the DNA fragment can be used to analyze DNA bending by EMSA. The principle of the EMSA-based, DNA-bending assay stems from the fact that a protein–DNA complex with a bend in the middle of the DNA fragment migrates in a polyacrylamide gel slower than the one with a bend closer to one of the DNA ends [9, 12]. Although a rigorously derived equation is not available to calculate protein-induced DNA bending angles, the empirical equation

$$\cos\frac{\alpha}{2} = \frac{\mu_{\mathrm{M}}}{\mu_{\mathrm{E}}}$$

can be used to estimate the bending angles from EMSA for rigid DNA fragments, where α is the DNA bending angle and μ_{M} and μ_{E} are the mobility of the protein–DNA complexes at the center and the end of the DNA fragment, respectively. As pointed out by Thompson and Landy [12] and also by Kim et al. [9], the mobility of a rigid DNA fragment is inversely proportional to the end-to-end distance. In this case, the mentioned-above equation can be easily derived (Fig. 1).

In 1989, Adhya and coworkers constructed a series of pBend vectors to facilitate the study of protein-induced DNA bending by using circular permutation assays [9]. Among them was pBend2 which could be used to generate a large number of DNA fragments of identical length with different positioning of a protein-binding site. A detailed procedure for using pBend2 and other pBend vectors has been described in the original paper [9] and has also been reviewed recently [13]. In this chapter, I will focus on presenting a detailed protocol for protein-induced DNA-bending assays using a pBend2 derivative, pBendAT, and using PCR to generate a large number of DNA fragments of identical length with protein-binding sites at variable positions within the fragments. pBendAT (Fig. 2; [14]) was derived from pBend2 and carries a 230 bp DNA insert containing five pairs of restriction–endonuclease recognition sites, which can be used to produce a set of DNA fragments of identical length but different in the positioning of the

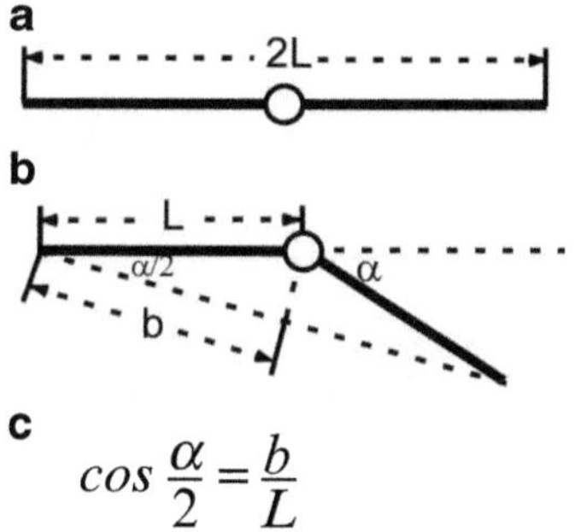

Fig. 1 The DNA bending angle is inversely proportional to the end-to-end distance of a rigid DNA molecule. (**a**) A linear, rigid DNA molecule containing a DNA-binding site in the center of the molecule. (**b**) A rigid DNA molecule with a bend in the center. (**c**) The bending angle (α) is related to the end-to-end distance of the DNA molecule. *2L* and *2b* are the end-to-end distances for the linear and the bent molecules, respectively. α is the bending angle

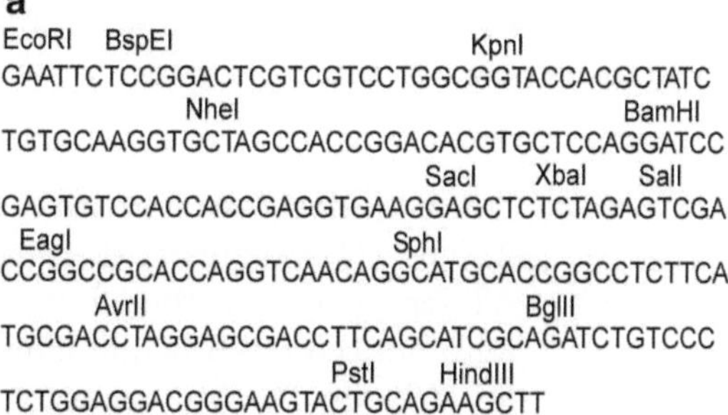

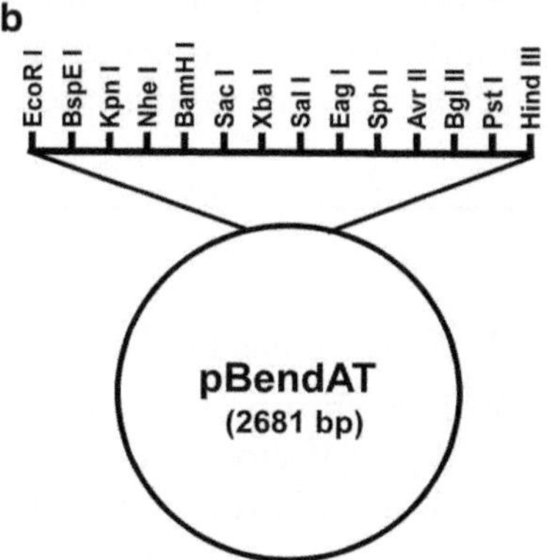

Fig. 2 (**a**) The nucleotide sequence of the 230-bp fragment of pBendAT between *Eco*RI and *Hin*dIII sites. The restriction enzyme sites for *Eco*RI, *Bsp*EI, *Kpn*I, *Nhe*I, *Bam*HI, *Sac*I, *Xba*I, *Sal*I, *Eag*I, *Sph*I, *Avr*II, *Bgl*II, *Pst*I, and *Hin*dIII are shown. (**b**) The restriction map of pBendAT

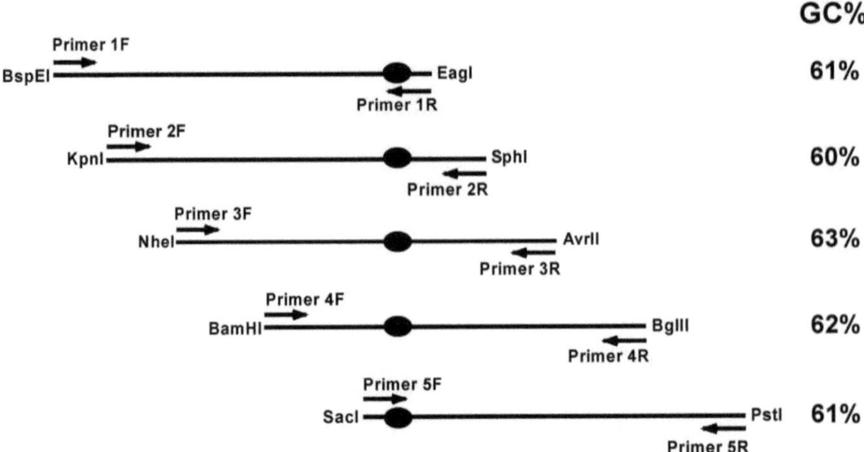

Fig. 3 The five DNA fragments of identical length, which can be produced either by restriction enzyme digestions or by PCR amplification and used in the DNA-bending experiments, are shown. The GC content is indicated on the right. The ovals indicate the DNA-binding sites. The following are the DNA sequences for the primers: primer 1F, 5′-TCGTCGTCCTGGCGGTAC-3′, 1R, 5′-GCGGCCGGTCGACTCTAG-3′, 2F, 5′-ACCACGCTATCTGTGCAA-3′, 2R, 5′-TGCCTGTTGACCTGGTGC-3′, 3F, 5′-AGCCACCGGACACGTGCT-3′, 3R, 5′-AGGTCGCATGAAGAGGCC-3′, 4F, 5′-TCCGAGTGTCCACCACCG-3′, 4R, 5′-TCTGCGATGCTGAAGGTC-3′, 5F, 5′-ACCGAGGTGAAGGAGCTC-3′, 5R, and 5′-TCCAGAGGGACAGATCTG-3′

DNA-binding sites relative to the DNA fragment ends. The 230 bp DNA sequence does not contain more than two consecutive AT base pairs and therefore is particularly useful for studying protein-induced DNA bending by the DNA-binding proteins recognizing AT-rich DNA sequences. Alternatively, PCR amplification can be used to generate a similar set of DNA fragments of identical length (Fig. 3), which greatly facilitates the production of these DNA fragments for the subsequent EMSA assays [14].

2 Materials

2.1 Cloning of the DNA-Bending Site of a DNA-Binding Protein into the XbaI or SalI Site of pBendAT

1. *E. coli* strains DH5α, Top10, and FL#401 (Top10/pBendAT).
2. Luria broth (LB) medium. Dissolve 10 g of Bacto tryptone, 5 g of Bacto yeast extract, and 10 g of NaCl in 1 L of dH_2O. Adjust pH to 7.5 with 5 M NaOH and autoclave.
3. SOC medium. Dissolve 20 g of Bacto tryptone and 5 g of Bacto yeast extract in 800 mL of dH_2O. Add 2 mL of 5 M NaCl, 2.5 mL of 1 M KCl, 10 mL of 1 M $MgCl_2$, and 10 mL of 1 M $MgSO_4$. Add dH_2O to 900 mL and autoclave the medium. Then add 100 mL of filter-sterilized 0.2 M glucose.
4. LB agar plates containing 50 μg/mL ampicillin. Dissolve 10 g of Bacto tryptone, 5 g of Bacto yeast extract, 10 g of NaCl, and 20 g of agar in 1 L of dH_2O and autoclave the medium.

Let it cool to ~55 °C, add 500 μL of 100 mg/mL filter-sterilized ampicillin, and pour it into Petri dishes (~25 mL/10 cm dish).

5. QIAprep Spin Miniprep Kit.
6. Ampicillin stock solution (100 mg/mL). Dissolve 1 g of ampicillin in 10 mL of dH_2O and sterilize by filtration through 0.2 μm membrane. Store 500 μL aliquots at −20 °C.
7. Synthetic oligonucleotides containing the designated DNA-bending sites with the ends compatible with the cloning sites (*Xba*I or *Sal*I) of pBendAT.
8. Restriction enzymes. SalI (20 U/μL) and *Xba*I (20 U/μL). 10× buffers for the restriction enzymes are provided by the vendors (*see* **Note 1**).
9. A Cary 50 UV–Visible Spectrophotometer and a cuvette for small volume samples (40–50 μL).
10. 0.5 M EDTA pH 8.0. Dissolve 93.05 g of EDTA (disodium salt) in 400 mL of dH_2O and adjust pH to 8.0 by using 5 M of NaOH and a pH meter. Bring the volume to 500 mL using a graduated cylinder.
11. 50× TAE buffer. Dissolve 121 g Tris base in 250 mL dH_2O and add 28.6 mL acetic acid and 50 mL of 0.5 M EDTA, pH 8.0. Bring the volume to 500 mL in a graduated cylinder by adding dH_2O.
12. Acetic acid.
13. 5× TBE buffer. Dissolve 27 g of Tris base and 13.75 g of boric acid in 400 ml of dH_2O and add 50 mL of 0.5 M EDTA, pH 8.0. Bring the volume to 500 mL in a graduated cylinder by adding dH_2O.
14. Calf intestinal alkaline phosphatase (CIP) at 10 U/μL (New England Biolabs).
15. T4 DNA ligase with 10× T4 DNA ligation buffer (New England Biolabs).
16. 1× DNA annealing buffer: 20 mM Tris–HCl, pH 8.0, 1 mM EDTA, and 100 mM NaCl.
17. T4 polynucleotide kinase with 10× T4 Polynucleotide Kinase Reaction Buffer (New England Biolabs).
18. 20 mM ATP. Dissolve ~120 mg of ATP in 10 mL of 40 mM Tris base solution. To measure the concentration, dilute the solution 1,000 times, take OD_{259} using a Cary 50 UV–Visible Spectrophotometer, and calculate the concentration with the extinction coefficient of 15.4 mM^{-1} cm^{-1}. Bring the volume to the final volume to make the 20 mM of ATP by adding 40 mM Tris base.
19. Agarose.
20. A submarine agarose mini-gel system.

21. A UV transilluminator.
22. A gel documentation system or a digital camera.
23. 6×gel loading dye. Dissolve 4 g of sucrose, 25 mg of xylene cyanol, and 25 mg of bromophenol blue in 10 mL of dH_2O. Store at 4 °C.
24. DNA size markers: 1 kb DNA ladder or λ DNA *Hin*dIII digest for 1 % agarose gels. 50 bp or low molecular weight DNA ladders for polyacrylamide gels.
25. Ethidium bromide solution (10 mg/mL). Dissolve 100 mg of ethidium bromide in 10 mL of dH_2O with stirring overnight. Store away from light at room temperature.
26. An air shaker and a 37 °C incubator.
27. 30 % acrylamide/Bis (29:1) solution. Dissolve 145 g of acrylamide and 5 g of bis-acrylamide in 350 mL of dH_2O. After the solutes are completely dissolved, adjust the volume to 500 mL using a graduated cylinder. Filter the solution using a 0.2 μm filter.
28. *N*,*N*,*N*′,*N*′-Tetramethylethylenediamine (TEMED).
29. 10 % ammonium persulfate (APS). Dissolve 1 g of ammonium persulfate in dH_2O to make a total volume of 10 mL. Store the solution at 4 °C. Make it fresh every 2 weeks.
30. A vertical polyacrylamide mini-gel system.
31. QIAquick Gel Extraction Kit (Qiagen).
32. Razor blades or scalpels.
33. 10 % (v/v) glycerol. Autoclave and store at room temperature.
34. 0.1 cm electroporation cuvettes.
35. Gene Pulser apparatus for electroporation (BioRad).

2.2 Generation of a Set of DNA Fragments for EMSA

1. Design five sets of DNA primers to amplify the five DNA fragments of identical length containing the protein-bending site (Fig. 3; *see* **Note 2**). Please notice the fact that primers overlapping the protein-bending site are specific for the designated protein-binding sequence. For instance, the primers 1R and 5F in Fig. 3 are unique for each protein-binding site. Other primers can be used for any DNA sequences cloned into the *Xba*I or *Sal*I site of pBendAT.
2. *Pfu* DNA polymerase or another high fidelity DNA polymerase with a corresponding reaction buffer.
3. 100 mM dNTPs at 25 mM of each dNTP.
4. pBendAT containing the DNA-bending site at 10 ng/μL.
5. 0.2 mL thin-wall PCR tubes.
6. A thermal cycler.

7. QIAquick Gel Extraction Kit (Qiagen).
8. QIAquick PCR Purification Kit (Qiagen).
9. Qiagen Plasmid Midi Kit or Plasmid Maxi Kit (Qiagen).
10. Restriction enzymes: *Bsp*EI, *Eag*I, *Kpn*I, *Sph*I, *Nhe*I, *Avr*II, *Bam*HI, *Bgl*II, *Sac*I, and *Pst*I.
11. Sterile 17 mm × 100 mm polypropylene tubes.

2.3 DNA End Labeling Using [γ-^{32}P]ATP

1. [γ-^{32}P]ATP, 1 mCi at 3,000 mCi/mmol.
2. T4 Polynucleotide Kinase with 10× reaction buffer (New England Biolabs).
3. Sephadex G-25 or G-50 spin columns (GE Healthcare).
4. A microcentrifuge.
5. Fuji FLA 3000 analyzer.
6. A phosphor storage screen.
7. Science Lab software.

2.4 Analysis of Protein–DNA Complexes by EMSA

1. Purified DNA-bending proteins, such as *E. coli* CRP or LacI, or the mammalian high mobility group protein AT-hook 2 (HMGA2), or your protein of interest.
2. 10× DNA-binding buffer: 200 mM Tris–HCl, pH 8.0, 2 M NaCl, 5 mM EDTA, 10 mM DTT, 5 mM $MgCl_2$, and 50 % glycerol. To make 10 mL of 10× DNA-binding buffer, add 2 mL of 1 M Tris–HCl, pH 8.0, 100 μL of 0.5 M EDTA, pH 8.0, 100 μL of 1 M DTT, ~2 mL of dH_2O, and 1.17 g of NaCl to a 15 mL conical tube. Vortex the tube until NaCl is dissolved. Then add 5 mL of glycerol (6.3 g) into the conical tube and bring the volume to 10 mL by adding dH_2O.
3. A vertical polyacrylamide gel system for larger gels.
4. A gel drier.
5. A power supply.
6. X-ray film and cassettes.

3 Methods

3.1 Cloning a DNA-Bending Site of a DNA-Binding Protein into the XbaI or SalI Site of pBendAT

1. In a 15 mL tube, grow *E. coli* strain FL#401 (*E. coli* Top10 containing pBendAT) in 5 mL of LB containing 50 μg/mL ampicillin at 37 °C overnight.
2. Purify pBendAT from 3 to 4 mL of the overnight cell culture using QIAprep Spin Miniprep Kit (*see* **Note 3**).
3. After the purification step, run a 1 % agarose mini-gel in 1× TAE or 1× TBE buffer for both supercoiled and *EcoRI-linearized* pBendAT to verify the size of the plasmid. To make a 1 % agarose

mini-gel (0.5 cm thick, 7 cm wide, and 10 cm long), add 0.3 g of agarose to 29.4 mL of dH_2O and 0.6 mL of 50× TAE buffer in a 100 or 200 mL flask. Heat the mixture in a microwave oven until the agarose is completely melted. Allow the molten agarose to cool to ~55 °C and pour the gel into a UV-transparent gel tray (7 cm wide and 10 cm long) with a comb inserted. Let the gel solidify at room temperature.

4. Estimate the DNA concentration using UV spectroscopy by measuring OD_{260} in a Cary 50 UV–Visible Spectrophotometer using a cuvette for 40 μL of solution (1.0 OD_{260} of dsDNA = 50 μg/mL = 0.075 mM (bp)).
5. In a 1.5 mL microcentrifuge tube, digest 2–5 μg of pBendAT in 50 μL of 1× *Xba*I or 1× *Sal*I buffer using *Xba*I (40 U) or *Sal*I (40 U) in the presence of 20 U of calf intestinal alkaline phosphatase (CIP) at 37 °C for 1 h.
6. Run a 1 % agarose gel in 1× TAE buffer to separate the linearized pBendAT from the undigested pBendAT.
7. Stain the gel in 0.5 μg/mL of ethidium bromide solution (ethidium bromide solution is diluted from ethidium bromide stock solution (10 mg/mL)).
8. Use a razor blade or a scalpel to quickly excise a gel slice with the *Xba*I- or *Sal*I-digested pBendAT under UV light. Purify the DNA using QIAquick Gel Extraction Kit or a similar gel extraction kit.
9. Estimate the DNA concentration using UV spectroscopy by measuring OD_{260} in a Cary 50 UV–Visible Spectrophotometer using a cuvette for 40 μL of solution. Run a small aliquot of DNA on a 1 % agarose gel in 1× TAE buffer to verify that enough linearized DNA has been purified.
10. Dissolve a pair of complementary DNA oligonucleotides that carry a designated DNA-bending site in dH_2O at 100 μM for each oligonucleotide. Phosphorylate 150 pmol of each oligonucleotide in a single 50 μL reaction containing 1× T4 polynucleotide kinase buffer, 1 mM of ATP, and 10 U of T4 polynucleotide kinase. Incubate the reaction at 37 °C for 30 min.
11. Anneal the oligos by heating the tube to 98 °C for 10 min in a 4 L beaker filled with hot water and allowing the water to cool down slowly, usually overnight.
12. Set up a 10 μL ligation reaction in 1× T4 DNA ligase buffer containing 1 mM of ATP and 200 U of T4 DNA ligase. Keep the total DNA concentration between 1 and 10 ng/μL, i.e., 10–100 ng of total amount of DNA. The molar ratio of insert to the vector should be 3–10 (3–6 are optimal for the ligation reaction). Incubate the ligation reaction at 16 °C overnight.

13. To make "competent" cells for electroporation, grow *E. coli* DH5α or Top10 cells to early exponential phase (OD_{600} ~0.5) and then wash the *E. coli* cells with autoclaved 10 % glycerol three times. Resuspend cells in a final volume of 3–4 mL in ice-cold 10 % glycerol, if you started with 1 L of LB culture. The cell titer should be about 1–3 × 10^{10} cells/mL.
14. Add 0.5–1 μL of a DNA sample to 40 μL of *E. coli* cells and transfer the mix into a chilled 0.1 cm electroporation cuvette. Electroporate the *E. coli* cells using an electroporation apparatus, such as Gene Pulser apparatus (set the Gene Pulser apparatus at 1.80 kV when using 0.1 cm cuvettes).
15. Add 1 mL of SOC medium into the cuvette. Transfer the cell suspension to a 17 mm × 100 mm polypropylene tube and incubate at 37 °C for 1 h. Plate the transformed cells on LB agar plates containing 100 μg/mL of ampicillin and incubate at 37 °C overnight or until colonies appear.
16. Pick ten *E. coli* colonies and grow each of them in 5 mL of LB containing 100 μg/mL of ampicillin at 37 °C overnight. Purify plasmid DNA from the overnight cultures using QIAprep Spin Miniprep Kit and run a 1 % agarose gel in 1× TAE for both supercoiled and *Eco*RI-linearized plasmid to verify the size of the plasmids.
17. Make a 20 % polyacrylamide mini-gel (0.75 mm thick, 8.2 cm wide and 7.5 cm long). First, assemble the Glass Cassette and Casting Stand for pouring the gel. Then make 10 mL of 20 % acrylamide solution: 6.67 mL of 30 % acrylamide/Bis (29:1) solution, 0.2 mL of 50× TAE, and 3.045 mL of H_2O. Add 80 μL of 10 % APS and 5 μL of TEMED just before pouring and mix thoroughly. Pour the acrylamide solution into the preassembled glass cassette. Insert a comb. Let the gel solidify at room temperature.
18. Digest the plasmids (~2–4 μg) using a restriction enzyme, *Xba*I or *Sal*I, and run a 20 % PAGE in 1× TAE buffer to screen for clones containing the designed DNA-bending site after the gel is stained in 0.5 μg/mL ethidium bromide and visualized under UV light.
19. Sequence across the cloned insertion to confirm the sequence of the designed DNA-bending site in the plasmids.

3.2 Generating a Set of DNA Fragments for EMSA

3.2.1 Generating a Set of DNA Fragments by PCR

1. Set up five PCR reactions (in 50 μL each) each using one of the five pairs of primers: 1F and 1R, 2F and 2R, 3F and 3R, 4F and 4R, and 5F and 5R. Each reaction should include 5 μL of 10× *Pfu* reaction buffer, 0.4 μL of dNTPs (stock solution, 100 mM, 25 mM for each dNTP), 1 μL of pBendAT (stock solution, 10 ng/μL), 1 μL of a forward primer (stock solution, 20 μM), 1 μL of a reverse primer (stock

solution, 20 μM), 2.5 U of *Pfu* DNA polymerase, and 40.6 μL of H_2O. Use the following protocol for the PCR reaction: for the initiation step, heat the PCR reaction at 95 °C for 3 min; in the second step, run 20 cycles of the denaturation at 95 °C for 30 s, the annealing at 55 °C for 30 s, and the extension at 72 °C for 1 min; in the final step, perform an extension at 72 °C for 10 min.

2. Make an 8 % polyacrylamide mini-gel (0.75 mm thick, 8.2 cm wide, and 7.5 cm long). First, assemble the Glass Cassette and Casting Stand. Second, make 10 mL of 8 % acrylamide solution: 2.67 mL of 30 % acrylamide/Bis (29:1) solution, 0.2 mL of 50× TAE, and 7.045 mL of H_2O. Add 80 μL of 10 % APS and 5 μL of TEMED just before pouring. Pour the acrylamide solution into the preassembled glass cassette. Insert a comb. Let the gel solidify at room temperature.
3. Run the PCR reactions from the **step 1** on the 8 % polyacrylamide gel in 1× TAE buffer to verify the success of the PCR and also to determine whether the DNA-bending site carries an intrinsic curvature (*see* **Note 4**).
4. Make a 2 % agarose mini-gel (0.5 cm thick, 7 cm wide, and 10 cm long). Add 0.6 g of agarose to 29.4 mL of dH_2O and 0.6 mL of 50× TAE buffer in a 100 or 200 mL flask. Heat the mixture in a microwave oven until the agarose is melted. Pour the gel into the UV-transparent gel tray (7 cm wide and 10 cm long) with a comb inserted. Let the gel solidify at room temperature.
5. Run the five PCR products on the 2 % agarose gel in 1× TAE buffer and purify the DNA fragments using QIAquick Gel Extraction Kit. Alternatively, the PCR products can be purified directly by using QIAquick PCR Purification Kit. Estimate the DNA concentrations by measuring OD_{260} in a Cary 50 UV–Visible Spectrophotometer using a cuvette for 40 μL of solution.

3.2.2 Generating a Set of DNA Fragments by Restriction Enzyme Digestion

1. Purify a large quantity of pBendAT carrying a DNA-bending site (pBendAT-X) using Qiagen Plasmid Midi Kit or Plasmid Maxi Kit from 100 to 250 mL of *E. coli* overnight culture grown in LB with 100 μg/mL of ampicillin (*see* **Note 3**).
2. In five 1.5 mL microcentrifuge tubes, set up 50 μL of buffered solution for each tube to digest pBendAT-X using five pairs of restriction enzymes: *Bsp*EI (20 U) and *Eag*I (20 U), *Kpn*I (20 U) and *Sph*I (20 U), *Nhe*I (20 U) and *Avr*II (10 U), *Bam*HI (20 U) and *Bgl*II (20 U), and *Sac*I (20 U) and *Pst*I (20 U) at 37 °C for 1 h to produce five DNA fragments of identical length.
3. Run the digests on a 2 % agarose gel in 1× TAE buffer to separate the five DNA fragments of identical length from the backbone

of the pBendAT-X vector. Purify these DNA fragments using QIAquick Gel Extraction Kit.

4. Estimate the DNA concentration by measuring OD_{260} in a Cary 50 UV–Visible Spectrophotometer using a cuvette for 40 μL of solution and run an 8 % polyacrylamide gel electrophoresis to determine whether the DNA-bending site carries an intrinsic curvature (if the mobility of the five free DNA fragments is identical, there is no intrinsic curvature for the protein-binding site (*see* **Note 4**)).

3.3 DNA End Labeling Using [γ-^{32}P]ATP

1. Set up five labeling reactions as following: use ~50 pmol of 5′ termini in a 50 μL reaction containing 1×T4 Polynucleotide Kinase Buffer, 50 pmol of [γ-^{32}P] ATP (100 μCi), and 20 U of T4 Polynucleotide Kinase. Incubate the reaction mixtures at 37 °C for 1 h.
2. Purify ^{32}P-labeled DNA fragments using Sephadex G-25 or G-50 columns.
3. Estimate the DNA concentration using UV spectroscopy by measuring OD_{260} in a Cary 50 UV–Visible Spectrophotometer as described above. Alternatively, run an 8 % polyacrylamide gel electrophoresis in 1×TAE buffer by loading an aliquot of each DNA sample before and after the purification step and determining the DNA concentration by comparing the DNA samples using a phosphor storage screen, Fuji FLA 3000 image analyzer, and Science Lab software for data analysis.

3.4 Analysis of Protein–DNA Complexes by EMSA

3.4.1 Quantitative EMSA to Determine K_d of a DNA-Binding Protein

The K_d of a protein binding to DNA corresponds to the protein concentration at which 50 % of the DNA is bound by the protein if the DNA concentration used in the assays is much lower than the K_d. Establishing the K_d value is required to determine the protein concentration for the DNA-bending assays.

1. Incubate 0.1–1 nM of ^{32}P-labeled DNA fragments (the *Nhe*I–*Avr*II fragment in Fig. 4) in a 50 μL of 1×DNA-binding buffer, with the increasing concentrations of your DNA-binding protein at 22 °C for 1 h (*see* **Note 5**).
2. Make an 8 % native polyacrylamide gel in 0.5×TBE buffer (0.75 mm thick, 16 cm long, 20 cm wide). First, assemble the Glass Cassette and Casting Stand. Then make 100 mL of 8 % acrylamide solution: 26.7 mL of 30 % acrylamide/Bis (29:1) solution, 10 mL of 5×TBE, and 62.45 mL of dH_2O. Add 800 μL of 10 % APS and 50 μL of TEMED just before pouring the gel. Pour the acrylamide solution into the preassembled glass cassette. Insert a comb. Let the gel solidify at room temperature.
3. Load the protein–DNA complexes on the polyacrylamide gel and run it in 0.5×TBE buffer on ice for 30–60 min at 19–38 V/cm to separate free and protein-bound DNA (*see* **Note 6**).

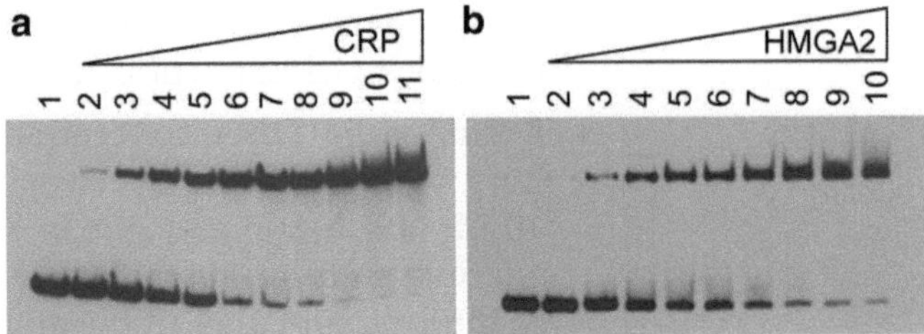

Fig. 4 Analysis of protein–DNA binding by EMSA. 0.2 nM of ^{32}P-labeled DNA fragments containing DNA-binding sites for CRP (**a**) or HMGA2 (**b**) was incubated with increasing concentrations of the corresponding protein in 50 μL of 1 × DNA-binding buffer. The autoradiograms or autoradiographs of the ^{32}P-labeled FL-SELEX1 are shown. (**a**) Lane *1* is a protein-free sample. In addition to the ^{32}P-labeled DNA fragment, the reaction mixtures in lanes *2–11* contain 0.1, 0.25, 0.5, 1, 1.5, 2.0, 3.0, 5.0, 10.0, and 20.0 nM of CRP (as a dimer), respectively. (**b**) Lane *1* is a protein-free sample. In addition to the DNA fragment, lanes *2–10* also contain 1, 2, 5, 10, 20, 50, 100, 200, and 500 nM of HMGA2, respectively

4. Disassemble the glass plates and dry the gel in a gel drier.
5. Use autoradiography or a phosphor storage screen with a Fuji FLA 3000 image analyzer to visualize the DNA bands.
6. Quantify the bands corresponding to the free and protein-bound DNA.
7. Determine the ratio (R) between the protein-bound DNA and the total DNA (the sum of the bound and unbound DNA).
8. Determine the K_d of the DNA-binding protein (Fig. 4) using the following equation:

$$R = \frac{a + b + K_d - \sqrt{(a + b + K_d)^2 - 4ab}}{2a}$$

where a and b represent the total DNA and total protein concentration, respectively.

3.4.2 Analysis of Protein-Induced DNA Bending

1. Set up five reactions: mix each of the five ^{32}P-labeled DNA fragments (0.2 nM) in a 50 μL of 1× DNA-binding buffer with the DNA-binding protein at the concentration equal to its K_d for the DNA (*see* Subheading 3.4.1). For example, 1 nM for CRP and 10 nM for HMGA2 are the concentrations used in the DNA-bending assays (Figs. 5 and 6). Incubate the reactions at 22 °C for 1 h.
2. Make an 8 % native polyacrylamide gel in 0.5× TBE buffer (0.75 mm thick, 16 cm long, 20 cm wide). First, assemble the Glass Cassette and Casting Stand. Then make 100 mL of 8 % acrylamide solution: 26.7 mL of 30 % acrylamide/Bis (29:1) solution, 10 mL of 5× TBE, and 62.45 mL of dH_2O. Add 800 μL of 10 % APS and 50 μL of TEMED just before pouring the gel.

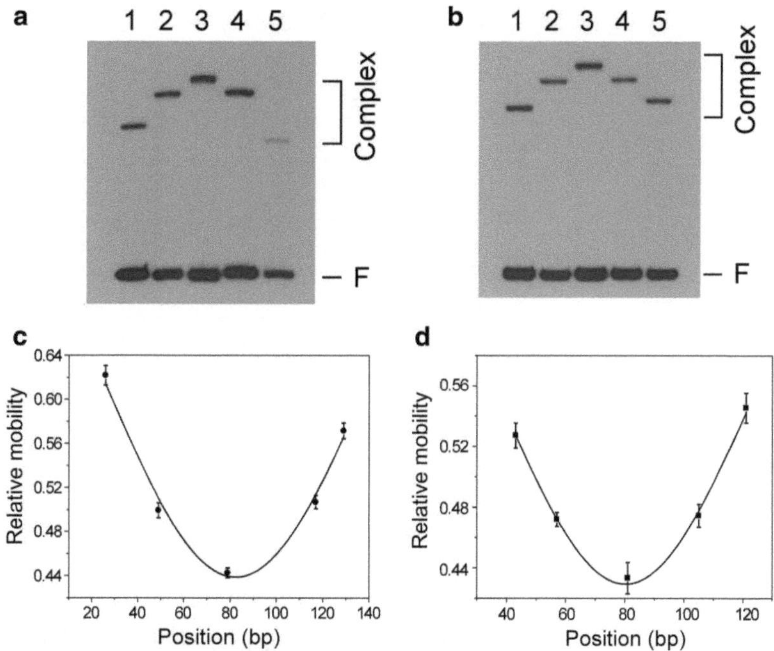

Fig. 5 DNA-bending induced by CRP. DNA-bending assays were carried out as described in Subheading 3. After the binding of CRP to the permutated DNA fragments in the presence of 20 μM of cAMP, an 8 % polyacrylamide gel was used to separate the protein-bound and free DNA fragments. The autoradiograms of ^{32}P-labeled DNA fragments are shown. The DNA fragments in the bottom of the gels are free DNA and those in the upper part are protein–DNA complexes. Labels: F is the free DNA and complex stands for the protein–DNA complex. (**a**) The CRP fragments of the *E. coli lac P1* promoter produced by digestion of plasmid pBendAT-CRP with the following pairs of restriction enzymes: *Bsp*EI–*Eag*I (lane *1*), *Kpn*I–*Sph*I (lane *2*), *Nhe*I–*Avr*II (lane *3*), *Bam*HI–*Bgl*II (lane *4*), and *Sac*I–*Pst*I (lane *5*). (**b**) The CRP fragments of *E. coli lac P1* produced by using PCR amplification as described in Fig. 3. (**c**) and (**d**) are plots of the relative mobility of the CRP–DNA complexes vs. the location of the CRP-binding site within the 149 bp fragment of pBendAT for the CRP fragments generated by restriction digestions of pBendAT-CRP and PCR amplifications of pBendAT-CRP, respectively. The standard deviations are also shown

Pour the acrylamide solution into the preassembled glass cassette. Insert a comb. Let the gel solidify at room temperature.

3. Load the protein–DNA complexes onto the polyacrylamide gel and run the gel in 0.5× TBE for 60 min at 19–38 V/cm to separate free DNA and the protein–DNA complexes of different mobilities.
4. Dry the gels in a gel drier.
5. Use autoradiography or a phosphor storage screen with a Fuji FLA 3000 image analyzer to visualize the DNA bands.
6. Measure the distance between the well and the position of the protein–DNA complexes for the *Nhe*I–*Avr*II fragment and the

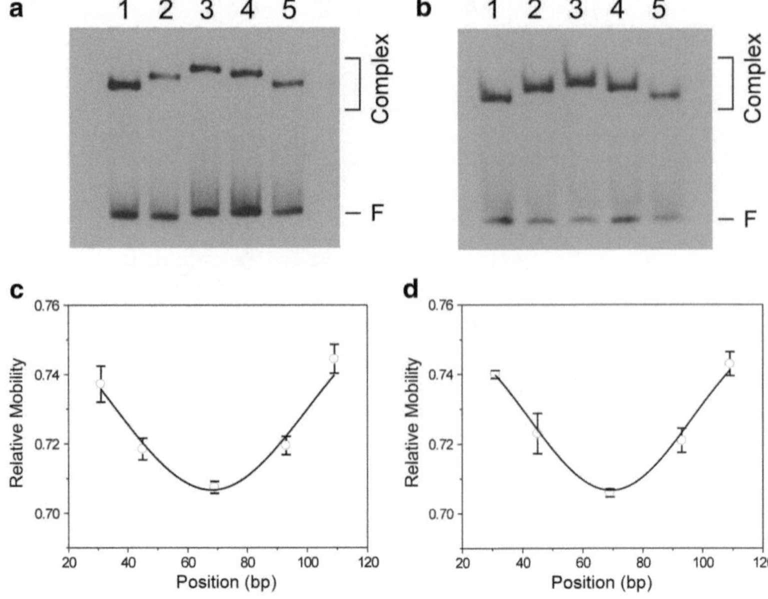

Fig. 6 DNA bending induced by HMGA2. DNA-bending assays were carried out as described in Methods. After the binding of HMGA2 to the DNA fragments of identical size generated from pBendAT-SELEX1 (**a**) or pBendAT-SELEX2 (**b**) by PCR amplification, an 8 % polyacrylamide gel was used to separate the protein-bound and free DNA fragments. The autoradiograms of ^{32}P-labeled DNA fragments are shown. The DNA fragments in the bottom of the gels are free DNA and those in the upper parts are protein–DNA complexes. Labels: F is the free DNA and complex refers to the protein–DNA complex. (**c**) and (**d**) are the plots of the relative mobility of HMGA2–DNA complexes vs. the location of the HMGA2-binding site within the 135 bp fragment for pBendAT-SELEX1 or pBendAT-SELEX2, respectively. The standard deviations are also shown

*Sac*I–*Pst*I fragment, respectively (lanes 3 and 5, respectively, of Figs. 5 and 6).

7. Calculate the μ_M and μ_E and the DNA bending angles using the equation:

$$\cos\frac{\alpha}{2} = \frac{\mu_{\mathrm{M}}}{\mu_{\mathrm{E}}}.$$

4 Notes

1. We purchase restriction enzymes, Calf intestinal alkaline phosphatase, T4 DNA ligase, and T4 Polynucleotide Kinase from New England Biolabs, Inc. All buffers are provided by the vendor.

2. Although we used five pairs of DNA primers in our DNA-bending experiments, it is possible to design and synthesize many more pairs of primers for different scenarios.
3. Alternatively, alkaline lysis method can be used to purify plasmid DNA [15].
4. The length of the PCR products is 115 bp plus the length of the protein-binding sequence. If a set of rigid, rodlike DNA fragments of identical length containing a DNA-bending site carries an intrinsic curvature or a bend, the DNA fragment with a curvature or a bend in the middle of the DNA fragment migrates in a polyacrylamide gel slower than the one with a bend closer to one of the DNA ends.
5. Regarding the range of protein concentrations used in the quantitative EMSA assays, we rely on the published K_d values for well-characterized DNA-binding proteins. For a new DNA-binding protein, we recommend to run a few EMSA assays and vary the protein concentrations by a few orders of magnitude, e.g., from nM to μM range. It is very important to determine the K_d before the DNA-bending assays. Firstly, the quantitative EMSA assays help to determine if a DNA-binding protein is still active. Secondly, the quantitative EMSA assays provide the optimal protein concentration for a specific binding and bending of the DNA sequence.
6. Since the 1× DNA-binding buffer contains 5 % glycerol, we did not use a loading dye in the protein–DNA samples. However, we add 1× loading dye (without DNA) to an empty lane to track the progress of sample migration in the gel.

References

1. van D V, Verrijzer CP (1993) Bending of DNA by transcription factors. Bioessays 15:25–32
2. Chen B, Xiao Y, Liu C, Li C, Leng F (2010) DNA linking number change induced by sequence-specific DNA-binding proteins. Nucleic Acids Res 38:3643–3654
3. Kuznetsov SV, Sugimura S, Vivas P, Crothers DM, Ansari A (2006) Direct observation of DNA bending/unbending kinetics in complex with DNA-bending protein IHF. Proc Natl Acad Sci USA 103:18515–18520
4. Matthews KS, Nichols JC (1998) Lactose repressor protein: functional properties and structure. Prog Nucleic Acid Res Mol Biol 58: 127–164
5. Lewis M, Chang G, Horton NC, Kercher MA, Pace HC, Schumacher MA, Brennan RG, Lu P (1996) Crystal structure of the lactose operon repressor and its complexes with DNA and inducer. Science 271:1247–1254
6. Lawson CL, Swigon D, Murakami KS, Darst SA, Berman HM, Ebright RH (2004) Catabolite activator protein: DNA binding and transcription activation. Curr Opin Struct Biol 14: 10–20
7. Harman JG (2001) Allosteric regulation of the cAMP receptor protein. Biochim Biophys Acta 1547:1–17
8. Schultz SC, Shields GC, Steitz TA (1991) Crystal structure of a CAP–DNA complex: the DNA is bent by 90 degrees. Science 253: 1001–1007
9. Kim J, Zwieb C, Wu C, Adhya S (1989) Bending of DNA by gene-regulatory proteins: construction and use of a DNA bending vector. Gene 85:15–23

10. Wu HM, Crothers DM (1984) The locus of sequence-directed and protein-induced DNA bending. Nature 308:509–513
11. Crothers DM, Gartenberg MR, Shrader TE (1991) DNA bending in protein–DNA complexes. Methods Enzymol 208:118–146
12. Thompson JF, Landy A (1988) Empirical estimation of protein-induced DNA bending angles: applications to lambda site-specific recombination complexes. Nucleic Acids Res 16:9687–9705
13. Zwieb C, Adhya S (2009) Plasmid vectors for the analysis of protein-induced DNA bending. Methods Mol Biol 543:547–562
14. Chen B, Young J, Leng F (2010) DNA bending by the mammalian high-mobility group protein AT hook 2. Biochemistry 49: 1590–1595
15. Sambrook J, Maniatis T, Fritsch EF (1989) Molecular cloning a laboratory manual, 2nd edn. Cold Spring Harbor Laboratory, Cold Spring Harbor, NY

Chapter 19

Using PCR Coupled to PAGE for Detection and Semiquantitative Evaluation of Telomerase Activity

Laura Gardano

Abstract

Telomerase is the enzyme that extends the chromosome ends, thereby contributing to eukaryotic cell genome stability. Telomerase is expressed in the majority of cells that have an unlimited proliferation such as stem cells and cancer cells. The increased interest in telomerase in cancer research, challenged by the low cellular abundance of the enzyme, has led to the development of a reliable and at the same time very sensitive approach to detect telomerase activity. The telomeric repeat amplification protocol (TRAP) represents an easy and rapid method for detection of telomerase activity in cells. A non-telomeric TS primer is extended by telomerase in the first step followed by the PCR amplification of the products. The PCR step renders this protocol very sensitive to detect telomerase activity at the single cell level making it compatible with the analysis of tumor samples. When run on a polyacrylamide gel, the PCR product is a characteristic ladder of bands due to the repetitive nature of telomeric DNA sequence. The densitometric analysis of the ladder allows the TRAP assay to be used for comparative quantification of telomerase activity in different samples.

Key words Telomerase activity, Telomeres, Telomerase detection

1 Introduction

Telomeres are structures at the end of chromosomes that protect them from end-to-end fusion and erosion [1]. They are represented by a repeated sequence of DNA, TTAGGG in vertebrates, associated with specific proteins that stabilize the folding of this particular portion of the chromosome [2]. The length of the telomeres is maintained by the enzyme telomerase, a reverse transcriptase that copies into DNA a small portion of the RNA template stably associated with it [3]. Cells lacking telomerase experience telomere shortening due to the DNA end-replication problem which prevents the very end of the chromosome being replicated [4]. Most somatic cells do not possess sufficient telomerase activity to guarantee the maintenance of telomere length; hence, they have a limited number of cell divisions before going into replicative senescence [5]. However, immortal cells such as stem cells do

Svetlana Makovets (ed.), *DNA Electrophoresis: Methods and Protocols*, Methods in Molecular Biology, vol. 1054, DOI 10.1007/978-1-62703-565-1_19, © Springer Science+Business Media New York 2013

possess telomerase activity, thus attenuating telomere loss due to replication. More generally, cells that replicate in an unlimited fashion must possess a mechanism of telomere maintenance. In fact, in 85 % of cancers, telomerase is reactivated, whereas in the remaining 15 %, alternative mechanisms of telomere lengthening must take place to guarantee the unlimited proliferation [6]. The presence of telomerase in the majority of cancer cells and its absence in somatic cells makes it an ideal target for cancer therapy [7].

The interest in telomerase as therapeutic target necessitated a rapid technique to detect very low levels of telomerase activity. The telomeric repeat amplification protocol (TRAP), described in Kim et al. (1994), represented a valid substitution to previous methods of telomerase detection, such as primer extension that required a much larger amount of cells to be able to detect telomerase activity [8]. In the standard assay described in 1985, a telomeric primer was extended by telomerase in the presence of radioisotope-labelled (α-^{32}P)-GTP. The products of extension were extracted with phenol/chloroform and ran on a sequencing gel. The results were visualized by autoradiography with a relatively long exposure time [3, 9].

The introduction of the TRAP assay rendered the analysis of telomerase activity faster and easier. In the original TRAP assay, a cell lysate is incubated with a non-telomeric oligonucleotide (TS) that was extended by telomerase (Fig. 1, step 1). In a second step, a PCR is performed to amplify the products generated by telomerase using the TS primer as a forward primer (Fig. 1, step 2). Initially, the reverse primer CX contained essentially the complementary sequence of three telomeric repeats containing a mismatch T/A to decrease the formation of primer dimers and staggered products. When resolved by PAGE, the product of a TRAP assay shows a characteristic six-nucleotide step ladder of DNA molecules representing the elongation of the TS primer by telomerase (Fig. 1, step 3). In addition, as control for the PCR step, an internal telomerase assay standard (ITAS) was included in the assay to be used also as normalizer. It was generated by the amplification of the 150 bp DNA fragment of rat myogenin by the same primers TS and CX [10]. Thus, the combination of mild detergent-based cell lysis and the PCR amplification rendered the TRAP assay very sensitive to detect telomerase in just tens of cells [8].

In 1996, the TRAP-eze™ protocol was made commercially available as a kit from Oncor, Inc [10]. This modification of the original protocol used as the reverse primer the sequence RP (*see* **Note 1**) that reduces even more the primer dimer formation. In addition, this sequence starts with few nucleotides that are not telomeric and not complementary to the telomeric sequence so that they generate PCR products that are not 3′-end extendible when the primers are annealed in a staggered manner to the telomerase products. This way, the formation of artifacts during the PCR step is greatly reduced. Additionally, the reverse primer mix in the

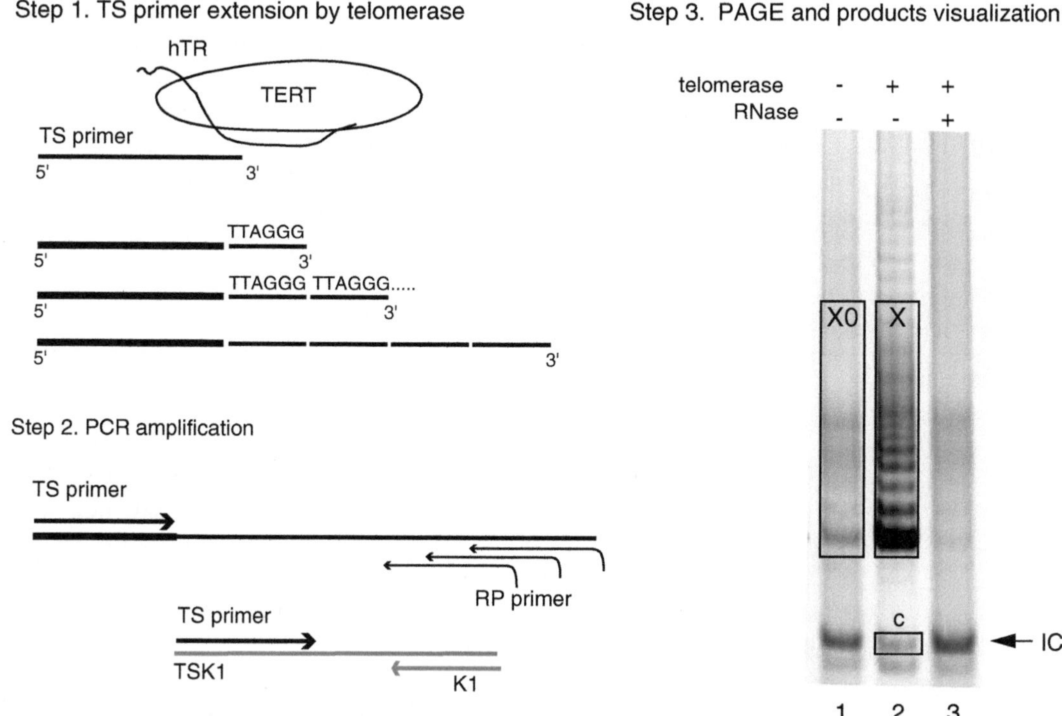

Fig. 1 A schematic of the three steps of the TRAP assay. In the *step 1*, telomerase (TERT) extends the TS primer copying the template sequence provided by telomerase RNA (hTR). In the *step 2*, the product of telomerase-dependent extension is PCR amplified using the primers TS and RP that contains a short sequence that does not anneal to telomeric products. The reverse primer mix also contains a 36 bp template TSK1 amplified by the primer TS and the reverse primer K1 that generate the internal PCR control (IC). In the *step 3*, polyacrylamide gel electrophoresis and SYBR green staining are used to visualize the typical 6-nucleotide step ladder of DNA products from the PCR step. A typical TRAP assay gel is shown. The assay was performed on rabbit reticulocyte lysate (RRL) used to reconstitute human telomerase activity in vitro. The TRAP-eze kit (Millipore) was used to perform the assay. The samples tested were as follows: *lane 1*, water; *lane 2*, 1 μL of RRL expressing human telomerase; and *lane 3*, negative control and RRL expressing human telomerase treated with RNase. The 36 bp internal control (IC) is indicated by the *arrow* on the gel. The *boxes* indicated on the gel as X, X0, and C were drawn to evaluate the amount of telomerase activity as described in Subheading 3.5

kit also contains a 36 bp oligonucleotide (TSK1) and a second primer (K1) used in combination with TS to generate the internal control PCR product (IC). Because only the primer TS is shared between the telomerase amplification products and the IC, problems related to primers competition are reduced compared to the original ITAS [11]. The TRAP-eze™ kit is indeed cost beneficial because it has higher sensitivity, is more reliable, and guarantees more quantitative linearity over the standard TRAP assay. In addition, all its components are carefully quality controlled and do not need to be tested before use [10]. However, one drawback of this kit is that the internal control (IC) is less sensitive to PCR inhibitors than the original 150 bp ITAS used in the standard TRAP protocol [10]. This might pose a limitation to the use of the TRAP-eze kit to detect telomerase in tumor samples because of the risk of false negatives [10]. In the following pages, the TRAP assay protocol is described according to the TRAP-eze method. No hot start PCR is required and the use of the RP primer compared to CX limits the formation of primer dimers increasing the efficiency of telomerase detection. In the following protocol, CX and RP primer sequences are both indicated and can be used to perform the assay with the considerations mentioned (*see* **Note 1**).

2 Materials

Because telomerase contains a stably associated RNA component indispensable for its activity, all the precautions used for RNA handling must be observed when performing TRAP assay, i.e., work on clean surfaces, change gloves often, use filter tips, and use only RNase- and DNase-free tubes and tips. RNase contamination will result in negative TRAP results because of the degradation of the telomerase RNA component.

2.1 Cell Lysis

1. PBS: 137 mM NaCl; 2.7 mM KCl, 10 mM Na_2HPO_4; 1.8 mM KH_2PO_4.
2. Pestle or grinder if TRAP assay is performed on tissue extract.
3. 1× CHAPS lysis buffer:

   10 mM Tris–HCl, pH 7.5, 1 mM $MgCl_2$, 1 mM EGTA, 0.1 mM benzamidine; 5 mM β-mercaptoethanol, 0.5 % CHAPS (w/v), 10 % glycerol. Store at −20 °C.
4. RNAse inhibitor at working concentration of 100–200 U/μL (Protector RNase inhibitor, Roche).
5. Reagent for protein assay, e.g., Bradford reagent.

2.2 TRAP Assay

1. 10× TRAP Reaction Buffer:
   200 mM Tris–HCl, 15 mM $MgCl_2$, 630 mM KCl, 0.5 % Tween-20 (v/v), 10 mM EGTA. Aliquot and store at −20 °C (*see* **Note 2**).
2. 50× nucleotides mix, 2.5 mM each dATP, TTP, dCTP, dGTP. Store the dNTP mix at −80 °C.
3. TS primer sequence: 5′-AATCCGTCGAGCAGAGTT-3′ (*see* **Note 3**). Prepare primer stock at a concentration of 10 μM in DNase-free, RNase-free water.
4. Reverse primer: RP: 5′-GCGCGG[CTTACC]$_3$CTAACC-3′ (*see* **Note 1**). Prepare primer stock at a concentration of 10 μM in DNase-free, RNase-free water.
5. Thermocycler.

2.3 Polyacrylamide Gel Electrophoresis

1. Vertical gel running system, with glass plates, spacers and comb. For best resolution, use gel ~15 cm × 15 cm and thickness of 1.0 mm.
2. 10× TBE: 890 mM Tris base, 890 mM boric acid, 20 mM EDTA.
3. For a volume of 60 mL of gel solution: 14 mL acrylamide/bis-acrylamide (29:1), 3.5 mL 10× TBE, 0.450 mL 10 % ammonium persulfate (*see* **Note 4**), 0.045 mL TEMED.
4. A 10× loading dye is added to the samples before running the PAGE: 0.25 % bromophenol blue, 0.25 % xylene cyanol FF, 30 % glycerol in water.
5. Power supply (>500 V).

2.4 Gel Staining

1. SYBR green™ (Invitrogen).
2. Glass dish.
3. 0.6× TBE, dilute from 10× TBE.

2.5 Gel Visualization and Analysis

1. Acquisition system with UV filter and CCD imaging system.
2. ImageJ (available on the web) or similar software for the analysis.

3 Method

Before starting, add RNase inhibitor to CHAPS 1× at a concentration of 100–200 U/μL.

3.1 Extract Preparation

Cell extract can be prepared from cells (cell culture or blood samples) or from solid tissue. Choose the appropriate protocol from the two alternatives provided below.

3.1.1 Preparation of Cell Extract from Adherent Culture

1. Wash adherent cells grown with 1× PBS once (*see* **Note 5**).
2. Add enough PBS to cover the surface of the plate and scrape the cells to resuspend them in PBS (*see* **Note 6**).
3. Pellet cells at 1,500 × *g* for 5 min.
4. Resuspend the pellet in 1× CHAPS lysis buffer by adding 200 μL of buffer per 10^5–10^6 cells.
5. Lyse cells on ice for 30 min.
6. Centrifuge at 12,000 × *g* for 20 min at 4 °C.
7. Transfer the supernatant into a fresh tube.
8. Determine the protein concentration.

3.1.2 Preparation of Cell Extract from Solid Tissue

1. Wash 40–100 mg of the tissue with 1×PBS.
2. Homogenize the tissue with a pestle or by grinding.

3. Add 1× CHAPS, 200 μL per 40–100 mg of tissue.
4. Lyse cells on ice for 30 min.
5. Centrifuge at 12,000 × *g* for 20 min.
6. Transfer the supernatant into a fresh tube.
7. Determine the protein concentration with Bradford or other suitable method.

3.2 Preparation of Negative and Positive Controls

1. Telomerase negative control. Telomerase is heat sensitive; thus, a negative control can be easily obtained by heating an aliquot of the cell extract prepared in the previous step for 15 min at 95 °C. Alternatively, an aliquot of the cell extract can be treated with RNase A at a concentration of 4 μg/μL for 15 min at room temperature to digest telomerase RNA.
2. Telomerase positive control. A positive control can be a telomerase positive cell extract, for example, a HeLa cell lysate or similar that has to be treated as described in Subheading 3.1.1 (*see* **Note 7**).
3. No cell lysate control. Include a TRAP assay sample with no cell extract present; instead 2 μL of the CHAPS buffer used to lyse the cells is added. This reaction should not produce any DNA product unless the reagents are contaminated and therefore the validity of the assay is impaired.

3.3 TRAP Reaction

Given the sensitivity of this assay, less than 1 μg of total protein or extract from as few as 30 cells is sufficient to detect telomerase activity. If the cell lysate gives a yield of about 100 ng/μL of total protein content, then a volume of 2 μL is enough to be assayed in the TRAP assay.

1. Prepare the TRAP reaction in a volume of 50 μL in RNase-free PCR tubes by adding:

   x μL of cell lysate.

   5 μL of 10 × TRAP reaction buffer.

   1 μL of 50 × dNTP mix.

   1 μL of TS primer (*see* **Note 1**).

   1 μL of reverse primer (*see* **Note 1**).

   PCR water up to 49.5 μL.
2. Incubate the reaction mix at 30 °C for 30 min. During this step, telomerase extends the TS primer (*see* **Note 8**).
3. Inactivate telomerase by heating the reaction at 94 °C for 2 min.
4. Add 0.5 μL of Taq polymerase.

5. Start the PCR program set to run 27–30 cycles of:

   94 °C for 30 s.

   50 °C for 30 s.

   72 °C for 1 min.

6. While the PCR is running, prepare polyacrylamide gel as described in the next section.

3.4 PAGE: Gel Preparation and Electrophoresis

1. Assemble clean glass plates and spacers in a gel casting mold, according to manufacturer's recommendations.
2. For one 15 cm × 15 cm gel of 1 mm thickness, prepare 60 mL of 10 % non-denaturing polyacrylamide in 0.6× TBE. Mix 15 mL of acrylamide to bis-acrylamide (29:1) with 3.6 mL of 10× TBE and 41.4 mL of water.
3. Add 450 μL of 10 % APS (freshly prepared) and 45 μL of TEMED. Mix well and pour the gel mix between the glass plates (*see* **Note 9**). Insert a comb to make a desired number of wells. Let the gel to polymerize for ~ 1 h (*see* **Note 10**).
4. When the PCR run is finished, stop the PCR machine and take out your samples. Add 5 μL of 10× blue loading dye to each sample and mix.
5. Carefully remove the comb from the polymerized gel and place the gel into the gel running apparatus as for electrophoresis. Fill the gel chambers with 0.6× TBE.
6. Load 20–25 μL of TRAP samples per each well.
7. Run the gel at constant voltage of 1.2 V/cm^2 for about 2.5 h. If the loading dye is prepared with both xylene cyanol and bromophenol blue, let the darker marker (bromophenol blue) run out of the gel and stop the electrophoretic run when xylene cyanol is about halfway into the gel.
8. To stain the gel with SYBR green™, dilute the dye 10,000 times in 0.6× TBE (*see* **Note 11**). Prepare enough staining solution to cover the gel (100 mL is enough for a 15 cm × 15 cm gel). Leave the gel in the SYBR green™ solution for 30 min and proceed with the visualization and scanning (*see* **Note 12**).
9. After the incubation with SYBR™ green, the gel should be rinsed one time with water before scanning. Place the gel on the clean scanning glass surface and take note of the coordinates of the area of the gel that will have to be selected on the software that controls the scanning. Set the wavelength of scanning to 473 nm for SYBR green™. The lane image of a typical TRAP sample should appear as a ladder of six-nucleotide steps.

3.5 Semiquantification of Telomerase Activity

Wherever possible, it is best to have several dilutions of the sample to be assayed in order to verify the linearity of the reaction and the lowest level of detection. The TRAP-eze kit is supplied with a positive control (TSR8) (*see* **Note 13**) that can be used as a reference for the quantification. The method to estimate telomerase activity is to calculate the ratio of the intensity of the ladder and the IC band using densitometry (*see* **Note 14**).

1. Scan the gel using a Gel doc or similar scanner with UV filter.
2. Using ImageJ, draw a rectangle around the telomeric products (X) from the first band as shown in Fig. 1, step 3 (*see* **Note 15**).
3. Draw a rectangle of the same area also on the lane with buffer only assay (X0).
4. Draw a rectangle to include the IC band (C).
5. Rectangles over the TSR8 lane, R for the ladder and C_R for the IC.
6. Telomerase activity (TA) can be estimated by:
   $TA = [(X - X0)/C]/[(R - X0)/C_R$ (*see* **Note 16**).

4 Notes

1. In the original TRAP assay, the reverse primer used was CX (sequence 5′-(CCCTTA)$_3$CCCTAA-3′) [8]. A number of modifications of this protocol were later introduced to solve some of the issues with the original protocol, e.g., the formation of primer dimers [12]. To overcome this issue, the sequence of the CX already contains a mismatch T/A to decrease the primer dimer formation. However, the RP sequence eliminates the primer dimers formation and other PCR artifacts even more efficiently and also improves the resolution. To obtain the IC band, 0.01 amol of the 36 bases sequence Kl 5′-AATCCGTCGAGCAGAGTTAAAAGGCCG AGAAGCGAT-3′ and 0.1 μg the specific reverse primer TSKl 5′-ATCGCTTCTCGGCCTTTT-3′ can be added to the TRAP reaction [11].
2. It is a good practice to aliquot all the reagents to be used in the TRAP assay, i.e., dNTPs, primer stocks, and TRAP buffer, to avoid repetitive freeze-thaw cycle. For example, make aliquots sufficient to test ten samples and thaw an aliquot prior to use.
3. These sequences of the TS and RP primers are the ones supplied by the TRAP-eze™ kit. If the kit is not used and the primers are custom made, the highest degree of quality for both synthesis and purification has to be requested, e.g., HPLC purification.
4. Ammonium persulfate 10 % solution should be prepared fresh, although it is possible to store it for short periods of time at 4 °C.

5. Because of the small number of cells that are needed for the TRAP assay, cells can be grown on 6-well plates to have about 10^5–10^6 cells to be lysed. Even smaller surface, such as 12-well plates, can be employed reducing the volume of CHAPS buffer used to lyse the cells.
6. Scraping is preferred to trypsinization to detach adherent cells, because trypsin can inhibit telomerase activity.
7. A pellet of telomerase positive cells to be used as positive control in the TRAP assay can be stored at −80 °C. The TRAP-eze™ kit is supplied with a cell pellet of 10^6 cells to be lysed in 200 μL of CHAPS buffer and aliquoted. Before use, dilute this cell lysate 20 times in CHAPS buffer and use 2 μL in the TRAP assay. Aliquots of telomerase positive cell lysates can be stored for several months at −80 °C. Because a few microliters are enough for this assay, it is recommended to prepare aliquots of not more than 10 μL to be stored.
8. For the extension step, the recommended temperature is 30 °C, and the time is 30 min. This reaction can be set in a thermocycler. Alternatively, the reaction can be performed at room temperature and the inactivation of telomerase in a thermocycler at 94 °C for 2 min.
9. Because the stacking gel is not required, the gel plates used to cast the gel can be kept horizontally while pouring the gel solution as well as during the gel polymerization. Be careful to clean of any acrylamide solution that might have leaked on the external side of the glasses because it might result in permanent residual of dried polymerized acrylamide.
10. If the excess of gel solution is poured into a beaker, it can be used to check the progression of gel polymerization.
11. To increase the sensitivity of the TRAP assay, it is recommended to perform the assay using radioisotope-labelled TS primer. In this case, end label the TS primer as follows: incubate 2.5 μL of γ-^{32}P-ATP (3,000 Ci/mmol, 10 mCi/mL), 10 μL of TS primer, 4 μL 5× polynucleotide kinase buffer (supplied with the enzyme), 0.5 μL T4 polynucleotide kinase, and 3 μL distilled water at 37 °C for 20 min followed by 5 min incubation at 85 °C to inactivate the enzyme. Clean the labelled primer from unincorporated nucleotides using mini-column prepacked with Sephadex G-50 (GE Healthcare). Use 2 μL of this TS primer in the TRAP reaction. Follow the rest of the protocol as described until the end of the run of the PAGE. Then dry the gel with a vacuum pump system and expose to an X-ray film or a phosphor storage screen. Develop the film or scan the screen using phosphorimager. In the case of phosphorimager, the analysis is more quantitative as the amount of the signal in a band is strictly proportional to the amount of ^{32}P.

12. The steps are darker at lower molecular weights and faint progressively towards higher molecular weights in relation to their abundance. If darker and thicker bands appear on the gel in the region of higher molecular weights, these are most likely signs of PCR artifacts and telomerase activity cannot be accurately evaluated. Also, if a ladder appears in the buffer only lane, the assay has to be repeated entirely with fresh solutions (primer stocks, TRAP buffer, dNTPs, etc.) to exclude contaminations.
13. TSR8 is a TS primer extended with eight telomeric repeats that can be used as a positive control and used as reference for semi-quantitative analysis of telomerase activity.
14. Make sure that the IC band is always present. It should have similar intensity in the different wells throughout the gel. Because of the competition for Taq polymerase of the primer TS used to generate the telomeric ladder and the IC band, when the sample contains too much telomerase, i.e., too many telomeric products, the IC band may not be visible. In this case the experiment should be repeated using several dilutions of the sample in order to visualize clearly the IC band that can then be used as normalizing factor.
15. Because the results of a TRAP assay appear as a ladder, it is arbitrary how to draw the rectangle around telomerase products. The important element is to keep the size of the rectangle equal for all the lanes that have to be compared. For example, it is a good criterion to set the rectangle size on the lane where the ladder has the highest number of visible steps because, most likely, these represent the highest telomerase activity.
16. Alternatively, if telomerase activity has to be compared among the samples within a given experiment, it is not necessary to perform a TRAP assay on the TSR8. Rather draw rectangles on the ladder of lane 1 (X1), the IC on lane 1 (C1), the ladder of lane 2 (X2), and the IC of lane 2 (C2) and continuing up to n samples (Xn and Cn). The relative TRAP activity can be represented in the form of a graph where the data points are represented by (X1/C1) - (X0 - C0); (X2/C2) - (X0/C0); ...(Xn/Cn) - (X0/C0), being X0 and C0 the ladder and the IC of the TRAP assay done on the buffer only, respectively [13].

References

1. Blackburn EH (1991) Structure and function of telomeres. Nature 350(6319):569–573
2. Blackburn EH (1991) Telomeres. Trends Biochem Sci 16(10):378–381
3. Greider CW, Blackburn EH (1985) Identification of a specific telomere terminal transferase activity in Tetrahymena extracts. Cell 43(2 Pt 1):405–413
4. Olovnikov AM (1973) A theory of marginotomy. The incomplete copying of template margin in enzymic synthesis of polynucleotides and biological significance of the phenomenon. J Theor Biol 41(1):181–190
5. Hayflick L, Moorhead PS (1961) The serial cultivation of human diploid cell strains. Exp Cell Res 25:585–621

6. Shay JW, Bacchetti S (1997) A survey of telomerase activity in human cancer. Eur J Cancer 33(5):787–791
7. Buseman CM, Wright WE, Shay JW (2012) Is telomerase a viable target in cancer? Mutat Res 730(1–2):90–97
8. Kim NW et al (1994) Specific association of human telomerase activity with immortal cells and cancer. Science 266(5193):2011–2015
9. Morin GB (1991) Recognition of a chromosome truncation site associated with alpha-thalassaemia by human telomerase. Nature 353(6343):454–456
10. Holt SE, Norton JC, Wright WE, Shay JW (1996) Comparison of the telomeric repeat amplification protocol (TRAP) to the new TRAP-eze telomerase detection kit. Methods Cell Sci 18:237–248
11. Kim NW, Wu F (1997) Advances in quantification and characterization of telomerase activity by the telomeric repeat amplification protocol (TRAP). Nucleic Acids Res 25(13):2595–2597
12. Saldanha SN, Andrews LG, Tollefsbol TO (2003) Analysis of telomerase activity and detection of its catalytic subunit, hTERT. Anal Biochem 315(1):1–21
13. Gardano L et al (2011) Native gel electrophoresis of human telomerase distinguishes active complexes with or without dyskerin. Nucleic Acids Res 40(5):e36

Index

Svetlana Makovets (ed.), *DNA Electrophoresis: Methods and Protocols*, Methods in Molecular Biology, vol. 1054,
DOI 10.1007/978-1-62703-565-1, © Springer Science+Business Media New York 2013

D

E

P

R

S

MIX
Papier aus verantwortungsvollen Quellen
Paper from responsible sources
FSC® C105338

If you have any concerns about our products,
you can contact us on
ProductSafety@springernature.com

In case Publisher is established outside the EU,
the EU authorized representative is:
Springer Nature Customer Service Center GmbH
Europaplatz 3, 69115 Heidelberg, Germany

Printed by Libri Plureos GmbH
in Hamburg, Germany